INTRODUCTION TO
MATHEMATICAL IDEAS

THE COVER: Sculpture, *Complementary Forms,* by Ruth Vollmer, 1967. *Courtesy Betty Parsons Gallery.* Photograph by H. Landshoff.

David G. Crowdis
and
Brandon W. Wheeler

DEPARTMENT OF MATHEMATICS
SACRAMENTO CITY COLLEGE

INTRODUCTION TO MATHEMATICAL IDEAS

McGraw-Hill Book Company

NEW YORK ST. LOUIS SAN FRANCISCO TORONTO LONDON SYDNEY

INTRODUCTION TO
MATHEMATICAL IDEAS

Library of Congress Catalog Card Number 68-27505

14705

1234567890 HDMM 7543210698

PREFACE

The pressures of completing studies in economics, art, music, history, psychology, or education leave the student little time for the exploration of other subjects, but mathematics is important to these subjects and to most others. Most students are asked to use (or, in the case of education candidates, to teach) mathematics after leaving college. This text is designed for such students who take only one or two college mathematics courses.

Most students who find themselves in a general education mathematics course may have limited background in the subject. For this reason the authors assume that the reader knows relatively little about mathematics, having had possibly only an introduction to algebra and geometry in high school. However, the reader who has a stronger background in mathematics will find topics here that challenge him and hold his interest.

The topics in this text were selected for three reasons. First, they represent important areas of mathematics that can be understood by students with limited mathematical preparation. Second, they form what the authors feel is a reasonable progression of topics that hold the interest of students, especially those who may have found mathematics less than exciting in the past. Third, they are indicative of the diversity and usefulness of mathematics. As a natural starting point we selected the one topic a student is sure to be familiar with: numerals and numeral systems. This choice has several advantages. It shows the student that he really knows more about mathematics than he thinks, having already mastered a relatively complex numeral system. It also introduces him to some of the history of mathematics and, at the same time, shows him the importance and power of symbolism as an intellectual tool.

As the second of the five major areas considered, we chose to examine the digital computer. An important reason for this choice involves the mathematical concepts upon which these machines are based. If a student can understand the nature and limitations of these machines, he is less likely to accept the mystery all too often surrounding them. The discussion of computers in Chapters 3 and 4 is not necessary to an understanding of the material in the chapters which follow, and thus can be delayed for later presentation or even omitted. Furthermore Chapters 3 and 4 are self-contained; neither an actual computer nor extensive knowledge of computers on the part of the instructor is required. Naturally, if students can see a computer at work, they benefit from the experience. The programs in Chapter 4 are written with Fortran II-D in mind but will run on any computer system that can compile Fortran II or Fortran with Format programs.

It is the close relationship between the structure of symbolic logic and simple electrical systems that makes it possible to construct a modern computer system, and Chapters 5 and 6 examine the nature of symbolic logic and its companions, argument and proof. These chapters also show which parts of reasoning are mechanical and thus lend themselves to machine-like processing, and which are not mechanical and thus require human thinking. The separation of the mechanical from the nonmechanical is an underlying precept of the axiomatic method.

Any study of logic brings up questions related to symbolic model building. The fundamental modeling material of mathematics is the set. In Chapter 7 sets are presented alone, as a model of logic, and as a basis for the operations on real numbers.

At this point, students should be ready to examine a formal mathematical system, both to solidify the ideas of proof from Chapters 5 and 6 and to illustrate the value of a completely formal approach to a subject.

Many formal systems are available to use as examples, notably euclidean geometry and the real-number system. However, the example used here is Boolean algebra, introduced in Chapter 8. Boolean algebra is radically different from systems the student may have encountered in the past; thus he is not asked to prove statements that he has long known to be true. Moreover, Boolean algebra is the abstraction of the elements of logic and sets he has just finished studying, and is the basic symbolic tool in computer design. Chapter 8 is the most difficult chapter for most students.

As the final topic, we sought one that could hold the interest of the nonmathematics student but at the same time bring together some of the general ideas considered in the earlier chapters, including model building and the development of calculational methods. The twin subjects of probability and statistics seemed a natural choice. They demonstrate the power of model building and show that mathematics is not simply a game or a tool for scientists and engineers, but that it also has direct applications to everyday activities.

We would like to thank Molly Crowdis for editing the original manuscript, Charles Slater for his encouragement and assistance in preparing this work, Susanne Shelley for her research and technical advice, and our friends and colleagues at Sacramento City College for the encouragement they gave us during the writing of this book.

David G. Crowdis
Brandon W. Wheeler

CONTENTS

INTRODUCTION TO MATHEMATICAL IDEAS

A Babylonian clay tablet, CA. *2300* BC, *recording salary payments for temple services.*
COURTESY THE METROPOLITAN MUSEUM OF ART,
ACQUIRED IN EXCHANGE WITH THE J. PIERPONT MORGAN LIBRARY, 1911.

chapter one

NUMBERS, NUMERALS, AND SYMBOLS

In these pages we present some of the ideas underlying the subject of mathematics. You should not find them difficult; you have already encountered them, worked with them, and, in many cases, mastered them. The difference is that we view them as mathematical concepts rather than as individual and unrelated mathematical calculations. For example, we all use logic, but only few of us know that formal logic is a branch of mathematics and that many of its ideas were first studied by mathematicians as they struggled with their own problems of mathematical analysis. In Chap. 5 we examine how, in mathematics, we reduce logical concepts to their essence and thus gain insight into the processes of reasoning.

Another example of mathematical ideas used in everyday problems is found in Chap. 9, where we examine probability and statistics. We have all acted on the basis of chance or have wondered: What are the chances? We examine these ideas and attempt to establish numerical procedures for using probability. As we study these ideas, we hope that a transfer takes place, so that the patterns of thinking involved find their way into your everyday activities. We do not attempt to teach you many skills; our main goal is to change your point of view and improve your understanding of mathematics. We also show you some of the relationships of mathematics to man's other activities.

We hope that you will be pleased by how much you already know. We require only that you can add, subtract, multiply, and divide whole numbers. If you are able to find the product of 375 and 48, you will have no difficulty with what follows, at least as far as any prior skill is concerned.

1-1 SYMBOLS

To begin our discussion, let us note the simple but often overlooked fact that there is a great difference between an object and the various symbols used to identify it. Although this may seem obvious, it is often a student's failure to note and use this difference that causes difficulty in his study of mathematics. The selection of a useful symbol to stand in place of an object is one of man's most powerful intellectual tools, and although there are many good examples in mathematics, the best is language—man's most impressive collection of symbols. The marks on this page are only symbols agreed on and used by man as he exchanges his thoughts with other men. Another simple example of the useful selection of a symbol is one's name. We often hear people complain that soon we are not going to have names; we shall simply have numbers. Although we know that they are commenting on the loss of identity in a social sense, they do not understand the difference between objects and their symbols. First, they do not mean numbers, as we shall see, but numerals; and second, from a logical point of view a set of numerals is no less a name than a collection of English letters.

In exploring the man-made world of mathematics, we shall take advantage of the object-name game, sometimes studying the objects themselves, and other times the special properties of the names we assign to them. If we were students of history, we might study the history of California, but if we were interested in linguistics, we might study the history of the word California.

Our only detailed study of the object-name game will be the names we use for numbers. These names are called numerals. Before doing so, however, we might consider why certain symbols become attached to certain objects. The most common reason is tradition. For most of us, the reasons for using a particular symbol are lost in time. We can often recover them if we explore the records of the past. For example, our symbol + is probably a shortened form of the Latin word *et* (and). Another example is our symbol = for the idea of equality. This symbol was first used in "The Whetstone of Witte," a text on algebra written in 1557 by the English mathematician Robert Recorde, who felt that nothing could be more equal than two parallel lines.

Another reason for the selection of a symbol is that the symbol reminds the reader of the object in some descriptive manner or is keyed to a phrase that describes the object. This is often what the past reveals when we look for the original reason for the adoption of what is now a traditional symbol. A humorous example of this is the selection of the name of one of the first modern computers, developed by one of the world's great mathematicians, the late John von Neuman. He called the machine the Mathematical Analyzer, Numerical Integrator and Computer, a name highly descriptive of its nature and properties. The humor lies in the abbreviated name of this mechanical brain—MANIAC. Our use of the word *object* is not limited to physical objects but includes ideas as well. The numerals 1 and 0 are symbols for objects in this class, because they stand for nonmaterial numbers. The numeral 1 reminds us of a single thing, and 0, enclosing an unmarked area, has the feeling of a void. This void or empty feeling also seems to be the reason for the use of the word *zero*, which probably comes from the Latinized form *zephirum* of the Arabic *sifr*, which is a translation of the Hindu word for empty or void, *sunya.*

Modern technology has adopted many descriptive words and symbols to stand for ideas. The word *program*, used to identify the sequence of steps a computer executes in solving a problem, because it is similar to the program, or sequence of events, followed in a play or concert, is one such word. Although we could examine many other examples of this process of the creation and development of symbols, we shall note only that today's descriptive symbols will probably become tomorrow's traditional ones.

Printing has slowed down, but not stopped, the ancient tradition of each writer's creating and using his own symbols for the concepts he is working with. However, in many of the rapidly changing areas of science and mathematics, scholars often fail to communicate, because they misunderstand the definition or assigned meaning of symbols. The use of different symbols by different authors for the same thing will remain a partial barrier to the interchange of ideas between experts and laymen.

Some mathematicians think that mathematics is limited to the study of symbols and their properties, without any other relation to reality. We do not agree; we think that mathematics cannot be reduced to symbol juggling but must be considered in relationship to the physical world, because almost all mathematics has been devised to symbolize and describe it.

Let us now assume that you are aware of an undefined concept called "numbers," and let us consider some of the ways man has chosen to symbolize it. We know that there are primitive civilizations unaware of this concept, people who cannot identify the similarity between two groups of different objects with the same number of objects in each group. We shall not explore this point. Instead, we shall begin our discussion of numerals, the symbols for numbers, with an awareness of the concept of numbers, but without a system for handling them, and speculate on how the present state of affairs came to be.

1-2 PRIMITIVE BEGINNINGS

Man's records of the past indicate that the first use of numerals in a definite system took place approximately 5500 years ago. But before this, man had some kind of primitive number sense; he could differentiate between a group of three objects and a group of four, without counting. During this period of his development, man discovered that he could compare the size of two collections of things by pairing them off, one object from one collection with one object from the other. A primitive tribal chief who wished to keep track of the number of members in his tribe might keep a sack of stones, one stone for each member. When a tribe member died, one stone would be thrown away; when a birth occurred, one stone would be added to the collection. Such a scheme is referred to, by modern man, as "one-to-one correspondence," or tallying. If, while pairing off the objects in two collections, we find that we use up both sets of collections at the same time, we say that the two collections match, or are in a one-to-one correspondence.

Although controlled experiments have shown that some animals have a number sense, only man has gone on to the notion of number symbols. Simple tallying was probably the earliest method of keeping track of number size. One-to-one correspondence is the usual way a child is introduced to the concept of numbers. As we look around us we see this idea in many forms. We observe that a play was attended by 375 persons, not because we counted them, but because 375 tickets were collected—one per person. We know we are buying a dozen objects, not because we counted them, but because the case they came in has a dozen places for objects and it is full. How often have we chosen sides for a game by pairing off one person for each side? All these examples illustrate the use of one-to-one correspondence. Indeed, as we shall see, this one basic idea can be used, together with some others about collections of things and how they work, to define and examine almost all the ideas of mathematics discovered and invented by man.

1-2 EXERCISES

1. Give two examples of symbols for objects where the reason for using the symbol has been lost.

2. Give two examples of symbols for objects where the reason for assigning the symbol is common knowledge.

3. When is one collection said to be in one-to-one correspondence with another?

4. A company hires a man to distribute free samples house to house. What is the easiest way for them to keep track of the number of houses he visits, assuming he is honest?

5. How might one-to-one correspondence be used to simplify the counting of a large number of dimes?

6. Can the number of empty chairs in a classroom be counted to determine the number of absentees on any given day? Explain.

7. What property is shared by two groups of objects that can be put into one-to-one correspondence?

8. Consider the following quotation from Lewis Carroll's "Through the Looking-Glass . . .":

> "The name of the song is called 'Haddocks' Eyes'."
> "Oh, that's the name of the song, is it?" Alice said, trying to feel interested.
> "No, you don't understand," the Knight said, looking a little vexed. "That's what the name is *called.* The name really is 'The Aged Aged Man'."
> "Then I ought to have said, 'That's what the *song* is called'?" Alice corrected herself.
> "No, you oughtn't: That's quite another thing! The *song* is called 'Ways And Means': but that's only what it's *called*, you know!"
> "Well, what is the song, then?" said Alice, who was by this time completely bewildered.
> "I was coming to that," the Knight said. "The song really is 'A-sitting On A Gate': and the tune's my own invention."

Identify two cases where Alice has confused objects and names.

1-3 SIMPLE GROUPING

Although tallying is an excellent method of keeping count of small collections, it loses its appeal as the numbers increase in size. Consider counting a flock of sheep by keeping a pebble for each one. If the flock increased to thousands, the shepherd would have to carry around a ton or so of rocks, even if they were small. To decrease the pile of rocks, a clever shepherd might let one stick represent ten rocks. Then he could use one-tenth as many sticks as rocks, with a few pebbles to take care of the odd number of sheep. If the pile of sticks were too large, he might let a pine cone represent ten sticks (or 100 sheep) and an old leg bone ten pine cones (or 1000 sheep) and so on, as he wished. Now, if he had 7925 sheep in his flock, he would have to carry seven old leg bones, nine pine cones, two sticks, and five rocks, instead of 7925 rocks. He might not be aware of it, but he has invented a system of numerals modern man would refer to as a "simple-grouping" system. The shepherd still has some problems, however. He has to be careful not to confuse his symbols. It would be disastrous to confuse pine cones and sticks, that is, hundreds with tens. Also, if one or two of his sticks were inadvertently broken, he might be led to believe that he had 200 more sheep than were actually in his flock. Lastly, carrying around a group of objects, such as sticks, rocks, and so on, might, at times, be inconvenient. These problems were overcome by a written, or "picture," notation, along with the development of writing materials, such as papyrus and parchment. With these inventions, records could be easily kept, transported, and stored.

One of the earliest examples of a simple-grouping system is Egyptian hieroglyphics (see Fig. 1-1). This system was based on a scale of 10, with distinct symbols for 1, 10, 100, 1000, 10,000, 100,000, and 1,000,000. Multiples of these numbers were indicated by repetitions of them. The symbol for 5000 was five of those for 1000, the symbol for 90 was nine of those for 10, and so on. Any number could be expressed by repeating

OUR NUMBER	HIEROGLYPHIC SYMBOL	OBJECT REPRESENTED
1	𓏺	VERTICAL STAFF
10	𓎆	HEEL BONE
100	𓍢	SCROLL
1000	𓆼	LOTUS FLOWER
10,000	𓂭	POINTING FINGER
100,000	𓆐	BURBOT FISH
1,000,000	𓁨	MAN IN ASTONISHMENT

FIG. 1-1 EGYPTIAN NUMERALS.

the symbols the required number of times. Therefore

$7925 = 7000 + 900 + 20 + 5$

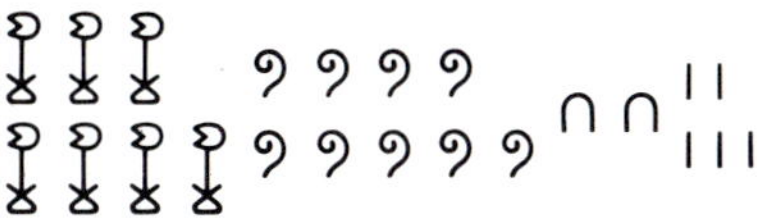

Note that although the larger quantities are usually grouped to the left, this is not necessary in a simple-grouping system.

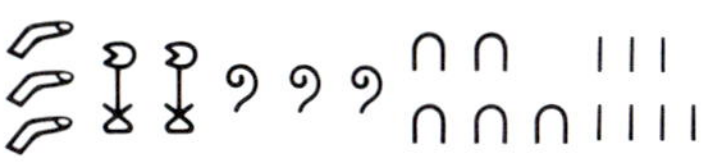

could be written

In this system the order of symbols has little relevance. The shepherd of a few paragraphs above could rearrange his rocks, sticks, and so on, and still be sure that the count of his flock was accurate. However, we cannot always rearrange our symbols that represent numbers. Try interchanging the present-day numerals that represent your age! Unless you are 22, 33, or a multiple of 11, it won't work.

Hieroglyphics has two obvious drawbacks. The first is that the numerals are awkward. It takes a relatively long time to draw the symbol for 1,000,000. In fact, I and ∩ are the only numerals that are easily drawn. The second is the number of symbols required for very large numbers. Writing 9,999,999 is a tedious job. Some early civilizations, when confronted by these difficulties, devised schemes to overcome them. One such scheme is the "multiplicative-grouping" system.

1-4 MULTIPLICATIVE GROUPING

Some simple-grouping systems evolved into a multiplicative-grouping system, in which symbols are adopted for digits 1, 2, 3, . . . , 9 and for 10,

1	一	6	六	10	十
2	二	7	七	100	百
3	三	8	八	1000	千
4	四	9	九		
5	五				

FIG. 1-2. TRADITIONAL CHINESE-JAPANESE NUMERALS.

100, and 1000. To write thirty, for example, the symbol for three is placed in front of the symbol for ten, instead of writing the symbol for ten three times, as in simple grouping.

The traditional Chinese-Japanese numeral system is of the multiplicative-grouping kind (see Fig. 1-2). Numerals in this system are written vertically. For example,

$50 = 5 \times 10 =$ 五
十

$231 = 2 \times 100 + 3 \times 10 + 1 =$

二 百	2×100
	+
三 十	3×10
	+
一	1

$7469 = 7 \times 1000 + 4 \times 100 + 6 \times 10 + 9 =$

七 千	7000
	+
四 百	400
	+
六 十	60
	+
九	9

Another multiplicative-grouping system with which you are more familiar is our written English numeral system. When we say, or write, five thousand, four hundred, and fifty-two, we mean $5 \times 1000 + 4 \times 100 + 5 \times 10 + 2$, clearly an example of multiplicative grouping. Although the principle is clearly seen in the *five thousand*, the *fifty* is a shortened form of five tens; in a dictionary, you would find *fifty* so defined. The words *twenty, thirty,* and so on, are such shortened forms. As you can see, we still make extensive use of a multiplicative-grouping system.

Multiplicative systems are more efficient than simple-grouping systems, because it is not necessary to write a single symbol many times. But are they the most efficient system?

1-4 EXERCISES

1. Translate the following Egyptian numerals into ours.

(*a*) ∩∩∩|| ∩∩∩||| (*b*) 𓆼∩ (*c*) 𓆐∩∩∩∩||| 𓆐∩∩∩∩||||

2. Translate the following into Egyptian hieroglyphics.

(*a*) 489 (*b*) 3058 (*c*) 1499 (*d*) 90,004
(*e*) 443,829 (*f*) 2,623,467 (*g*) 494,211 (*h*) 4,329,575

3. Translate the following traditional Chinese-Japanese numerals into ours.

(*a*) 七十五 (*b*) 八百九 (*c*) 三千四百二十一 (*d*) 九千六十

4. Translate the following into the traditional Chinese-Japanese.

(*a*) 11 (*b*) 777 (*c*) 4008 (*d*) 9710

5. Perhaps the most common simple-grouping numeral system is the Roman, given below.

1	I
5	V
10	X
50	L
100	C
500	D
1000	M

Example: MMMCCXXVII = 3227

The Roman system has a subtraction principle. A smaller numeral placed to the left of a larger one is subtracted from it. For example,

IX = 10 − 1 = 9 but XI = 10 + 1 = 11

XC = 100 − 10 = 90 but CX = 100 + 10 = 110

Translate the following:

(*a*) CCCXXXVIII (*b*) CCXCI
(*c*) MDI (*d*) MMCXLIX
(*e*) MMCDXLVIII

6. Translate the following into Roman numerals:

   (*a*) 489 (*b*) 3058 (*c*) 1499 (*d*) 3538 (*e*) 3002

7. Why do not the Roman and traditional Chinese-Japanese systems have a zero symbol?
8. Give an example of a simple-grouping system still in use.
9. As previously stated, symbols in most simple-grouping systems can be written in any order. Is there any practical reason for writing them in ascending or descending order?
10. Let the elements of a multiplicative system be our digits 1, 2, 3, . . . , 9 and the letters $a = 10$, $b = 100$, $c = 1000$, and $d = 10{,}000$. Translate the following into this multiplicative system:

    (*a*) 29 (*b*) 777 (*c*) 4209 (*d*) 87,035

11. Translate the number from Exercise 10 into a simple-grouping system that uses the same symbols as Exercise 10.

1-5 POSITIONAL NUMERAL SYSTEMS

Positional numeral systems may be a natural outgrowth of multiplicative systems, although there is no historical evidence to prove this. Recall that, in the multiplicative systems, larger-valued numerals precede smaller-valued ones. In the traditional Chinese-Japanese system, thousands are written above hundreds, hundreds above tens, and units are always written lowest. Again, when we write numbers in English, we use a definite order—hundreds to the right of thousands, thousands to the right of ten thousands, and so forth. In the traditional Chinese-Japanese system, we could eliminate the symbol for hundreds by agreeing that, if a digit symbol appeared in the third position from the bottom, it would mean so many hundreds. In the same way, a numeral in the fourth position from the bottom could mean so many thousands, although the symbol for thousand did not appear. This is the essence of positional numeral systems. Only symbols for digits are necessary. Larger numbers are indicated by positioning the digits according to a previously agreed-on arrangement. The obvious example of such a system is the one you are most familiar with. When we write 268, the 2, in the third position from the right, means two hundreds; the 6, in the second position, six tens; and the 8, eight units, because it is on the right end. If we rearrange the 2, 6, and 8, we change the number completely, because the position of the digits determines the number as much as the particular digits written.

In all the numeral systems mentioned so far, counting is done by arranging objects in groups of ten, then in groups of ten times ten (hundreds), then ten times ten times ten (thousands), and so on. We call this a "base-10" system. However, a base of 10 was not always used. Natives of Queensland used a base of 2 and counted thus: one, two, two and one, two twos, and so on. Some African Pygmies use a base of 2 and 4 combined and count: a, oa, ua, oa-oa for 1, 2, 3, 4 and oa-oa-a, oa-oa-oa, oa-oa-ua for 5, 6, 7.

The first historical example of a positional numeral system is that of the Mayas of Mexico and Central America. There is evidence that this system existed at least 500 years before any comparable European or Asiatic system. Besides being the first positional system, the Mayan system was revolutionary in another aspect. It was the first to use a zero, or absence symbol (such a place-holding symbol is needed to avoid confusion). For example, if 207 were written 2 7, with a blank space as a position in the numeral denoting no tens, we might confuse it with 27 or 2007 (2 7) or even with 20,007 (2   7). The Mayan system was not a decimal, or base-10 system, but vigesimal, or base-20. The Mayan symbols are shown in Fig. 1-3. Some students, seeing Fig. 1-3, say that the Mayan is not a positional but a simple-grouping system with ___ standing for five and · for one. However, we shall see that these are only the basic symbols for digits, like our own 0, 1, 2, . . . , 9, and that position is indeed the key element of the Mayan system. The numbers from 0 to 19 are symbolized by pebbles (the dots) and sticks (the lines). Multidigit numbers are written vertically, as in the traditional Chinese-Japanese system. The lowest posi-

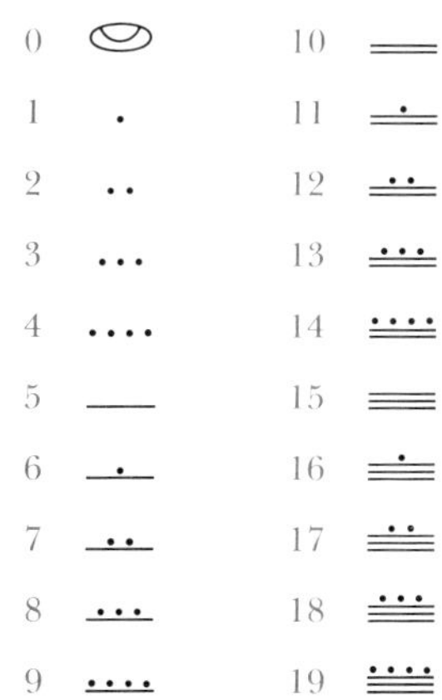

FIG. 1-3 MAYAN DIGITS.

tion indicates units (1 to 19), the second lowest position multiples of the base 20, and so on. For example, we write 46 as

The upper row of •• means 2 twenties, and [Mayan numeral] below it means 6 units, or $2 \times 20 + 6$. We write 68 as

The upper row of ••• means 3 twenties, and [Mayan numeral] below it means 8 units, or $3 \times 20 + 8$.

We must be careful, when writing numbers larger than 19 in the Mayan system, to leave space between positions, or confusion will result. For example, [Mayan numeral] $= 6$ and [Mayan numeral] $= 25$. More examples of Mayan numerals follow:

 $= \begin{matrix} 2 \times 20 \\ + \\ 0 \end{matrix} = \begin{matrix} 40 \\ + \\ 0 \end{matrix} = 40$

 $= \begin{matrix} 12 \times 20 \\ + \\ 2 \end{matrix} = \begin{matrix} 240 \\ + \\ 2 \end{matrix} = 242$

 $= \begin{matrix} 17 \times 20 \\ + \\ 16 \end{matrix} = \begin{matrix} 340 \\ + \\ 16 \end{matrix} = 356$

In our system, the first right-hand position denotes units, the second tens, and so on. The Mayan has an analogous placement, with the lowest position denoting units, and the next twenties. In our system, 10×10, or hundreds, appears in the third position. We might then expect that the Mayan had 20×20, or four hundreds, in the third position, but we would be wrong. The third position denotes multiples of 20×18 or 360. It is believed that the Mayan set up their system like this because there were 360 days in the Mayan year.

In a positional system, the positions need not be graduated equally. The only requirement is that those who use it agree on the meaning of each position. In other words, equally graduated positions are a convenience, not a necessity. Happily, all other positional values in the Mayan system are obtained by multiplying the previous positional value by 20.

Examples of the three-digit Mayan numerals follow:

$$\begin{array}{ccccc} 1 \times 360 & & 360 & & \\ + & & + & & \\ 0 \times 20 & = & 0 & = & 361 \\ + & & + & & \\ 1 & & 1 & & \end{array}$$

$$\begin{array}{ccccc} 3 \times 360 & & 1080 & & \\ + & & + & & \\ 1 \times 20 & = & 20 & = & 1105 \\ + & & + & & \\ 5 & & 5 & & \end{array}$$

$$\begin{array}{ccccc} 6 \times 360 & & 2160 & & \\ + & & + & & \\ 2 \times 20 & = & 40 & = & 2210 \\ + & & + & & \\ 10 & & 10 & & \end{array}$$

$$\begin{array}{ccccc} 16 \times 360 & & 5760 & & \\ + & & + & & \\ 6 \times 20 & = & 120 & = & 5892 \\ + & & + & & \\ 12 & & 12 & & \end{array}$$

If we wish to write very large numbers, we increase the number of positions: the fourth position represents 360 × 20, or 7200s; the fifth 7200 × 20, or 144,000s; the sixth 144,000 × 20, or 2,880,000s; and so on. More examples follow:

$$\begin{array}{ccr} 2 \times 7200 & & 14{,}400 \\ 6 \times 360 & = & 2160 \\ 3 \times 20 & & 60 \\ 6 & & \underline{6} \\ & & 16{,}626 \end{array}$$

$$\begin{array}{ccr} 2 \times 2{,}880{,}000 & & 5{,}760{,}000 \\ 0 \times 144{,}000 & & 0 \\ 10 \times 7200 & = & 72{,}000 \\ 12 \times 360 & & 4{,}320 \\ 1 \times 20 & & 20 \\ 5 & & \underline{5} \\ & & 5{,}836{,}345 \end{array}$$

The Mayan system is the most sophisticated of those we have discussed so far, excluding our own. Could it be made more sophisticated? The fact that the third position gives multiples of 360 instead of 400 is disconcerting, but not fatal. Using a base as large as 20 makes the multiplication table have 20 numbers to a side. This means that the table is four times as large as the one you memorized in elementary school. Perhaps the real lack of sophistication is the form of the digits, $\cdot$ to $\overset{\cdot\cdot\cdot\cdot}{\equiv}$. The system would be better if distinct symbols were used for them. Then one would not have to worry about whether $\overset{\cdot}{\text{—}}$ is 6 or 25, and the memorizing of addition and multiplication facts would be greatly facilitated.

1-5 EXERCISES

Translate the following Mayan numerals into ours.

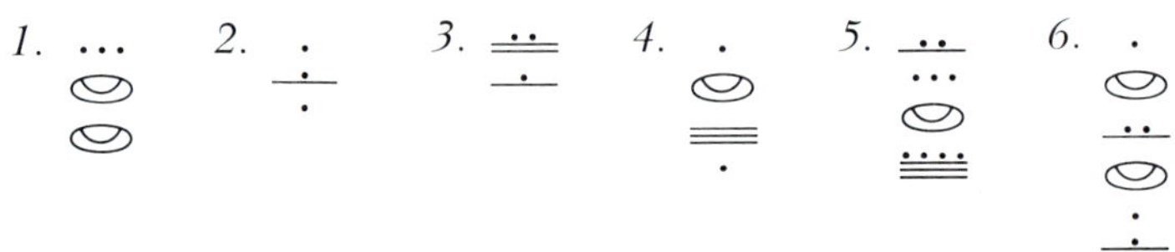

Translate into the Mayan system.

7. 489	*8.* 3058	*9.* 1459	*10.* 90,004
11. 443,829	*12.* 2,623,467	*13.* 494,211	*14.* 2,889,575

15. Suppose a primitive tribe used the following system to count: oh, ah, so, say, say-oh, say-ah, say-so, ...

(*a*) What base does this tribe use?
(*b*) Write the next four numbers in the system.
(*c*) How would they write the number 13?
(*d*) Does this tribe have a symbol for zero?

1-6 BABYLONIAN NUMERALS

From the preceding sections, you might think that mankind began with one-to-one correspondence, proceeded to simple grouping, to multiplicative grouping, and finally to positional numeral systems, without any great difficulty. Not so. There were many transitional numeral systems and there were many complete failures. About 4000 years ago the Babylonians used a positional system that was much like the Mayan but not so efficient. This system is known as "cuneiform." It used only two symbols: / for one and < for ten. Both numerals were made by pressing a stylus into wet clay and so could be easily and quickly produced. Numerals were also shortened by using a subtraction principle whose symbol was /‾ . For example,

53 = 5 × 10 + 3 = <<<<< ///

57 = 6 × 10 − 3 = <<<<<< /‾///

Most of the population of Babylonia, at that time, used only simple grouping, because they did not need large numbers. Similarly a Maya who never had to count past 19 would not know that his numeral system was positional. Only a few scholars of that time worked with large numbers, and so there was practically no standardization in the symbols or notation used. Each scholar varied the method to suit himself. However, the general pattern of symbolizing large numbers is much like that of the Mayas. The system is positional, like the Mayan, and the individual digits are formed by grouping < (ten) and / (one). The base is 60; that is, cuneiform numerals are sexagesimal. It is thought that a base of 60 was used because the Babylonians worshiped 60 gods. Also there is no inconsistency, as in the Mayan system. Each positional value is sixty times that of the previous one.

The values of the positions for a base of 60 are:

Fifth position	*Fourth position*	*Third position*
216,000 × 60 = 12,960,000s	3600 × 60 = 216,000s	60 × 60 = 3600s

Second position	*First position*
60s	Units

A number written in cuneiform might look like

</ <<// = 11 × 60 + 22 = 660 + 22 = 682

This is a two-digit number. The first digit on the right is <<//, and the second is </. Because this is a positional system, the second digit means 11 sixties, or 11 × 60 = 660. The first position is units, and there are 22 of them. The number

/// <<<<<//// <</‾// = 3 × 3600 + 54 × 60 + 18

= 10,800 + 3240 + 18 = 14,058

is a three-digit number. The positional values are units, 60s, and 3600s. The subtraction principle is used in the units position. Cuneiform numerals contain two flaws. First, a careless Babylonian could run the digits together so as to confuse numbers. For example, does <<// mean 22 or 1202? Second, there is no absence symbol. This caused great trouble. It was sometimes necessary to read an entire text to decide what a particular symbol meant. For example, / could mean 1 or 60 or 3600, depending on whether it was thought to be in the first, second, or third position. More examples follow:

</ /// ≤≤ over << < $= 11 \times 216{,}000 + 3 \times 3600 + 40 \times 60 + 10$
$= 2{,}376{,}000 + 10{,}800 + 2400 + 10$
$= 2{,}389{,}210$

/ ≤ over < // ≤ over </̅/̅/̅ / $= 1 \times 12{,}960{,}000 + 20 \times 216{,}000 + 2 \times 3600 + 18 \times 60 + 1$
$= 12{,}960{,}000 + 4{,}320{,}000 + 7200 + 1080 + 1$
$= 17{,}288{,}281$

1-6 EXERCISES

Translate the following Babylonian numerals:

1. << //

2. << // <<< /

3. /// ≤≤ over << // over // <<<// </̅/̅/̅

4. / </ </ </ </

5. ≤≤ over << ≤≤ over << // over // <<</̅/̅/̅ <<< ///

Translate the following into cuneiform.

6. 62	*7.* 661	*8.* 14,295
9. 544,753	*10.* 3612	*11.* 216,001

12. What difficulty did you encounter in Exercises 10 and 11?

13. Give an example of counting in base 60 that is used today.

1-7 HINDU-ARABIC NUMERALS

The numerals we use every day are called Hindu after the people who invented them and Arabic after those who carried them to Europe. The digits 1, 2, 3, . . . , 9 were used for approximately 400 years before zero and place value were introduced. As late as 1500, Europeans were trying to decide whether to adopt Hindu-Arabic numerals or to keep Roman numerals. The Hindu-Arabic system has much to recommend it. It is positional, and addition, subtraction, multiplication, and division are easily carried out in such a system. Also, its digits are easily distinguished from one another, which was not so in the Mayan system. Finally, a base of 10 is more tractable than the Mayan base of 20. As you may have noted, almost all the numeral systems discussed have a base of 10. It is assumed that man chose a base of 10 because he had 10 fingers and originally used them for counting. Although a base of 10 is an important part of our civilization, its choice is quite arbitrary. In fact, other bases have practical value. For example, a base of 2 is used in computers and satellites, because little equipment, relatively, is needed to transmit data in this base. Eggs are counted and sold in dozens and gross, that is, in a base of 12. *Dozen* is another word for twelve, and *gross* means 12 × 12.

1-8 VARIATIONS OF HINDU-ARABIC

The most important characteristics of the Hindu-Arabic system are the pattern and symbols it uses, and not its base of 10. Let us consider the pattern a moment. The first position on the right indicates units, the second multiples of 10, or multiples of the base. The third position gives multiples of 100, which equals 10 × 10, or base times base. The fourth position gives the number of thousands, which equals 10 × 10 × 10, or base times base times base. Then, for any base, the following gives the value of each position:

Fifth position	*Fourth position*	*Third position*	*Second position*	*First position*
Base × Base × Base × Base	Base × Base × Base	Base × Base	Base	Units

Let us see how this serves us in writing numerals in bases other than 10. (We shall use a subscript to indicate the base. Thus 345_7 means that we are using a base of 7. If no subscript is used, then the base is the usual 10.) If we count in base 5, we actually collect in groups of five and use 0, 1, 2, 3, and 4 as digits. Thus 11_5 means 1 five plus 1 unit, or six, in base 10. A slightly more complicated example is 24_5, which equals 2 fives plus 4 units, or $2 \times 5 + 4 = 14$. Again, 102_5 means one 5×5, or 25, plus zero fives plus 2 units, or $25 + 0 \times 5 + 2 = 27$. Similarly

$$342_5 = 3 \times 25 + 4 \times 5 + 2 = 75 + 20 + 2 = 97$$
$$1000_5 = 1 \times 5 \times 5 \times 5 + 0 \times 25 + 0 \times 5 + 0 = 125$$
$$3241_5 = 3 \times 125 + 2 \times 25 + 4 \times 5 + 1 = 375 + 50 + 20 + 1 = 446$$

In the "binary," or base-2, system the only digits used are 0 and 1. The second rightmost position in any numeral gives the number of twos, the third (base $\times$ base) the number of fours, the fourth (base $\times$ base $\times$ base) the number of eights. Thus

$$10_2 = 2 + 0 \times 1 = 2$$
$$101_2 = 1 \times 4 + 0 \times 2 + 1 = 5$$
$$1111_2 = 1 \times 8 + 1 \times 4 + 1 \times 2 + 1 = 15$$

Note that base 5 has no single symbol for five, for we write $10_5 = 5$. In base 2, there is no single symbol for two, because $10_2 = 2$. In fact, for any base b, $10_b = b$. Also, one must refrain from saying "ten base five" for 10_5. The word *ten* is an English word meaning this many:  . One should say "one zero base five" for 10_5. Similarly 341_7 is said "three four one base seven," not "three hundred and forty-one base seven."

However, there is a base that traditionally has words associated with its positions. This is base 12. The numeral $11_{12} = 12 + 1 = 13$ could be read one dozen and one. Similarly 75_{12} is seven dozen and five, or 89. Also $101_{12} = 12 \times 12 + 0 \times 12 + 1 = 145$ is one gross and one, and 243_{12} is two gross and four dozen and three, or 339. Here we have indicated groups of twelve by the word *dozen* and numerals in the third position, that of 12×12, by the word *gross*. Let us call twelve gross a "supergross." Thus $144 \times 12 = 1728$ is a supergross. More to the point, however, is the fact that the fourth position in base twelve gives multiples of $12 \times 12 \times 12 = 1728$, or supergross. Then 3452_{12} could be read three supergross, four gross, five dozen, and two.

Base 12 (in fact, any base larger than 10) contains a pitfall that catches many students. How does one write ten in base 12, since $10_{12} = 12$ and $9_{12} = 9$? Recall that base 10 contains digits 0, 1, 2, . . . , 9 and base 5 contains digits 0, 1, 2, 3, 4. In both cases, the largest digit is one less than the base being used. Base 12 also follows this pattern. The last digit, then, is eleven, but it cannot be written 11_{12}, since $11_{12} = 12 + 1 = 13$. So digits for bases larger than 10 include new single symbols for ten and larger numbers, up to but not including the base. Thus a duodecimal (base 12) numeral system might have the digits 0, 1, 2, 3, 4, 5, 6, 7, 8, 9, *t*, *e*. Various numbers could be written thus:

$1e_{12} = 12 + 11 = 23$
$5t4_{12} = 5 \times 144 + 10 \times 12 + 4 = 720 + 120 + 4 = 844$
$ee_{12} = 11 \times 12 + 11 = 143$
ete_{12} could be read *e* gross, *t* dozen, and *e*

Knowing that there are as many bases as numbers, one wonders which base is best. We are prejudiced in favor of base 10. However, probably no base is best. Base 2 does not tax one's memory with many symbols, but large numbers can have very large numerals. A base of 40 would allow numbers to be written compactly, but no one could remember the multiplication table. However, any base near 10 could be used as well as 10 for both counting and computation.

1-8 EXERCISES

1. Count to 10 in base-2 notation.
2. Write the numerals for 65 using:

(*a*) Base 2
(*b*) Base 5
(*c*) Base 12

3. Change 132_5 to base-10 notation.
4. Change 1011_{12} to base-10 notation.

Change to the indicated base.

5. $724 = (\quad\quad)_4$ *6.* $114 = (\quad\quad)_3$ *7.* $191 = (\quad\quad)_{12}$
8. $19 = (\quad\quad)_{12}$ *9.* $110111_2 = (\quad\quad)_{10}$
10. $10101010101_2 = (\quad\quad)_{10}$
11. Let 12 = one dozen, 12 dozen = one gross, and 12 gross = one supergross. Translate each of the following into an appropriate duodecimal numeral.

(*a*) Six dozen and six
(*b*) Nine gross

(*c*) Eleven gross, seven dozen, and four
(*d*) Five supergross and three
(*e*) Eleven supergross and ten dozen
(*f*) Ten supergross, eleven gross, eleven dozen, and ten

12. Twenty is sometimes called a "score." Write a numeral for four score and seven in:

(*a*) The vigesimal system
(*b*) The Mayan system

13. Write nineteen in the vigesimal system.

14. Use the symbols 0, *a, b, c,* and *d* to represent the decimal numbers 0, 1, 2, 3, and 4, respectively, and *x, y,* and *z* to represent the decimal numbers, 5, 25, and 125. Write 327 in:

(*a*) A simple-grouping system.
(*b*) A multiplicative-grouping system
(*c*) A positional system

15. Use the symbols 0, *a, b,* and *c* to represent the decimal numbers 0, 1, 2, and 3, respectively, and *w, x, y,* and *z* to represent 4, 16, 64, and 256. Write the decimal number 327 in:

(*a*) A simple-grouping system
(*b*) A multiplicative-grouping system
(*c*) A positional system

REVIEW EXERCISES

1. What is the difference between a number and a numeral?

2. Why is it impossible for 3472 to be a base-7 numeral?

3. Translate: 

4. Translate: 

5. Translate: 

6. Translate: 六千九百五

7. Translate: MMCDXLIX

8. Translate 3004 into: (*a*) Egyptian numerals; (*b*) Babylonian numerals; (*c*) Roman numerals; (*d*) Mayan numerals; (*e*) traditional Chinese-Japanese numerals.

9. Count to 15 using base-3 numerals.

10. $e3t_{12} = (\qquad)_{10}$

SELECTED REFERENCES

Eves, Howard: "An Introduction to the History of Mathematics," Holt, Rinehart and Winston, Inc., 1959.

The first part of this text presents the history of numeral systems in a very readable manner. However, the rest of the work requires a knowledge of calculus.

Holme, Roger W.: The Philosopher's Alice In Wonderland, *Antioch Review,* Summer, 1959.

This presents the usual interpretation of Exercise 8, Sec. 1-2.

Newman, James R.: "The World of Mathematics," vol. 3, pp. 1886–1890, Simon and Schuster, Inc., New York, 1956.

This mathematical anthology covers nearly every phase of mathematics and relates mathematics to other disciplines. The special reference is to another interpretation of that part of "Through the Looking-Glass . . ." used in Exercise 8, Sec. 1-2.

Wren, F. Lynwood: "Basic Mathematical Concepts," McGraw-Hill Book Company, New York, 1965.

Here is a source of examples of using the historical numeration systems and of problems in other bases.

A nineteenth-century Japanese abacus.
Beads above the bar have the value 5; those below, the value 1.
Only beads pushed toward the bar are counted.
COURTESY ALADDIN HOUSE, NEW YORK.

chapter two

COMPUTATION

In Chap. 1 we examined some of the ways man has devised to symbolize numbers. In the present chapter we examine some of the processes he has devised to use these symbols and thus the objects they stand for.

2-1 THE OPERATION OF ADDITION

ADDITION HISTORICALLY

Addition has been known by many other names: aggregation, composition, collection, assembly, joining, and summation. A form of the word *summation* is still used in the phrases *to do a sum* and *what is the sum?* Our word *addends,* for terms to be added, is a shortened form of the Italian *numeri addendi,* numbers to be added. In antiquity, *product* was used for the result of any operation, most often, however, for addition and multiplication. Today, of course, it is used for multiplication only, and *sum* means the result of addition.

The earliest symbol for addition is found in an Egyptian papyrus written around 1600 B.C., called the Ahmes papyrus for the man who wrote it. Legs walking forward  were used for addition, and legs walking backward  for subtraction. This is a very good description of the operation. To add 17 and 5, begin at 17 and walk (count) forward five numbers. Diophantus, the third-century Greek algebraist, as well as some Hindu mathematicians, used juxtaposition to show addition. They placed numbers next to one another with no sign between. We do this today with mixed fractions. For example, $2\frac{1}{2}$ means 2 plus $\frac{1}{2}$. If a child is $2\frac{1}{2}$ years old, he is 2 years plus 6 months.

Early Italian arithmetic was concerned with commerce, as in other societies. The operation of addition was first used to indicate that sacks were too full or not full enough. If a sack was too heavy, a + was written on it, if too light a −. Thus the Italians wrote the word *plus* between numbers to be added. It was a contraction of the word *surplus*. Later they shortened it even more to *p*, and finally to differentiate between the operation of addition and the letter of the alphabet, $\overline{p}$, was adopted for addition. There is some confusion concerning the origin of +. A Hindu manuscript uses a cross that is like + to mean subtraction. It is generally believed that + comes from the Latin *et*. Many German manuscripts of around 1450 used *et* for addition. It was written in an abbreviated form as &, which is very close to +. The plus and minus signs first appeared in print in Widman's "Arithmetic" about 1500. He was discussing the weight of boxes as being 30 lb + 2 oz or 30 lb-2 oz (the actual units Widman used were not pounds or ounces). All the signs and symbols for arithmetic operations seem so familiar to us that we often forget that they are only the symbols we have grown up with. Our brief discussion of their history serves to remind us that they, too, are arbitrary, and any other symbols would do.

ADDITION PROCESSES

As we learned in Chap. 1, counting by using a positional numeral system has many advantages over simple-grouping and multiplicative-grouping systems. Although the invention of place value led to a simple and efficient method of symbolizing numbers, it had an even greater effect on computation. The numeral systems used by ancient societies were not readily adaptable to arithmetic operations. In fact, some form of abacus was mostly used to do computation. The operation of an abacus is based on a positional system and place value. This means, then, that ancient societies used a computation machine incorporating place value when their official numeral system did not. The following examples of computation show how tedious addition is in nonpositional systems. In each of them we add 257 and 396. Our system is shown with each example to compare its ease of operation.

HIEROGLYPHICS

257
396
653

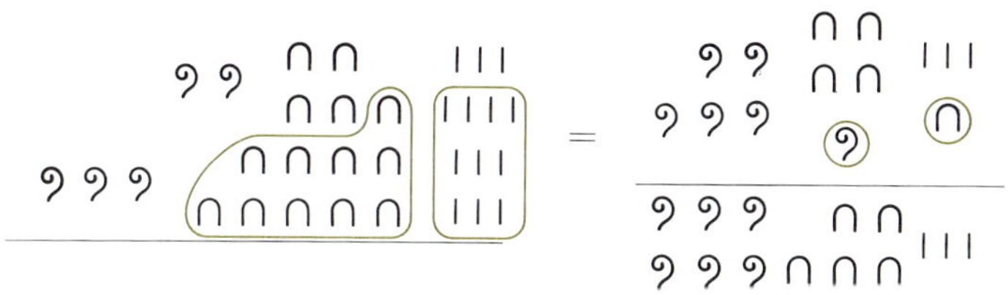

The addition is done in two steps. First we change 10 staffs | to the equivalent 1 heel bone, and then 10 heel bones to 1 scroll. This is all the simplification possible, and the answer is written by reordering all the remaining symbols. Note that when we rewrite 13 staffs as $\overset{|\,|\,|}{\cap}$, the operation is similar to our adding $7 + 6 = 13$. Here we have mentally exchanged 10 ones for 10. In hieroglyphics, addition seems to follow a natural plan. The main difficulty is the awkwardness of the symbols.

ROMAN NUMERALS

257	C C L V I I	=	C C L I I
396	C C C X C V I		C C C C I
653			D C L I I I

Addition is not difficult with Roman numerals. It is complicated somewhat if the subtraction principle is used. Otherwise, it is like hieroglyphics. In this example, the circled part reduces to C, since V + V = X and XC + X = C $(90 + 10 = 100)$. The answer follows, since five C's equal D.

CHINESE-JAPANESE SYSTEM

257
396
653

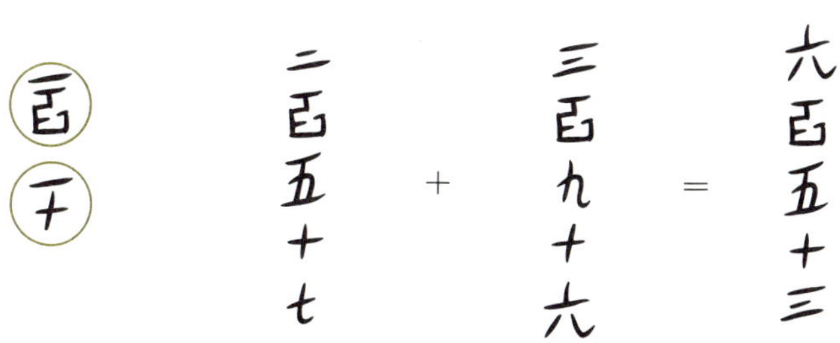

In this system note that 七 + 六 = $\genfrac{}{}{0pt}{}{十}{三}$. In other words, we write 三 (three) and carry 十 (ten). Add $\genfrac{}{}{0pt}{}{一}{十} + \genfrac{}{}{0pt}{}{五}{十} + \genfrac{}{}{0pt}{}{九}{十} = \genfrac{}{}{0pt}{}{1}{10} + \genfrac{}{}{0pt}{}{5}{10} + \genfrac{}{}{0pt}{}{9}{10} = \genfrac{}{}{0pt}{}{\genfrac{}{}{0pt}{}{1}{100}}{\genfrac{}{}{0pt}{}{5}{10}} =$ $\genfrac{}{}{0pt}{}{百}{\genfrac{}{}{0pt}{}{五}{十}}$. In the second position, we write $\genfrac{}{}{0pt}{}{五}{十}$ (5 tens) and we carry 百 (one hundred).

Add $\genfrac{}{}{0pt}{}{一}{百} + \genfrac{}{}{0pt}{}{二}{百} + \genfrac{}{}{0pt}{}{三}{百} = \genfrac{}{}{0pt}{}{1}{100} + \genfrac{}{}{0pt}{}{2}{100} + \genfrac{}{}{0pt}{}{3}{100} = \genfrac{}{}{0pt}{}{6}{100} =$ $\genfrac{}{}{0pt}{}{六}{百}$. In this case nothing has to be carried, and we write $\genfrac{}{}{0pt}{}{六}{百}$. The final answer appears above. Addition is similar to ours, the only difference being that an extra symbol has to be carried for 10 or 100 or 1000, as the case may be.

Addition can easily be performed in the Mayan system if one remembers that ___ equals ..... and knows the positional values.

MAYAN SYSTEM:

257
396
653

Adding across, beginning at the bottom, we have 33 units. Since the second position has a value of twenty, we carry 1 twenty from the 33 units and write $\overset{\cdots}{=}$ (thirteen). In the second position we add $\overset{\cdot\cdot}{=}$ and · plus the 1 twenty we carried, and we write $\overset{\cdot\cdot\cdot\cdot}{=}$. Nothing is carried to the third position, and so · is in the third position of the answer.

Our addition is like the Mayan. We must remember a few more facts, since each of our digits is a single symbol instead of a group of sticks and rocks. As mentioned before, the use of a single symbol for a digit prevents confusion in writing multidigit numbers. This easily compensates the task of little more memorization.

Over the centuries our method of addition has changed little. Hindus used two methods. One was like ours. The other, called "retrograde" or "inverse" addition, was done by beginning at the left and canceling numbers when they were no longer necessary:

```
 7529
 2530
  752
   61
 9762
1087
```

(In the printed sum, the 9 and 7 of 9762 are struck out.)

The answer is 10,872. To arrive at this answer:

1. Begin at the left and add 7 and 2. Write 9 below the line.
2. Add $5 + 5 + 7 = 17$. Write the 7 and add the 1 to the 9 from step 1; cross out the 9.
3. Add $2 + 3 + 5 + 6 = 16$. Write the 6 and add the 1 to the 7 from step 2; cross out the 7.
4. Add $9 + 0 + 2 + 1 = 12$. Write the 2 and add the 1 to the 6 from step 3; cross out the 6.
5. The numerals under the line and not crossed out constitute the answer.

The Arabs used the same method of addition as ours. The only variation was that they recorded the answer above the addends instead of below. In the sixteenth century Gemma Frisius introduced a method of addition that is still used in business today when it is necessary to add a long column of numbers. In this method each column is added separately, beginning at the right. Then these partial sums are added to get the final answer:

```
 9325
 8394
  721
   75
 ----
   15
  20
 13
17
-----
18515
```

The separate steps are easier to see when we write sums this ways. We do not have to carry, so that fewer errors occur.

2-1A EXERCISES

Translate and add 751 and 479 in:

1. Hieroglyphics.
2. Roman numerals.
3. Mayan numerals.
4. Traditional Chinese-Japanese numerals.
5. In which of the above four exercises is addition most like the familiar Hindu-Arabic addition? Why?

Translate and add 14,991 and 10,931 in:

6. Hieroglyphics.
7. Mayan numerals.
8. Use retrograde addition to add 2793 + 4281 + 6593 + 7140.

Add 9452 + 9984 + 699 using:

9. Retrograde addition.
10. The method of Gemma Frisius.
11. Make a step-by-step outline of the procedure you would follow in adding 498 and 752 using the everyday method.

ADDITION IN BASES OTHER THAN TEN

When the nature of positional numerals became fully understood, other methods of addition were dropped in favor of the one we now use. Ours is a decimal system in which each successive position has ten times the value of the previous one. If when adding a column of numbers, the sum becomes 10, we simply add 1 to the next column. For example,

$$\begin{array}{rcrcrcrcr}
795 & & & & 7\times 100 & + & 9\times 10 & + & 5\times 1 \\
\underline{847} & & & & 8\times 100 & + & 4\times 10 & + & 7\times 1 \\
\hline
 & & & & 15\times 100 & + & 13\times 10 & + & 12\times 1 \\
1642 & = & 1\times 1000 & + & 6\times 100 & + & 4\times 10 & + & 2\times 1
\end{array}$$

The sum $5 + 7 = 1 \times 10 + 2$. We write the 2 and combine the 1×10 with the other tens. One more than the sum of the tens column is 14 tens, which is 1 hundred and 4 tens. We write the 4 under the tens column and

+	0	1	2	3	4	5	6	7	8	9
0	0	1	2	3	4	5	6	7	8	9
1	1	2	3	4	5	6	7	8	9	10
2	2	3	4	5	6	7	8	9	10	11
3	3	4	5	6	7	8	9	10	11	12
4	4	5	6	7	8	9	10	11	12	13
5	5	6	7	8	9	10	11	12	13	14
6	6	7	8	9	10	11	12	13	14	15
7	7	8	9	10	11	12	13	14	15	16
8	8	9	10	11	12	13	14	15	16	17
9	9	10	11	12	13	14	15	16	17	18

FIG. 2-1 ADDITION FACTS FOR BASE 10.

+	0	1
0	0	1
1	1	10_2

+	0	1	2	3	4
0	0	1	2	3	4
1	1	2	3	4	10_5
2	2	3	4	10_5	11
3	3	4	10_5	11_5	12_5
4	4	10_5	11_5	12_5	13_5

FIG. 2-2 ADDITION FACTS FOR BASES 2 AND 5.

then add the hundreds. One more than this sum is 16 hundreds, which is 1 thousand and 6 hundreds, and the computation is completed. When actually doing addition, one is hardly aware of the various place values. One merely remembers various basic sums and the process of carrying. All addition facts can be given in a table. Figure 2-1 gives them for base 10. To use the table to add 7 and 8, locate 7 in the column on the left. The number in the same row as the 7 and under the column headed by 8 is the answer. So $7 + 8 = 15$.

In this manner we can give all the addition facts in any base. Figure 2-2 gives them for bases 2 and 5. To perform complicated addition in bases 2 and 5, we need to know exactly what to carry. Whenever the sum of a column equals or exceeds the value of the next position, 1 is added to the next column. If the sum of the first column is twice the place value of the second position, then 2 is carried to the next column. If it is three times, then 3 is carried, and so forth. However, as in base 10, one seldom dwells on the values of the positions. When the sum of a column equals the base being used, 1 is added to the next column. This becomes automatic, particularly if one is referring to a table of addition facts. For example,

$$\begin{array}{r} 23_5 \\ \underline{43_5} \\ 121_5 \end{array}$$

Referring to the addition facts for base 5, $3 + 3 = 11_5$, and we write 1 and carry 1 to the second column, as with 11 in base 10. $1 + 2 = 3$ and $3 + 4 = 12_5$. Since there is no third column, we write 12, and the answer is 121_5. More examples follow:

$$\begin{array}{r} 244_5 \\ 341_5 \\ \underline{402_5} \\ 2042_5 \end{array} \qquad \begin{array}{r} 11_2 \\ \underline{10_2} \\ 101_2 \end{array}$$

$$\begin{array}{r r c r}
 & & & \textit{Check in base 10} \\
 & 10011_2 & = & 19 \\
111_2 & 11101_2 & = & 29 \\
101_2 & 11111_2 & = & 31 \\
\underline{110_2} & \underline{10101_2} & = & \underline{21} \\
10010_2 & 1100100_2 & = & 100
\end{array}$$

The steps involved in these calculations in other bases are identical to those we use ordinarily; only the numeral combinations change. The process remains the same.

CASTING OUT NINES The usual method for checking addition in base 10 is to add in reverse order. If the first sum is found by adding from top to bottom, then the accuracy of the addition is checked by readding from the bottom up. A less precise but more easily used method is "casting out nines." To check addition by casting out nines, add the digits of the addends, discarding sums that equal 9. For example, 543,623 has an excess of nines equaling 5. This is seen by adding the digits $5+4+3+6+2+3$. The $5+4$ is discarded, since it equals 9, and so is the $3+6$. The $2+3$ is what remains after casting out nines, or the excess of nines. To check $19+12$:

$$\begin{array}{r l}
 & \textit{Excess of nines} \\
19 & 1 \\
\underline{12} & \underline{3} \\
31 & 4
\end{array}$$

Cast out nines from 19. The excess is 1. The excess of nines in 12 is 3, since $1+2=3$. Now add these two excesses: $1+3=4$. If the addition is correct, 4 should also be the excess of nines of 31, the result of the addition; that is, the sum of the excesses of nines of the addends should equal the excess of nines of the sum of the addends. Another example follows:

$$\begin{array}{r l}
 & \textit{Excess of nines} \\
\begin{array}{r} 275 \\ 593 \\ 224 \\ \underline{322} \end{array} & \left.\begin{array}{l} \left.\begin{array}{l} 5 \\ 8 \end{array}\right\}4 \\ \left.\begin{array}{l} 8 \\ \underline{7} \end{array}\right\}6 \end{array}\right\}1 \\
1414 & 1
\end{array}$$

We had to cast out nines in the addends until we reached a one-digit number. The excess of 5 and 8 is 4, since $5+8=13$ and $13-9=4$. For the same reason the excess of 8 and 7 is 6 and, finally, of 4 and 6 is 1. This checks with the excess of the answer.

A modification of casting out nines involves repeated addition of digits until a one-digit answer is obtained. For example,

Casting out nines		*Modification*			
2783	2	2783	$2 + 7 + 8 + 3 = 20$	$2 + 0 = 2$	2
4591	1	4591	$4 + 5 + 9 + 1 = 19$	$1 + 9 = 10$	$1 + 0 = 1$
2768	5	2768	$2 + 7 + 6 + 8 = 23$	$2 + 3 = 5$	5
10142	8	10142	$1 + 0 + 1 + 4 + 2 = 8$		8

Casting out nines works as an addition check for two reasons. First, if a number that 9 does not divide exactly is divided by 9, the remainder is the sum of the digits of the number. For example, $12 \div 9$ has a remainder of 3 and the sum of its digits $1 + 2 = 3$, and $25 \div 9$ has a remainder of 7 and $2 + 5 = 7$. Second, if two or more numbers are added and then the sum is divided by 9, the remainder of the sum equals the sum of the remainders of each of the addends divided by nine. For example, $12 + 19 = 31$:

$12 \div 9$	remainder 3
$19 \div 9$	remainder 1
$31 \div 9$	remainder 4

Thus the sum of the remainders of the addends, 12 and 19, equals the remainder of 31.

This discussion centers on remainders of division by 9. A small difficulty is found if the addends are exactly divisible by 9. The sum of the digits of numbers that are exactly divisible by 9 equals 9 or a multiple of 9. For example, 27 is divisible by 9 and $2 + 7 = 9$; for 81, $8 + 1 = 9$; and for 36, $3 + 6 = 9$. Since we have been finding remainders by adding the digits, this presents an inconsistency. When the remainder is 0, the digits add to 9. Therefore, to restore our discussion to remainders we must discard sums of digits that equal 9, since the remainders of such numbers are zero. Then, casting out nines is, in effect, checking addition by examining the remainders of division by 9. The sum of the remainders of the addends must equal the remainder of the answer, for addition to check.

Casting out nines is not an infallible check. For example, $19 + 12 = 22$ checks by casting out nines, although the correct answer is 31. Since casting out nines is checking by division, we could cast out any number we wished. We choose nine because it is easy to add the digits to find the remainder of division by 9. The fact that 10, 100, 1000, and so on, when divided by 9, leave a remainder of 1 is the basic reason that digits can be added to find the remainder. Since $10 \div 9$ has a remainder of 1, then 2 tens or $20 \div 9$ has a remainder of 2, and 3 tens or $30 \div 9$ has a remainder of 3, and so on. For a two-digit number made up of nonzero digits, such as 23, we can think of division as first dividing 20 then dividing 3, and combining the two answers: $20 \div 9 = 2$ with a remainder of 2, and $3 \div 9 = 0$ with a remainder of 3, and the final answer is $2 + 0$ with a remainder of $2 + 3$. Another more complicated example is 75. By casting out nines, the excess equals 3. If we divide 70 and 5 by 9, then $70 \div 9 = 7$ with a remainder of 7 and $5 \div 9 = 0$ with a remainder of 5. The answer of $75 \div 9$ must then equal $7 + 0 = 7$ with a remainder of $7 + 5 = 12$. Since 12 is larger than 9, we raise the quotient of $70 \div 9$ by 1, to 8, and reduce the remainder to 3. We have then found the remainder by adding the digits $7 + 5$ and throwing out a 9.

Let us use another divisor:

321	$321 \div 15 = 21$ with $r = 6$	$6 + 14 = 20$
674	$674 \div 15 = 44$ with $r = 14$	and $20 \div 15 = 1$ with $r = 5$ (circled)
995	$995 \div 15 = 66$ with $r = 5$ (circled)	

We see that the remainders are both 5, and we have checked the arithmetic by casting out fifteens. An easy divisor is 5, since the remainder upon division by 5 is easily found; $321 \div 5$ yields remainder 1, and $674 \div 5$ yields 4. The remainder is the last digit of the numeral if it is 0, 1, 2, 3, 4; and 5 less than the last digit if it is 5, 6, 7, 8, 9. This is because the only products possible with 5 as a factor end in 0 or 5.

2-1B EXERCISES

1. Explain the basis for carrying in addition.

Find the following sums in base 5.

2. 32_5
$\underline{12_5}$

3. 224_5
$\underline{333_5}$

4. 1034_5
$\underline{1034_5}$

5. 230_5
223_5
430_5
$\underline{324_5}$

6. 3433_5
3444_5
2012_5
$\underline{4343_5}$

Find the following sums in base 8.

7. $\begin{array}{r} 41_8 \\ 23_8 \\ 32_8 \\ \underline{12_8} \end{array}$ *8.* $\begin{array}{r} 6647_8 \\ 1043_8 \\ 7045_8 \\ \underline{325_8} \end{array}$ *9.* $\begin{array}{r} 3027_8 \\ 2104_8 \\ 5744_8 \\ \underline{1406_8} \end{array}$ *10.* $\begin{array}{r} 3461_8 \\ 2053_8 \\ 3276_8 \\ \underline{6043_8} \end{array}$

11. $\begin{array}{r} 4071_8 \\ 1036_8 \\ 741_8 \\ 2031_8 \\ \underline{32_8} \end{array}$

12. How is casting out sevens in base 8 related to casting out nines in base 10?

13. Check Probs. 7 to 11 by casting out sevens.

Find the following sums.

14. $\begin{array}{r} 111_2 \\ 111_2 \\ 111_2 \\ 101_2 \\ \underline{100_2} \end{array}$ *15.* $\begin{array}{r} 11011_2 \\ 10001_2 \\ 10101_2 \\ \underline{111_2} \end{array}$ *16.* $\begin{array}{r} 11111_2 \\ 11111_2 \\ 11101_2 \\ 111_2 \\ \underline{10101_2} \end{array}$

18. Construct a table of basic addition facts for base 12.

17. $\begin{array}{r} 101111_2 \\ 101_2 \\ 10_2 \\ 1_2 \\ 11101_2 \\ \underline{110111_2} \end{array}$

Find the following sums in base 12.

19. $\begin{array}{r} 26_{12} \\ 12_{12} \\ 51_{12} \\ \underline{30_{12}} \end{array}$ *20.* $\begin{array}{r} 62_{12} \\ 55_{12} \\ 93_{12} \\ \underline{48_{12}} \end{array}$ *21.* $\begin{array}{r} 305_{12} \\ 639_{12} \\ 795_{12} \\ \underline{84_{12}} \end{array}$ *22.* $\begin{array}{r} 503_{12} \\ 729_{12} \\ 3t4_{12} \\ eee_{12} \\ \underline{tlt_{12}} \end{array}$

23. $\begin{array}{r} eooe_{12} \\ 10te_{12} \\ 30ee_{12} \\ \underline{830t_{12}} \end{array}$ *24.* $\begin{array}{r} etet_{12} \\ teet_{12} \\ eooe_{12} \\ \underline{ttte_{12}} \end{array}$

25. Add 753 + 291 + 452 and check by:
   (*a*) Casting out fives.
   (*b*) Casting out nineteens.
26. What is wrong with checking base-10 addition by casting out fives?
27. Why is it not possible to check base-2 addition by casting out ones?

2-2 THE OPERATION OF MULTIPLICATION

MULTIPLICATION HISTORICALLY

The word *multiply* is from the Latin *multus* (many) and *plicare* (to fold). Multiplication is a shortened form of addition in which equal addends are "folded" together. Much to the pleasure of most students, the words *multiplier* and *multiplicand* are being dropped in favor of the word *factor*. This is being done because the order of multiplication is irrelevant. 4×3 is equivalent to 3×4, and it does not matter which is the multiplicand and which the multiplier.

The symbols for multiplication were slower to develop than those for addition, because addition was necessary for commerce. The early Italian mathematicians used no symbol for multiplication; they simply placed numbers next to one another. The cross was first used to indicate which parts of fractions were multiplied when dividing common fractions; that is,

$$\frac{3}{4} \times \frac{7}{8}$$

shows how one divides

$$\frac{3}{4} \div \frac{7}{8}$$

First multiply 3 and 8 (the upper left and lower right numbers); then divide by the product of 4 and 7 (the lower left and upper right):

$$\frac{3}{4} \div \frac{7}{8} = \frac{3}{4} \times \frac{7}{8} = \frac{3 \times 8}{4 \times 7} = \frac{24}{28} = \frac{6}{7}$$

Check by using the usual division process for common fractions:

$$\frac{3}{4} \div \frac{7}{8} = \frac{3}{4} \cdot \frac{8}{7} = \frac{3 \times 8}{4 \times 7} = \frac{24}{28} = \frac{6}{7}$$

The use of $\times$ as the symbol for multiplication of two whole numbers is attributed to the seventeenth-century English clergyman and mathematician William Oughtred. It is said that he died in a transport of joy on hearing the news of the restoration of Charles II. The English mathematician and logician Augustus De Morgan remarked to this, "It should be added, by way of excuse, that he was 86 years old." Some mathematicians, notably Liebniz, objected to × as a symbol for multiplication because it resembled x, the most often used symbol for an unknown in algebra. Liebniz was one of the first writers to make extensive use of the dot to signify multiplication. He also used the symbol $\frown$.

MULTIPLICATION PROCESSES

Little is known of multiplication in ancient societies. We shall discuss some of the multiplication systems which used Hindu-Arabic numerals and which evolved into the modern method.

The ancient Egyptians are believed to have used a method of multiplication that involved repeated doublings of one of the factors. This method is called "duplation," and the end result looks like this (we are multiplying 47 and 19):

1✓	19		19
2✓	38		38
4✓	76		76
8✓	152		152
16	304		608
32✓	608	$1 + 2 + 4 + 8 + 32 = 47$	$47 \times 19 = 893$

The 47 is ignored momentarily. The 19 is written with a 1 to its left. Both 1 and 19 are doubled, and the results are written below. Hence 2 and 38 appear below the 1 and 19. Then 2 and 38 are doubled, and the results are written below, yielding 4 and 76 respectively. Then 4 and 76 are doubled, and so on. To find the product of 47 and 19, check the numbers in the left column whose sum is 47; that is, $1 + 2 + 4 + 8 + 32 = 47$. Thus the answer is the sum of the numbers to the right of those which add to 47; that is, $19 + 38 + 76 + 152 + 608 = 893$.

It is easy to see the principle underlying duplation if one considers 47×19 to mean 47 nineteens: 38 is 2 nineteens, 76 is 4 nineteens, 152 is 8 nineteens, and 608 is 32 nineteens. Then when we add $19 + 38 + 76 + 152 + 608$, we are adding 1 nineteen plus 2 nineteens, plus 4 nineteens, plus 8 nineteens, plus 32 nineteens, which equals 47 nineteens—the required answer.

With a complete understanding of place value in a positional system, multiplication processes became more efficient and more like our modern

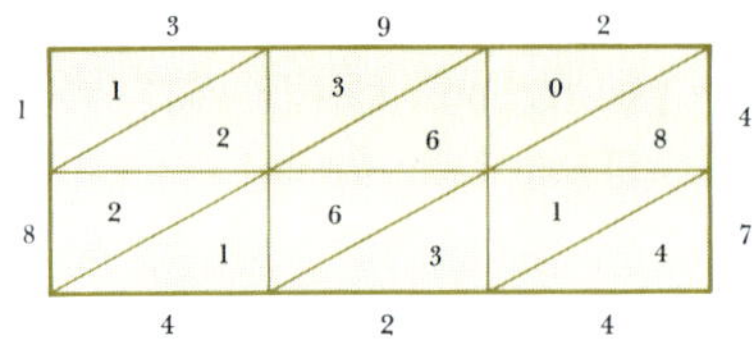

FIG. 2-3 GELOSIA MULTIPLICATION.

method. A very old method of multiplication is thought to have been invented in India. It is called "gelosia" because it uses a lattice design. A gelosia was a framework put in windows to protect ladies of the Arab world from the curious eyes of passers-by. The modern word *jalousie* is derived from gelosia because of its original protective purpose. Some modern authors call this method "galley multiplication." To multiply by gelosia (see Fig. 2-3), the two factors are placed on the outside of a rectangle. One factor is written on the top of the rectangle, the other vertically along the right side. Lines are drawn as in Fig. 2-3, which shows the multiplication 392×47. Each triangle thus formed is filled with a one-digit number. For example, the upper left corner contains 1 and 2, the result of multiplying 4 on the right by 3 on the upper left: $3 \times 4 = 12$. The 0 and 8 in the upper right are the result of the operation $2 \times 4 = 8$. The 0 is used to shown there are no tens. This position could just as well have been left empty. To get the final product, the digits are added diagonally. If a sum is 10 or larger, the second digit is carried to the next diagonal. For example, in the second diagonal from the right $8 + 1 + 3 = 12$; the 2 is written and the 1 is carried to the next diagonal $0 + 6 + 6 + 1 + 1 = 14$. Gelosia was a popular method of multiplication and probably would have remained so if the lattice had not been so difficult to print or draw.

The Hindus also used a method of multiplication that was much like retrograde addition. The Arabs called this method the "Hindu plan." To multiply 86×28 by this method, the factors are arranged as on the far left below:

								4
						3		~~3~~0
				7		2~~7~~6		2~~7~~~~6~~
		16		1~~6~~2		~~1~~~~6~~~~2~~		~~1~~~~6~~~~2~~8
28		28		28		28		28
86	next	86	then	86	now	86	finally	86

First multiply 8×2. Next the 16 is written above and to the left of 28. This is done because 16 is really $80 \times 20 = 1600$, without the zeros being written. Then $2 \times 6 = 12$. The 2 is written above the 2 of 28. The 1 is added to the 6 and 7 written above it, and then the 6 is crossed out. Now $8 \times 8 = 64$. The 4 is added to the top 2, making 6, and the 2 is crossed out. Add the 6 from the 64 to the 7, making 13. The 3 is written above the 7, and the 1 is carried, making the 2 on the left. The 1 and the 7 are crossed out. Finally multiply $6 \times 8 = 48$, record the 8, add the 4 to the 6, making 10, record the 0, and add the 1 to the 3. At the same time cross out 6 and 3. The answer is 2408.

MULTIPLICATION IN OTHER BASES

Multiplication in bases other than 10 is analogous to base-10 multiplication; the only difference is in the basic multiplication and addition facts. Figure 2-4 gives the multiplication facts for bases 5 and 2. To multiply $342_5 \times 4_5$:

$$
\begin{array}{r}
\text{(3)(1)} \\
3\;4\;2_5 \\
\underline{4_5} \\
3\;0\;2\;3_5
\end{array}
$$

Begin by multiplying $4_5 \times 2_5 = 13_5$. The 3 is written below the line, and the 1 is carried. Then $4 \times 4 = 31_5$. Adding the 1 that was carried gives $31_5 + 1 = 32_5$. The 2 is written below the line, and the 3 is carried. Then $4 \times 3 = 22_5$. Adding the 3 that was carried gives $22_5 + 3 = 30_5$. The 30 is written below the line, and the multiplication is complete. More examples follow:

$$
\begin{array}{r}
24_5 \\
\underline{31_5} \\
24 \\
\underline{132} \\
1344_5
\end{array}
\qquad
\begin{array}{r}
233_5 \\
\underline{42_5} \\
1021 \\
\underline{2042} \\
21441_5
\end{array}
\qquad
\begin{array}{r}
110_2 \\
\underline{11_2} \\
110 \\
\underline{110} \\
10010_2
\end{array}
\qquad
\begin{array}{r}
1011_2 \\
\underline{101_2} \\
1011 \\
\underline{10110} \\
110111_2
\end{array}
$$

×	0	1	2	3	4
0	0	0	0	0	0
1	0	1	2	3	4
2	0	2	4	11_5	13_5
3	0	3	11_5	14_5	22_5
4	0	4	13_5	22_5	31_5

×	0	1
0	0	0
1	0	1

FIG. 2-4 MULTIPLICATION TABLES OF BASES 5 AND 2.

The best way to check multiplication is to reverse the order of multiplication, that is, change multipliers. However, multiplication can easily be checked by casting out nines. The excesses of nines of the factors are multiplied, and the result equals the excess of nines of the product, if the factors have been correctly multiplied. For example,

	Excess		
257	5		
38	2		
2056			
711			
9766	1	$5 \times 2 = 10$	excess 1

The excess of nines of 257 is 5, and that of 38 is 2. Since this is a multiplication problem, 5 and 2 are multiplied; and the excess of nines of this product is 1. This checks with the excess of nines of the answer 9766. As in addition, casting out nines to check multiplication is not sure. If, in the above example, one assumes the answer to be 8866, it will check:

	Excess		
257	5		
38	2		
8866	1	$5 \times 2 = 10$	excess 1

2-2 EXERCISES

1. Multiply 56×32 by the duplation method.

2. Use gelosia to find the following products.

(*a*) 425×73 (*b*) 5482×503 (*c*) 8791×952

3. Multiply the following by the Hindu plan.

(*a*) 42×34 (*b*) 79×86

4. How is gelosia related to the modern method of multiplication?

5. Make a multiplication table for base 8.

6. Repeat Exercise 5 for base 12.

7. Multiply the following in base 8.

(*a*) 52_8 × 64_8 (*b*) 407_8 × 607_8 (*c*) 453_8 × 10_8 (*d*) 2735_8 × 461_8

8. Multiply in base 12 and check by casting out elevens.

(*a*)	(*b*)	(*c*)	(*d*)
te_{12}	eoe_{12}	$tete_{12}$	$etet_{12}$
$\underline{e9_{12}}$	$\underline{tot_{12}}$	$\underline{10_{12}}$	$\underline{8t7t_{12}}$

9. The table from Exercise 5 is symmetric about a line drawn from the upper left corner to the lower right corner. Why?

10. In "Through the Looking-Glass . . ." Alice is confused. To reorient herself, she starts to recite the things she knows: "Let me see; four times five is twelve, and four times six is thirteen, and four times seven is fourteen . . . oh dear. I shall never get to twenty at that rate." Why will Alice not get to twenty in six more multiplications?

2-3 THE OPERATION OF SUBTRACTION

SUBTRACTION HISTORICALLY

The earliest recorded subtraction symbol was found in the Ahmes papyrus of Egypt and denoted legs walking backward. Later the Babylonians used /‾ to indicate subtraction in their numerals. The Romans indicated subtraction by placing a smaller numeral to the left of a larger one: IX = 10 − 1 = 9. Later, in Italy, the letter *m* (for minus) was used as the subtraction symbol. To eliminate confusion, $\overline{m}$ was also used. Finally the − was used without the *m*. As with the plus sign, the first use of the minus sign was not for the subtraction of two numbers but to show that sacks were lighter than they should be.

The words *minuend* and *subtrahend* come from the Latin *numerus minuendus* (number to be diminished) and *numerus subtrehendus* (number to be subtracted). These words are rarely used today, even in elementary arithmetic. Subtraction, like addition, has been known by many names: diminution, extraction, detraction, and subduction.

SUBTRACTION PROCESSES

Subtraction is not so standardized as multiplication and addition. In buying groceries and conducting other business transactions, it is conceivable that you could encounter three different subtraction processes. All three were known in Europe as early as 1100. The first is the "make-change" method of subtraction. As in making change, we add from the number to be subtracted to the larger number. Thus to subtract 103 − 85, we start at 86 and count to 103; that is, 85 plus 15 make 100, and 3 more are 103. Thus 103 − 85 is 15 and 3, or 18.

We could also subtract 103 − 85 by the "equal-additions" method:

$$\begin{array}{r} 1\;0\;{}^{1}3 \\ \underline{{}_{9}\not{8}\;5} \\ 1\;8 \end{array}$$

First we add 10 to 103 and subtract 5 from 13. Then we add 10 to 85, calling it 95, and subtract 9 from 10. This method increases the top number by 10 and then removes the additional 10 from it by adding 10 to the number to be subtracted.

The third method is called "borrowing";

$$\begin{array}{r} \overset{9}{\not{1}\not{0}}\;{}^{1}3 \\ \underline{8\;5} \\ 1\;8 \end{array}$$

Here, 103 is regrouped into 9 tens and 13 units. Then 5 is subtracted from the 13 units and 8 tens are subtracted from the 9 tens.

The reason that subtraction has not settled on one process is that none of them is clearly superior to the others. If one is initially trained in any one of these methods, he can subtract about as quickly as one who is trained in either of the others. One of the authors of this book learned to subtract by equal additions, the other by borrowing. We have not been able to determine who is the faster subtractor.

SUBTRACTION IN OTHER BASES

Subtraction is defined in terms of addition and is said to be the "inverse" of addition. Thus $7 - 2 = 5$ because $5 + 2 = 7$. The same reasoning applies in other bases: $(10)_{12} - t = 2$ because $t + 2 = (10)_{12}$.

For most subtraction we use borrowing. For example, in base 5,

$$\begin{array}{r} \overset{3}{\not{4}}\;\overset{12}{\not{3}}\;{}^{1}1_5 \\ \underline{-2\;4\;2_5} \\ 1\;3\;4_5 \end{array} \qquad \begin{array}{r} 4\;3\;1_5 = 1\;1\;6 \\ \underline{2\;4\;2_5} = \underline{-7\;2} \\ 1\;3\;4_5 = \;\;4\;4 \end{array}$$

Subtraction in base 5 is similar to that in base 10. Borrowing or regrouping is done in the same manner as in base 10. One must be careful to remember the positional values of each base. In this example, we must first subtract 2 from 1. To do this, we borrow from the 3 and make it 2. Then we subtract 2 from 11_5. We are in base 5, and $11_5 - 2 = 4$. Next we subtract 4 from 2. To do this, we regroup again; the top number is now 3 twenty-fives and $12_5(7)$ fives. Subtracting 4 from 12_5, we get the second digit of the answer, 3. Finally $3 - 2 = 1$, and the subtraction is complete.

Another example, this time in base 7, follows:

5 12 12			
$\not6\,\not3\,\not3^{1}5_7$	$6\,3\,3\,5_7$	5	Check by casting out sixes
$2\,4\,5\,6_7$	$2\,4\,5\,6_7$	5	
$3\,5\,4\,6_7$	$3\,5\,4\,6_7$	0	$5 - 5 = 0$

In this example, since the subtraction is in base 7, we may check by casting out sixes. When we check subtraction by casting out, the excesses are subtracted, as in checking addition they are added and in checking multiplication they are multiplied.

A fourth method of subtraction was used in Europe around 1000. It was known as "subtraction by complements." By this method, to subtract $15 - 8$, we add 2 to 15 and then subtract 10. A method similar to this is used today in computers. It is called the "nines-complement" method. When used in base 10, it is only an interesting game. However, it is the method that many computers must use to do subtraction. The nines complement of any number is found by subtracting it from a number with the same number of digits all being nines. Thus the nines complement of 21 is $99 - 21$ or 78, and

Number	*Nines complement*
865	$999 - 865 = 134$
48,932	$99{,}999 - 48{,}932 = 51{,}067$
70,036	$99{,}999 - 70{,}036 = 29{,}963$

To subtract by the use of nines complement, consider the following:

$$\begin{array}{r} 6\,3\,4 \\ -3\,2\,5 \\ \hline \end{array}$$

Replace 3 2 5 by its nines complement (674):

$$\begin{array}{|r} 6\,3\,4 \\ 6\,7\,4 \\ \hline \end{array}$$

Now add and draw lines as shown:

$$\begin{array}{r|r} & 6\,3\,4 \\ & 6\,7\,4 \\ \cline{2-2} 1 & 3\,0\,8 \end{array}$$

Remove the 1 in the answer and add as shown:

```
     |6 3 4
     |6 7 4
  (1)|3 0 8
   └──→  1
     3 0 9
```

Thus the answer is 309, and this checks; $634 - 325 = 309$. We can see how this method works by examining what happens if we apply it to the subtraction of two general numbers, $a - b$. First, we add the nines complement of b to a; thus $a + (999 - b)$. By referring to the numerical example above, note that we removed 1 from the answer and then added 1 to the units position. The first 1 that was removed was in the thousands position, and, in effect, we subtracted 1000 and added 1; thus $a + (999 - b) - 1000 + 1$. But subtracting 1000 and adding 1 is the same as subtracting 999, so that the result is $a + (999 - b) - 999$. Thus we are adding 999 and subtracting 999 at the same time. The 999s cancel each other, and we see that the subtraction reduces to $a - b$. Therefore, the nines-complement method gives the same answer as any other subtraction process:

```
  7 6 4 8     complement of 5962 is 4037        |7 6 4 8
 -5 9 6 2                                       |4 0 3 7
                                             (1)|1 6 8 5
                                              └──→    1
                                                |1 6 8 6
```

and

```
 1 8 3 4 2    complement of 798 is 99201        |1 8 3 4 2
 -   7 9 8                                      |9 9 2 0 1
                                             (1)|1 7 5 4 3
                                              └──→      1
                                                |1 7 5 4 4
```

In the second example, we are subtracting numbers of unequal digits, and the smaller number must be supplemented with zeros so that it has the same number of digits as the larger one. Thus 798 is considered to be 00798, and its complement is $99999 - 00798 = 99201$.

We said that a computer used the complement method to do subtraction. You might wonder why it does not subtract by a more usual process, since it must subtract two numbers to get a number's complement. However, remember that computers deal in binary numbers, and to do subtraction, they use the "ones complement." The ones complement of a binary number is found by subtracting it from a number of equal digits all being ones:

Number	*Ones complement*
10_2	$11_2 - 10_2 = 01_2$
1100110_2	$1111111_2 - 1100110_2 = 0011001_2$
1010_2	$1111_2 - 1010_2 = 0101_2$

In the simplest case, the ones complement of 0 is 1, and the complement of 1 is 0. In a computer, 1 and 0 are indicated by a switch being on or off. Thus a computer can create the ones complement of any number by reversing the settings of all the switches that make up the number. Let us subtract 10111 − 111 by the complement method. The answer is clearly 10000:

$$\begin{array}{r} 1\,0\,1\,1\,1_2 \\ -\quad 1\,1\,1_2 \\ \hline \end{array} \qquad \text{complement of } 00111 \text{ is } 11000 \qquad \begin{array}{r|r} & 1\,0\,1\,1\,1 \\ & 1\,1\,0\,0\,0 \\ \hline \textcircled{1} & 0\,1\,1\,1\,1 \\ \hookrightarrow & 1 \\ \hline & 1\,0\,0\,0\,0 \end{array}$$

Therefore a computer that can add and create the ones complement can also subtract. It can also multiply using repeated additions and divide by using repeated subtractions.

2-3 EXERCISES

Subtract the following base-5 numbers. Check by casting out fours.

1. $\begin{array}{r} 331_5 \\ \underline{221_5} \end{array}$ *2.* $\begin{array}{r} 413_5 \\ \underline{242_5} \end{array}$ *3.* $\begin{array}{r} 102_5 \\ \underline{34_5} \end{array}$

4. $\begin{array}{r} 300_5 \\ \underline{123_5} \end{array}$ *5.* $\begin{array}{r} 303_5 \\ \underline{232_5} \end{array}$ *6.* $\begin{array}{r} 10000_5 \\ \underline{4444_5} \end{array}$

Find the following differences in base 12. Check by casting out *e*'s.

7. $\begin{array}{r} 6755 \\ \underline{2531} \end{array}$ *8.* $\begin{array}{r} 7853 \\ \underline{2446} \end{array}$ *9.* $\begin{array}{r} 9033 \\ \underline{2648} \end{array}$

10. 4*et*1 − 1*te*0

11. 4*et*1 − 3*tee*

12. 7*ttt* − 3*eee*

13. 100000 − 4*et*17

Find the following differences in base 2. Check by converting to base 10.

14. 111 − 11

15. 101 − 11

16. 1000 − 111

17. 10101 − 1010

Subtract the following base-10 numbers by the nines-complement method.

18. 654 − 421

19. 4000 − 732

20. 99432 − 15306

21. 35204 − 9911

22. 10000 − 9999

Subtract the following base-2 numbers by the ones-complement method.

23. 111 − 11

24. 101 − 11

25. 1000 − 111

26. 10101 − 1010

Subtract the following, using the complement method.

27. $9e49_{12}$ − 3541_{12}

28. $9e49_{12}$ − $43et_{12}$

29. 4000_5 − 1324_5

30. 4321_5 − 401_5

31. In Europe in the Middle Ages the most popular term for the result of subtraction was "the rest." Give two examples to show how this term could be used to indicate the result of subtraction.

2-4 THE OPERATION OF DIVISION

DIVISION HISTORICALLY

Even in the Middle Ages, division was considered difficult. It was thought that, if a student could do division, then he knew all arithmetic.

The symbol ÷ was used in Europe for subtraction. It is believed that it was first used to indicate division by Italian algebraists. However, the first time it appeared in print as a division symbol was in "Teusche Algebra" by Johann Heindrich Rahn, published in 1659.

DIVISION PROCESSES

The oldest division process was used by the Egyptians. It is called duplation and follows the same pattern as multiplication by duplation. For example, to solve $51 \div 8$:

1	8			
2	16✓	$51 \div 8 = 6$	remainder	51
4	32✓			-48
8	64			3

Write down 1 and the divisor 8, momentarily ignoring 51. Double 1 and 8. This gives 2 and 16. Then double 2 and 16, and so on. The doubling process continues until the last number under the divisor is larger than 51. Now check the numbers in the second column (headed by 8) that add to 51 or that add closest to 51 without going over—in this case, 16 and 32. The answer to the division is the sum of the numbers next to the ones checked: $2 + 4 = 6$. Thus $51 \div 8 = 6$. To find the remainder, add $16 + 32 = 48$ and subtract: $51 - 48 = 3$. The remainder is 3. This method is easily understood if we consider 16 as 2 eights and 32 as 4 eights. Then $16 + 32 = 48$ is the same as saying that 2 eights plus 4 eights are 48, or 6 eights are 48, and 3 more are 51. Hence $51 \div 8 = 6$ with remainder 3.

A method of division that was very suitable for use with an abacus or sand table was called "galley" or "scratch" division. It is reminiscent of retrograde addition and Hindu-plan multiplication and, like these processes, it is Hindu in origin. We shall divide $19432 \div 83$ by the galley method. First, we write 83 under the dividend:

```
1 9 4 ¦ 3 2 |
  8 3 ¦     |
```

The dotted line indicates that we shall not be concerned with 32 during the first phase of the division. The space to the right of solid line is the answer space. Second, divide 83 into 194:

```
194 ÷ 83 = (2)
1 9 4 ¦ 3 2 | (2)
  8 3 ¦     |
```

Then multiply 2×8 and subtract 16 from 19, $19 - 16 = 3$. The 3 is placed above the 19. We are through with 19 and 8, so cross them off:

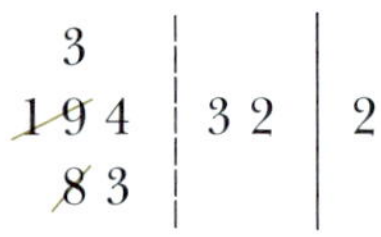

Now multiply 2 × 3 and subtract 6 from 34. We are through with 34 and 3, so cross them off. 34 − 6 = 28 and 28 is written above 34:

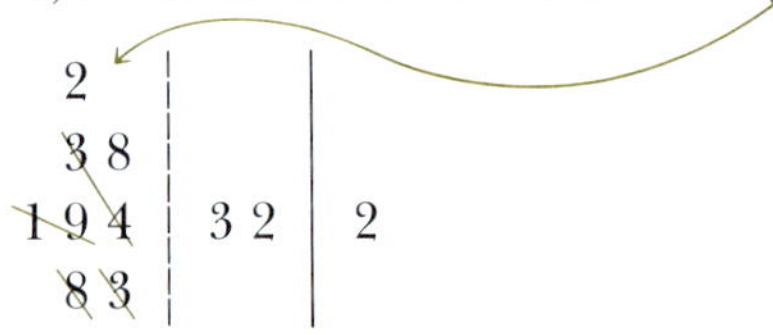

This completes the first phase. The next division by 83 involves 2832—the numbers that have not been crossed out. The next division is 283 ÷ 83, and we draw a dotted line to show that 2 is being momentarily ignored; we rewrite 83 next to the dotted line, making

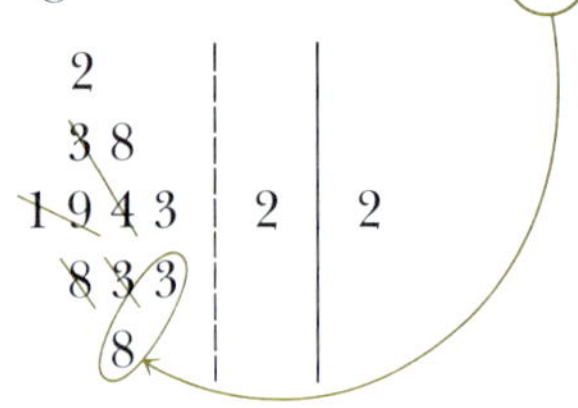

The partial quotient ($83\overline{)283}$) is 3. Write it next to 2:

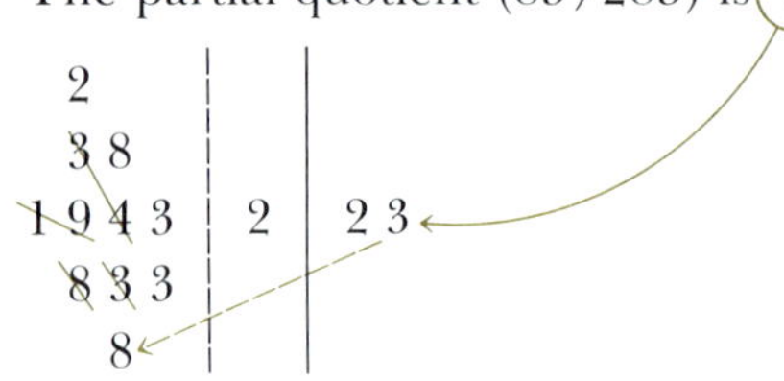

Multiply 3 × 8 and subtract 24 from 28. Then cross out the 8 and 28:

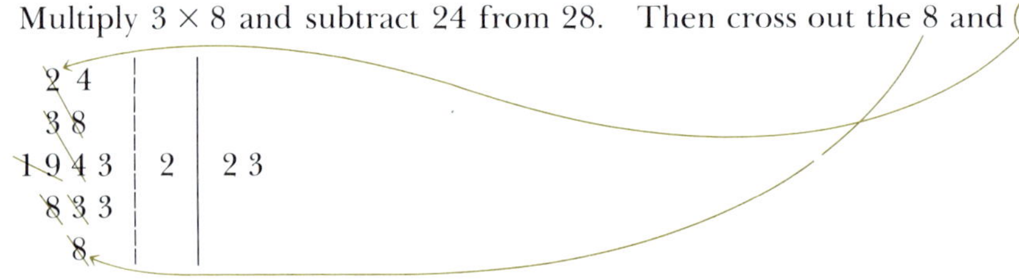

Now multiply 3 × 3 and subtract 9 from 43, write 34, and cross out 3 and 43. The second phase of division is complete. Finally, we have 342 as the only numerals not crossed out. When we divide this by 83, the division will be finished. The final (83) is now written next to the solid line:

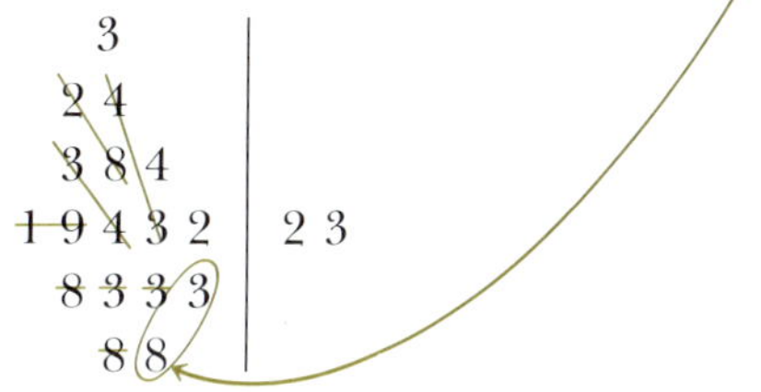

Divide 83 into 342, and write the answer next to 3 in the answer space:

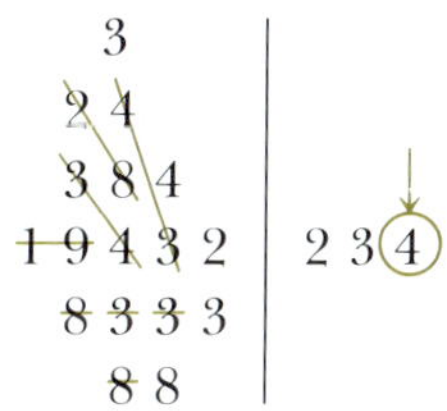

Multiply 8 × 4 and subtract 32 from 34, cross out 34 and 8, writing the (2). (34 − 32 = 2).

Lastly, multiply 4 × 3 and subtract 12 from 22. Cross out 22 and 3, and write the 10:

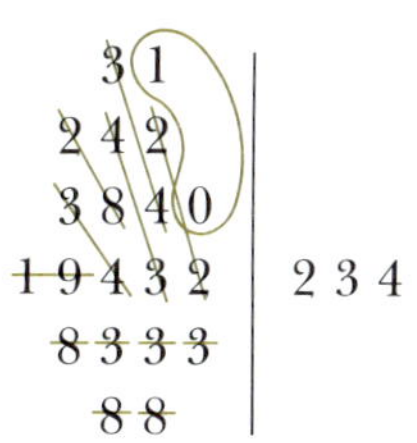

The 10 is the remainder, and 19432 ÷ 83 = 234 with remainder 10. Of course, the problem is normally not written over and over, as we have done it here. It is written only once. The entire problem and answer would appear as the last step of the example. The process is much simpler than it seems. Because subtraction is done mentally, it requires less writing than the usual process.

Another example of galley division, done in the normal way, is the following, in which we divide 17369 ÷ 24:

```
    1      |
   1 2     |
  3 5 8 7  |
1 7 3 6 9  |  7 2 3
  2 4 4 4  |
   2 2     |
```

17369 ÷ 24 = 723 with remainder 17.

Galley division was very popular in Europe until the eighteenth century. It is believed to have maintained its popularity because it is compact; it requires less writing than other processes, and it is adaptable to the abacus. Galley was the most popular method of division although our modern process began to evolve in the fifteenth century.

DIVISION IN OTHER BASES

Division is the opposite or inverse of multiplication and is defined in terms of multiplication. We say that $12 \div 3 = 4$ because $3 \times 4 = 12$ or that $29 \div 3 = 8$ with remainder 5 because $3 \times 8 + 5 = 29$. Generally $p \div q = a$ with remainder r if and only if $p = q \times a + r$. Thus p is called the dividend, q the divisor, and a the quotient. In the expression $12 \div 3 = 4$, then, 12 is the dividend, 3 the divisor, and 4 the quotient.

Our division process is based on a refined guess. We guess, using our experience with numbers and multiplication tables, and then check to see if our guess is accurate. Thus to divide 222 by 27,

```
      8
27 ) 222
```

we write 8 as the possible answer and then multiply to see if it is correct:

```
      8
27 ) 222
     216
     ---
       6
```

$8 \times 27 = 216$; thus there is a remainder of 6. If the remainder had been larger than 27, we should have had to guess again, because this would mean that there is at least one more 27 in the dividend. To divide 954 by 27, we note first that the quotient will have two digits. We proceed as follows:

First

$$\begin{array}{r} 3 \\ 27\overline{)954} \\ \underline{81} \\ 14 \end{array}$$

Second

$$\begin{array}{r} 35 \\ 27\overline{)954} \\ \underline{81} \\ 144 \\ \underline{135} \\ 9 \end{array}$$

The method is again based on guessing and checking. However, there is the added refinement that the first division involves only the 95 of 954. The reason that the first two digits are used can be seen by dividing $540 \div 27$. The answer is, of course, 20. However if, for some reason, we knew that the 0 and 20 would later have to be replaced, we could write the 20 as 2_ and fill in the blank later. In the same manner, when we first divide 27 into 954, the 3 in the answer is really 30 without the 0. The 81 under the 95 of 954 is actually 810 without the 0. We bring down the 4 because we are subtracting 810 from 954. Thus the division process has broken 954 into the sum $810 + 135 + 9 = (30 \times 27 + 5 \times 27) + 9$ and shown that there are 30 and 5 or 35 twenty-sevens in 954, with 9 units left over.

This process works with any positional numeral system; that is, it does not depend on the fact that we use base-10 numerals. Thus, to divide in base 5, proceed as above. Since division depends on remembering multiplication tables, we refer to Fig. 2-5. To divide $24_5 \div 4$,

$$\begin{array}{r} 3 \\ 4\overline{)24_5} \\ \underline{22_5} \\ 2 \end{array}$$

×	0	1	2	3	4
0	0	0	0	0	0
1	0	1	2	3	4
2	0	2	4	11_5	13_5
3	0	3	11_5	14_5	22_5
4	0	4	13_5	22_5	31_5

FIG. 2-5 MULTIPLICATION TABLE OF BASE 5.

We examine the multiplication table to see what times 4 gives 24_5, and we find $4 \times 3 = 22_5$ is as close to 24_5 as we can get; the answer is 3 with remainder 2. For a more complicated example, consider $342_5 \div 21_5$:

First

$$\begin{array}{r@{\,}l} & 1 \\ 21_5 & \overline{)\,342_5} \\ & \underline{21} \\ & 13 \end{array}$$

Second

$$\begin{array}{r@{\,}l} & 13 \\ 21_5 & \overline{)\,342_5} \\ & \underline{21} \\ & 132 \\ & \underline{113} \\ & 14_5 \end{array}$$

First, we divide 21_5 into the 34_5 of 342_5. The partial quotient is 1; we multiply and subtract as shown. Second, we divide 21_5 into 132_5. Here is where experience in multiplication is useful. To most of us the answer to $132_5 \div 21_5$ is not immediately obvious. However, after a little thought, 3 seems to be a likely trial quotient, since $2 \times 21_5$ is 42_5. Then we multiply $21_5 \times 3 = 113_5$ and subtract 113_5 from 132_5, leaving a remainder of 14_5. To check,

$$\begin{array}{r@{\,}l} & 8 = 13_5 \\ 11 & \overline{)\,97} \\ & \underline{88} \\ & 9 = 14_5 \end{array}$$

Checking is accomplished by changing 21_5 and 342_5 to base 10. Then $21_5 = 11$ and $342_5 = 97$. The quotient of $97 \div 11$ is 8 with remainder 9. This checks, since $13_5 = 8$ and $14_5 = 9$.

2-4 EXERCISES

Divide by duplation.

1. $88 \div 7$ *2.* $342 \div 10$ *3.* $7004 \div 23$

Divide by the galley method.

4. $3852 \div 14$ *5.* $14{,}357 \div 92$ *6.* $473{,}521 \div 321$

7. How do you check division by casting out nines? (*Hint:* consider how one checks division.)

Divide the following, and check by casting out nines.

8. $23\overline{)1327}$ *9.* $118\overline{)35{,}731}$ *10.* $805\overline{)953{,}211}$

Divide the following, and check by casting out the appropriate number.

11. $12311_5 \div 43_5$ *12.* $13232_4 \div 22_4$
13. $6029_{12} \div 3t_{12}$ *14.* $36336_7 \div 56_7$

Divide the following, and check by changing to base 10.

15. $10_2\overline{)10110_2}$ *16.* $110_2\overline{)1111_2}$
17. $101_2\overline{)101101101_2}$

18. If one number divides another evenly, the remainder is 0. In this special case $p \div q = s$ is an example of an even division. $p \div q = s$ if and only if $p = q \times s$. Because of this definition of division, mathematicians say that division by 0 is undefined. Why is this so; that is, why is $4 \div 0$ undefined?

2-5 CHANGING NUMBER BASES

Perhaps the most efficient method of changing bases is division. To understand this method, consider changing 482 to base 5. Remember that the positional values in base 5 are units, fives, twenty-fives, one hundred and twenty-fives, and so on.

$482 \div 5 = 96$ with remainder 2. This means that there are 96 fives in 482 and 2 units. Therefore, the base-5 number will end in 2 units. The 96 fives will be distributed to make fives, twenty-fives, and one hundred and twenty-fives; the other positional values of base 5. If we rearrange the 96 fives into groups of five, we find the number of twenty-fives in the 96 fives. To do so, we divide the 96 fives by 5. Thus $96 \div 5 = 19$ with remainder 1. This means that 96 fives contain 19 twenty-fives and 1 five. The 19 twenty-fives will be broken up into groups of one hundred and twenty-fives and twenty fives. Since 5 twenty-fives equal 125, we can divide the 19 twenty-fives by 5 to find the number of one hundred and twenty-fives it contains. $19 \div 5 = 3$ with remainder 4. The 19 twenty-fives can be thought of, then, as 3 one hundred and twenty-fives and 4 twenty-fives. Thus 482 contains 3 one hundred and twenty-fives, 4 twenty-fives, 1 five, and 2 units; that is, $482 = 3412_5$.

The actual process, in base 10, does not require an understanding of each step or even number bases or positional values. In using the division process to change bases, one simply divides by the new base repeatedly until the quotient equals zero. The remainders, in reverse order of their appearance, are the answer. Thus, to change 482 to base 5,

$$\begin{array}{rl}
0 & \\
5\,)\overline{3} & 4 \\
5\,)\overline{19} & 1 \\
5\,)\overline{96} & 2 \;\downarrow \\
5\,)\overline{482} & \qquad 482 = 3412_5
\end{array}$$

First write 482 and divide it by 5, writing the remainder on the right. Divide 96, then 19, then 3, by 5, each time recording the remainder on the right. The base-5 number is found by reading the remainders from top to bottom. To change 725 to base 8,

$$\begin{array}{rl}
0 & 1 \\
8\,)\overline{1} & 3 \\
8\,)\overline{11} & 2 \\
8\,)\overline{90} & 5 \;\downarrow \\
8\,)\overline{725} & \qquad 725 = 1325_8
\end{array}$$

To check,

$$\begin{aligned}
1325_8 &= 1 \times 512 + 3 \times 64 + 2 \times 8 + 5 \\
&= 512 + 192 + 16 + 5 = 725
\end{aligned}$$

To change 1630 to base 12,

$$\begin{array}{rr}
0 & 11 = e \\
12\,)\overline{11} & 3 = 3 \\
12\,)\overline{135} & 10 = t \;\downarrow \\
12\,)\overline{1630} & \qquad 1630 = e3t_{12}
\end{array}$$

For bases larger than 10, two-digit remainders must be changed to the single symbol of the base being used.

This method does not depend on a base of 10 and can be used to change from any base to any base. Thus, to change from one base to another, divide by the value of the base to which you wish to change. The answer will be the remainders, written in reverse order of their appearance, after they have been changed to their equivalent form in the new base. Thus, to change 11034_5 to base 10, we must divide by 10, which is 20_5:

	0	$R = 12_5 = 7$
third	$20_5 \overline{)\,12_5}$	$R = 11_5 = 6$
second	$20_5 \overline{)\,301_5}$	
	$\underline{20_5}$	
	101_5	
	$\underline{40_5}$	
	11_5	
	301_5	$R = 14_5 = 9$
first	$20_5 \overline{)\,11034_5}$	
	$\underline{110_5}$	
	34_5	
	$\underline{20_5}$	
	14_5	

$11034_5 = 769$

For the second division, we had to move 301 up, because division in base 5 is not so familiar to us as in base 10, and we could not "see" the quotient. We had to work it out by long division. To change 110110_2 to base 4, divide by $4 = 100_2$:

0	$R = 11_2 = 3_4$
$100\overline{)\,11}$	$R = 1_2 = 1_4$
$100\overline{)\,1101}$	$R = 10_2 = 2_4$
$100\overline{)\,110110}$	
$\underline{100}$	
101	
$\underline{100}$	
110	
$\underline{100}$	
10	*Thus:* $(110110)_2 = (312)_4$

To check, $110110_2 = (32 + 16 + 4 + 2) = 54$

$312_4 = 3 \times 16 + 4 + 2 = (48 + 4 + 2) = 54$

2-5 EXERCISES

Change the following base-10 numbers to the indicated base by division.

1. $724 = (\quad)_4$ *2.* $114 = (\quad)_3$ *3.* $191 = (\quad)_{12}$

4. $19 = (\quad)_2$ *5.* $2169 = (\quad)_{13}$

Make the following conversions of bases by division.

6. $1317_{10} \to (\qquad)_{12}$
7. $1011001_2 \to (\qquad)_5$
8. $1011001_2 \to (\qquad)_{10}$
9. $1011001_2 \to (\qquad)_{12}$
10. $2374_8 \to (\qquad)_{10}$
11. $2374_8 \to (\qquad)_5$
12. $2374_8 \to (\qquad)_2$
13. $2374_8 \to (\qquad)_{12}$

14. Why can't we change base-10 numbers to base-5 by dividing them by 2?

2-6 THE DIFFERENCE BETWEEN OPERATION AND PROCESS

In this chapter we have dealt mainly with some of the processes used to add, subtract, multiply, and divide. There are many more that could have been discussed. With most processes, as you probably noticed, you need have no understanding of exactly why they work. You need only know such things as when to carry and where to place certain intermediate results and when to add and so on. This mechanical method of finding results is important and is discussed fully in Chap. 3. Any method of finding the result of one of the arithmetic operations is based on the special characteristics of the operation. Let us look briefly at the arithmetic operations themselves.

Normally we deal with a number of operations we learned in our early training in arithmetic. We are introduced first to addition, then to subtraction, then to multiplication, and finally to division. We are told that they are all simply variations of the basic operation of addition, with multiplication being repeated addition, subtraction being the reverse of addition, and division being the reverse of multiplication. However, to a mathematician they mean something else. A mathematician says that these four operations are examples of "binary compositions," that is, mappings of $S \times S$ into S. Although this expression is meaningless to you now, it is only a compact way of saying, in the technical language of mathematics, that these operations are rules for associating a third number with a pair of numbers taken in a specific order. For example, the operation of division associates the numeral 5 with the pair 20 and 4, as in $20 \div 4 = 5$, and the numeral $\frac{1}{5}$ with 4 and 20, as in $4 \div 20 = \frac{1}{5}$. Note that the order of the pair of numbers makes a difference. Looking at addition, we note that 9 is associated with 5 and 4, as in $5 + 4 = 9$, and also with 4 and 5, as in $4 + 5 = 9$. The order of the pair in addition does not matter. Mathematicians say that the binary composition of addition on numbers is "commutative" (changing the order does not change the results) but division is not (the order of the pair must be considered).

If we went on at this point to study algebra, we might examine "associativity," another property of the operation of addition not shared by division. Associativity is the name given to the fact that, if we add 180 and 6 and then add 3 to the sum, we get the same result as if we added the 6 and 3 first and then added the sum to the 180. Using parentheses to mean do this first, we write this in the following mathematical sentence:

$(180 + 6) + 3 = 180 + (6 + 3)$ associativity of addition

If we try the same thing with division, we find that it does not have this property; that is, $(180 \div 6) \div 3$ equals 10—remember that $(180 \div 6)$ must be done first, then the result divided by 3; thus we have $180 \div 6 = 30$ and then $30 \div 3 = 10$—but $180 \div (6 \div 3)$ equals 90, because $6 \div 3 = 2$ and then $180 \div 2 = 90$. What we wish to point out here is that if a, b, and c are any numbers, then $a + b = b + a$ (commutative property) and $(a + b) + c = a + (b + c)$ (associativity property). Remember that parentheses mean do this first. We can use the letters in place of numbers because these are properties of the operations, not properties of or dependent on the way we write numerals; that is, we have seen that our rules for finding the sum, difference, quotient, and product of two numbers are based on the method we use for writing numbers, but the nature of the sum, product, and so on, depends on the nature of numbers themselves. If we say that $31 + 64 = 95$ or ∩∩∩| added to ∩∩∩∩∩∩|||| yields ∩∩∩∩∩∩∩∩∩|||||, the nature of the sum does not change simply because we write the problem in hieroglyphics rather than the more usual Hindu-Arabic.

All the processes for multiplying two numbers depend on a property of addition and multiplication called the "distributive law." Informally this property says that, when multiplying two numbers, such as 16×5, it is possible to multiply 16×3 and then 16×2 and add the two products:

$$16 \times 5 = 16 \times 3 + 16 \times 2 = 48 + 32 = 80$$

Recall that a process very similar to this is used in duplation. The reason that this is called the distributive law can be seen by writing the above in usual form:

$$16 \times (3 + 2) = 16 \times 3 + 16 \times 2$$

Another example follows:

$$3 \times (7 + 6) = 3 \times 7 + 3 \times 6 = 21 + 18 = 39$$

Letting a, b, c, represent any numbers, the distributive law is

$a \times (b + c) = a \times b + a \times c$

It states that, when multiplying one number by the sum of two others, it is possible to get the correct answer by multiplying the first number by each of the others and then adding the products. A phrase mathematicians sometimes use to describe this is that multiplication distributes over addition.

The rules for finding the results of the arithmetic operations are based on the fact that we use Hindu-Arabic as our method for symbolizing numbers. Other symbol systems use other rules. Indeed, we examined a number of other methods, different from the usual ones we all learned in school, for finding sums and products in Hindu-Arabic notation.

When mathematicians began to separate these ideas, that is, when they began to see the difference between the operation and the process for finding the result of the operation, they gained a greater insight into the workings of our number system.

In this chapter we used the fact that $a + b = b + a$ and $a \times b = b \times a$ and $(a \times b) \times c = a \times (b \times c)$ and $a \times (b + c) = a \times b + a \times c$. For if these properties of multiplication and addition did not hold, we should not be able to find sums and products using the processes we do. However, we did not examine how these procedures depend on these laws of arithmetic; rather, we looked only at procedures for calculating results. Also, in Chap. 1 we emphasized the difference between numbers and numerals. If we always noted this difference, this text would be full of statements like "When the number whose Hindu-Arabic symbol is 3 is added to the number whose Hindu-Arabic symbol is 5, the result is the number whose Hindu-Arabic symbol is 8." Instead, we use the usual "When 5 is added to 3, the result is 8," although we know that this is incorrect in the light of the discussion in Chap. 1. Actually we have been guilty of this already, but since most of us are used to dealing with the incorrect form, we continue to do so.

2-6 EXERCISES

1. Identify three sentences in the text where the word *number* is used incorrectly and two where it is used correctly.

2. Is multiplication associative? Commutative?

3. Is subtraction associative; that is, does $(a - b) - c = a - (b - c)$? (*Hint:* Let $a = 10$, $b = 6$, and $c = 2$, and try it. Remember that parentheses mean do this first.)

4. Show that division is not commutative.

5. Show that subtraction is not commutative.
6. Would the commutative law for addition hold if we added with Roman numerals? Explain.
7. Give two nonmathematical examples that differentiate between operation and process.
8. Give three examples of binary compositions of things other than numbers.
9. Why is it easier to show that division is not associative than to show that addition is?
10. Is addition distributive over multiplication? Give an example to verify your answer.

REVIEW EXERCISES

1. What is the base of the following example?

   $401 + 43 + 312 + 444 = 1422$

2. Multiply 36×759 by:

   (*a*) Duplation
   (*b*) Gelosia

3. Make a table of multiplication facts for base 4.
4. $709_{11} + 234_{11} + 189_{11} + 480_{11} = (\qquad)_{11}$.
5. Add by retrograde addition: $2642 + 573 + 86$.
6. $506_8 \times 327_8 = (\qquad)_8$. Check by casting out sevens.
7. Change 5266 to base 5 by division.
8. Change 1342 to base 6 by division.

SELECTED REFERENCES

Cajori, F.: "A History of Mathematical Notation," vol. I, The Open Court Publishing Company, La Salle, Ill., 1929.

Eves, Howard: "An Introduction to the History of Mathematics," Holt, Rinehart and Winston, Inc., New York, 1959.

Generally too complicated for the inexperienced mathematics student. However, the discussion of some historical processes is good.

Smith, David Eugene: "History of Mathematics," vol. II, Ginn and Company, Boston, 1925.

This is perhaps the best work on the mathematical operations. Here are presented many more of the now obsolete methods of adding, subtracting, multiplying, and dividing.

IBM
SYSTEM 360

chapter three

PROBLEM SOLVING WITH COMPUTERS

We have been using various methods to add, subtract, multiply, and divide. They almost all require the memorization of a series of steps. Only a few require complete understanding of the numeral system used. Any examination of methods leads eventually to better ways, one of which is to leave simple arithmetic to a machine.

3-1 ALGORITHM AND SIMILAR PROBLEM-SOLVING METHODS

Old words and symbols often take on new meaning with the passage of time when they become descriptive of new concepts. Such a word is "program." A dictionary may give several definitions of the word, the most common being a list of events, pieces, and so on, of a show or concert. The same dictionary may also give, as a less common definition, a plan of procedure. The purpose of this chapter is to explore the meaning of programs as a plan of procedure.

There are many ways of analyzing problem-solving methods; for we

A modern computer system.
COURTESY IBM CORPORATION.

wish to examine *methods,* not solutions. Many authors have explored the procedures, and many methods of attacking a problem have been devised. To examine how computers are used in problem solving, we break down problems into three segments: (1) the goal, (2) the process to be used in reaching the goal, and (3) the principles or underlying concepts on which the processes are based.

First, let us consider the goal. By goal we mean objective. It may be finding the product of two numbers, or an accurate forecast of tomorrow's weather, or landing men on the moon, or finding how much we should price an object to increase sales, or any other identifiable objective. The key element of any goal is identification.

Then, whatever the goal, we must next consider how to reach it, what principles or ideas permit the problem to be solved or the goal to be reached.

Finally, we must translate these principles into a concrete process, a set of steps to be carried out to reach the goal—a plan of action. For example, if our goal is to land a man on the moon and return him safely to earth, we must examine the laws of rocketry that could be used to build the necessary vehicle. The process, the physical translation of the underlying principles into action, is highly complex, involving the work of many thousands of persons and a cost of billions of dollars. For a simpler example, consider the arithmetic problem of finding the product of two numbers, such as 37 and 54. Here the underlying principles are not readily identifiable, but many different processes are available. In Chap. 2 we identified at least three processes that could be used to find the product of these two numbers. We should probably select the usual multiplicative process we learned in elementary school. Unless we have had advanced training in mathematics or our initial instruction was in new mathematics, we are probably not aware of the principles involved. Thus, in the two examples, the goals are clearly stated; however, in the first, the principle is clear but the process is not, and, in the second, it is the reverse.

We can identify the process normally used to find the product of two numbers as an "algorithm." In modern times, the word *algorithm* is used for any special process of solving a certain kind of problem. The word comes from *algorism* which means a systematic method of calculation with our Hindu-Arabic numeral system. The key point is that knowledge of both the process and the underlying principle is often not necessary for problem solving. 37 times 54 yields 1998, and to arrive at this result, one has only to follow a set of directions learned in the third or fourth grade. The goal can be reached in a mechanical manner.

Because process, principle, and goal can be considered separately, we may wonder which, if any, of them require intelligence to carry out their analysis. From what we have considered here it is clear that, many times, the carrying out of a process to reach a goal may be mechanical and does not require intelligent supervision. We then wonder whether we can build machines to release men from routine work and whether such machines exist now. We can, and they do. We have all seen desk calculators that can carry out multiplication. These machines do not, in general, find the product of two numbers by using the algorithm we are familiar with but, rather, an even simpler one involving the repeated addition of numbers. We shall see that more complex machines can solve more complex problems by algorithmic methods, even problems man cannot.

3-1 EXERCISES

1. Identify a goal or objective such that the method of reaching it is well known but the underlying principles are obscure.
2. Identify a goal or objective such that the underlying principles are clear but their translation into steps that can be acted on is complex.
3. Give examples of everyday problems that are solved by algorithmic procedures.
4. List four completely different machines or devices that release men from routine work.
5. Devise a program for counting a large number of dimes.
6. Devise an algorithm for finding the sum of two fractions.

3-2 PROGRAMMED APPROACH TO PROBLEM SOLUTION

If we examine the everyday methods of reaching specific goals, we find that many are based on algorithmic or programmed procedures; that is, in carrying out many activities, we follow programs which are based on concrete steps for reaching an objective and which do not require an understanding of the basic principles. For example, if you walked into a grocery store and picked up a box of cake mix, you might find something like this printed on the back of the box:

1. Heat oven to 350 degrees.
2. Grease pans generously.
3. Dust pans with flour.

4. Blend in large bowl:
   Cake mix
   $1\frac{1}{3}$ cups water
   2 egg whites
5. Beat 4 minutes at medium speed.
6. Bake at 350 degrees for 30 minutes.
7. Cool 10 minutes.
8. Invert cake pans on wire rack.

This set of directions is the program for converting the cake mix into a finished cake. To make the cake, we follow these directions one at a time, in order, until the job is completed. To understand and carry out this program, there is no need for any knowledge of the chemical, electrical, or mechanical principles involved. We do not need to know how the motor of the mixer converts electrical to mechanical energy or how the heat of the oven converts the mix into a cake. All that is required of us is that we understand and follow the directions. Let us consider a second example. On the market today are a number of kits for building things, from hi-fi sets to model airplanes—and even one for grooming the family dog. Each of these kits comes with a set of directions, a program for using it to complete the job. As with the cake mix, the person doing the work does not need to know the underlying principles but is required only to follow directions in the order they are presented. There are two factors, then, that are important in making and following such programs: the individual instructions and their order.

For a third example of a programmed problem solution, consider a modern assembly line for building an automobile. The assembly line is the physical translation of a program into action. If one wrote a step-by-step description of how the car was being put together, one would have a written program for the job. From the input of raw materials to the finished product each step must be carried out and, for the most part, in a certain order. The engine should be finished before the body, and so forth. Such a written program is kept, studied, and used by the automobile manufacturer and has usually been devised over the years through trial and error. Although trial and error is an acceptable process for working out a program when there is the time, under normal conditions a person working out such a program must analyze its goal and know how it will be followed and who or what will be following it. For example, if one wanted to write a set of directions for driving a car, he would have to know what kind of car it was, how it was equipped, and how much experience the driver had. Any such set of directions, or program, must be worked out similarly; that is, it must be made up of separate directions that are within the capabilities of the person or machine that is to carry them out. When a set of directions is being written, the writer must assume that they will be carried out exactly as written; for if a direction is unclear or incomplete, the whole project comes to a dead end—the goal unreached. If the sequence of events that are to take place is not in a carefully arranged order, it often happens that the desired goal will not be reached. Putting an automobile in gear does not necessarily produce any result if the engine is not running or if the brake is not off.

Although we can see the analogy of such processes to recipes, if the process is physical, we often do not see that similar parallels exist between recipes and many nonmaterial problem-solving procedures. We pointed out the cookbook nature of our usual addition and multiplication techniques. Division and subtraction methods are in the same class. As with the automobile, the procedures are based on the "make" and "model" of our numeral system. Because it is base-10 and positional we have sets of directions based on these facts. If our numeral system were of some other kind, the usual directions would be useless, and we should have to work out new ones.

As we consider the idea of problem solving by direction further, let us adopt the common name for such a set of directions—a program. Thus, any cake recipe is a program for making a cake; and a set of directions for driving a car, a program for driving a car. The word *program* has become widely used, and we commonly speak of computer programs and programmed learning.

A good method for showing the step-by-step breakdown of a problem is to use a device computer programmers call a "flow diagram." A flow diagram is a set of symbol blocks that indicate the step-by-step procedure in a problem solution. Figure 3-1 is the flow diagram for our cake mix.

Flow diagrams are of great value, because, with them, we can easily visualize the step-by-step nature of our problem-solving method. Figure 3-2 is a more complex flow diagram for finding the largest number in a column of numbers. Here we use two kinds of blocks to indicate two different kinds of functions. When giving a simple direction, we use a rectangle; those steps which require decisions are contained in diamond-shaped blocks. For more complex operations, other shapes are normally used. To follow a flow diagram one has simply to follow the directions of the arrows in order until the program is completed. In Sec. 3-3 we examine, in detail, a complete problem solution by program.

In both the flow diagrams (3-1, 3-2), we make certain assumptions about the person and equipment involved in carrying out the directions. For example, we assume that an oven with a temperature control is available; that the phrase *dust pans with flour* has some meaning to the person following the directions; that *minute* is understood to be a measure of time and not, for example, degrees of an angle; that, in Fig. 3-2, the word *largest* has some meaning; and so forth. Note that when the language of the

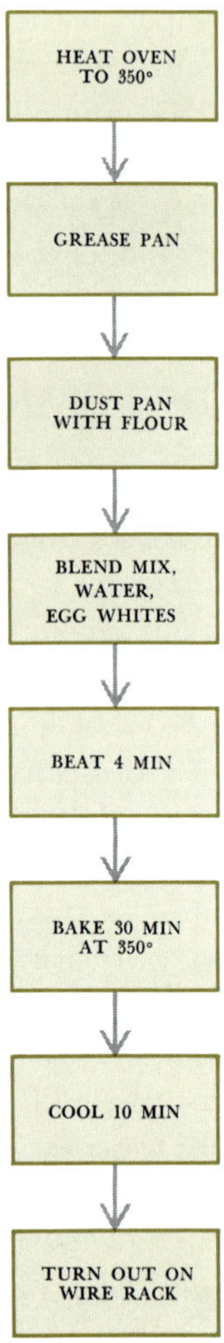

FIG. 3-1 FLOW DIAGRAM OF BAKING A CAKE.

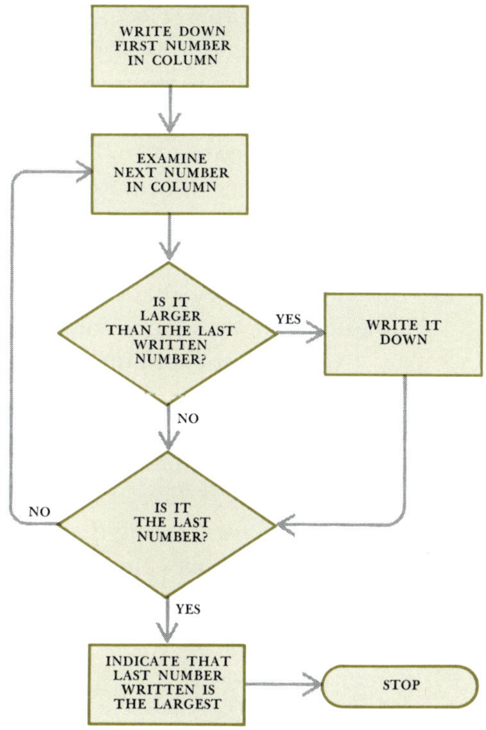

FIG. 3-2 FLOW DIAGRAM OF FINDING THE LARGEST NUMBER OF A SET.

instruction depends on the completion of some previous act, the instructions for this act have been given. For example, the direction in Fig. 3-2 that says "indicate that the last number written is the largest" does refer to a previous act, which was having some number written down. This direction does not make sense if taken out of context, because it refers to other parts of the set of directions.

One now wonders how detailed a set of directions or a program must be if it is to be useful. The amount of detail depends on who or what is going to be following the directions. Note that directions for a machine to follow must be simpler than those for a man.

3-2 EXERCISES

1. Make a flow diagram for a set of directions for making a cheese sandwich.

2. Assume that you have a machine that can read two cards, each with a number on it, and then eject a new card with the sum of the two numbers on it. You have 100 cards, each with a single number written on it; make a flow diagram showing how you would use the machine to find the sum of the 100 numbers.

3. In Exercise 2 it is necessary to know how the machine performs the addition of the numbers in order to use it in finding their sum? Explain.
4. Give examples of programs for doing jobs such that a person carrying out the program is usually not aware of the underlying principles.
5. List the steps, in order, that one follows in starting the engine and moving an automobile with an automatic transmission.
6. Repeat Exercise 5 for an automobile that does not have an automatic transmission.
7. What assumptions are you forced to make when doing Exercises 5 and 6?
8. Make a flow diagram that shows how you would add two 3-digit numbers, such as 498 and 752.
9. Write a program for locating the one counterfeit coin in a group of five coins. The counterfeit coin is of different weight from the four genuine coins and to weigh the coins you must use an assayer's balance. You must locate the counterfeit in three weighings.

3-3 SOLUTION BY PROGRAM

Although we have used the word *program* to describe any kind of set of directions, we want to consider how programs apply to computers. To gain understanding and to present an example in detail, let us consider the following problem.

The procedure for converting base-10 numerals to any other base was discussed in Chap. 2. The following is a program for converting a base-10 numeral to a base-7 numeral.

STEP 1 Call the number to be converted C, and on your paper write $C =$ ____, filling in the blank with the value of C.

STEP 2 Write $A = 1$ on your paper.

STEP 3 Write $B = 0$ on your paper.

STEP 4 Divide the value of C by 7 to the nearest whole number, and find a quotient and a remainder. For example, $25 \div 7 = 3$ with a remainder of 4.

STEP 5 Multiply the remainder from step 4 by the value of A.

STEP 6 Change the value of B by adding the result of step 5 to the present value of B.

STEP 7 Increase the value of A by multiplying it by 10.

STEP 8 If the quotient of step 4 is zero, go to step 11. Otherwise, go on to the next step.

STEP 9 Change the value of C to the quotient of step 4.

STEP 10 Go back to step 4.

STEP 11 Write __(1)__ written in base 7 is __(2)__, filling in blank 1 with the original base-10 numeral and blank 2 with the present value of B.

STEP 12 Return to step 1 to begin another conversion.

Let us now use these instructions to convert the base-10 numeral 67 to base 7.

STEP 1 $C = 67$.

STEP 2 $A = 1$.

STEP 3 $B = 0$.

STEP 4 $67 \div 7 = 9$ with remainder 4.

STEP 5 $4 \times 1 = 4$.

STEP 6 $B = 0 + 4 = 4$.

STEP 7 $A = 1 \times 10 = 10$.

STEP 8 9 is not equal to zero; therefore go on to step 9.

STEP 9 $C = 9$.

STEP 10 Go to step 4.

STEP 4 $9 \div 7 = 1$ with remainder 2.

STEP 5 $2 \times 10 = 20$.

STEP 6 $B = 4 + 20 = 24$.

STEP 7 $A = 10 \times 10 = 100$.

STEP 8 1 is not equal to zero; therefore go on to step 9.

STEP 9 $C = 1$.

STEP 10 Go to step 4.

STEP 4 $1 \div 7 = 0$ with remainder 1.

STEP 5 $1 \times 100 = 100$.

STEP 6 $B = 24 + 100 = 124$.

STEP 7 $A = 100 \times 10 = 1000$.

STEP 8 The quotient of step 4 is zero; therefore go to step 11.

STEP 11 67 written in base 7 is 124_7.

STEP 12 Go back to step 1.

Let us now check the results of this program:

$124_7 = 1 \times 49 + 2 \times 7 + 4 = 49 + 14 + 4 = 67$

Therefore, in this case, our program does correctly convert for us. We can make a flow diagram of this program by blocking in the steps. It might look like Fig. 3.3. We see that the steps involved do not require a knowledge of the goal or of the reason why the steps bring us to it; that is, anyone who can read the steps can carry them out, even if he has never heard of base 7 or different systems of numeration. Simply by following the directions, almost anyone can arrive at 67 written in base 7 is 124_7.

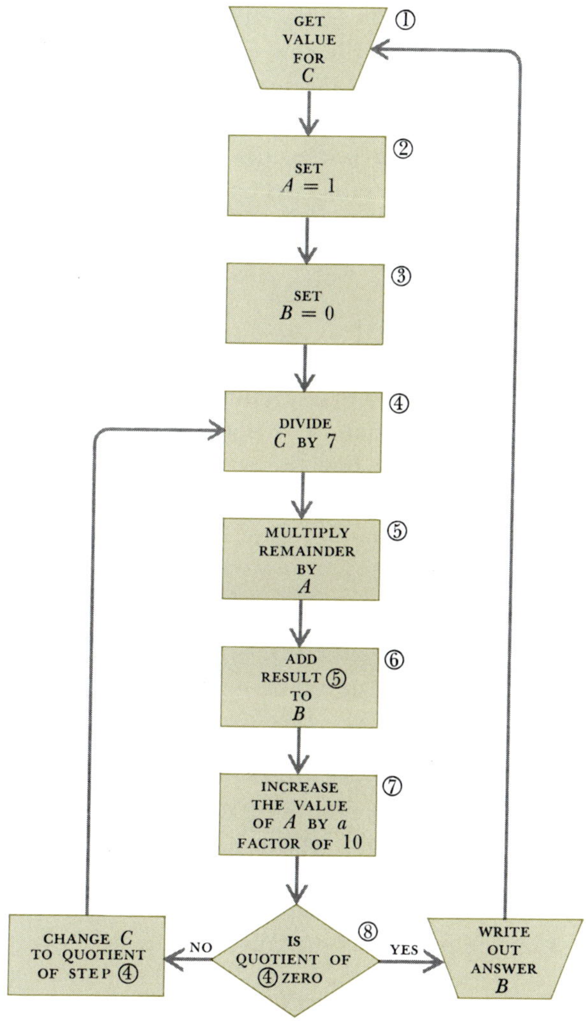

FIG. 3-3 FLOW DIAGRAM OF CHANGING FROM BASE 10 TO 7.

We have discussed how one develops such a problem. Usually, it requires some experience with this kind of problem solving and an understanding of the goal and the possible methods of reaching it. The program could be used to convert 13,956,852 to its base-7 representation of 226426350_7. However, it would require repetition of the various steps of the program nine times. If we had the time, this would be an acceptable procedure, easily followed by modern digital computers. They would be able to complete the conversion in 1 or 2 sec. In the next section, we consider what such a machine would have to be able to do. This is not to say that these machines are superior to man; man, without mechanical aids, cannot fly, or see events taking place thousands of miles away, or spend hours swimming underwater. Like the airplane, the television, and the scuba, the "math machine" or computer is only another of man's inventions.

3-3 EXERCISES

1. Convert 137 to a base-7 numeral, using the program.

2. Try the conversion of 13,956,852 to base 7, using the program.

3. We pointed out that the order of the instructions is important. One wonders whether any of the steps can be interchanged or whether they must remain fixed. Can the following changes be made without making the program unworkable?

 (*a*) Can steps 7 and 8 be interchanged?
 (*b*) Can steps 3 and 4 be interchanged?
 (*c*) Can steps 8 and 9 be interchanged?

4. Is there any way in which the program could be changed easily and used to convert numbers to base 5?

5. In the program some instructions are needed to place the remainders of division in their proper place. What steps of the program arrange these remainders, and what instructions would you replace these with, if you were instructing someone to change numerals to base 7?

6. Can the program be changed easily to convert numerals to base 12?

3-4 MACHINES TO EXECUTE PROGRAMS

If we designed a machine to carry out programs, such as those in Sec. 3-3, what should it be able to do? We do not have to break down our analysis to the actual mechanisms, but we must ask ourselves what functional parts such a machine should have. The main requirement of any such machine is that it has some method of storing information. It should be able to store instructions that are part of the program and to store data to be worked on. Let us call this storage a "memory unit." It is easy to see that the memory of our machine will be of little use unless information can be inserted and removed with ease. Therefore, our machine must have some method of reading and writing—"inputting" and "outputting" information. A fourth requirement of our machine is that it be able to do arithmetic, for a number of the steps of our program require arithmetic computation. Thus our computer must have an "arithmetic unit." Finally, all these units must have a boss, a supervisor, something to tell them when to act and what to act on; our machine must have a "control unit." We can summarize all this in a block diagram, as in Fig. 3-4, where arrows show the flow of information.

Figure 3-4 shows the flow of instructions and data from input to memory to arithmetic unit and then to output units—all under the supervision of the control unit. The parts of the machine do not have to be physically distinct; that is, they may be separated only when we consider the jobs they perform. If we built such a machine, we might use one part to do several jobs. For example, a typewriter could be used as both an input and an output unit. Another example is the control unit, which would have to function as memory in many problems, for it must hold instructions and then act on them.

There are many machines with such features. A modern car, for example, has some of them. The lock, gearshift, engine, and brakes all

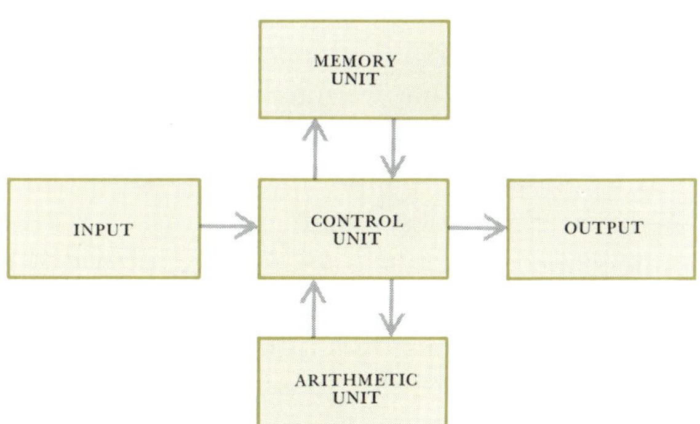

FIG. 3-4 FUNCTIONAL DIAGRAM OF A COMPUTER.

allow for instruction input as the driver, acting on the controls, tells these machines what to do. The controls of the car are input devices, which instruct and control its motion. For information and function output, there are the movement of the wheels and the readings of the various dials, gauges, and lights of the dashboard. The car's braking system is an excellent example of the functions we are considering. A modern power brake receives instructions from the driver through the brake pedal. It takes these directions and converts them into action, both in stopping the car and in actuating the brake lights.

Consider the simple desk adding machine, such as might be found in a modern office. It has all the features or units we have considered. For input it has its keyboard. For memory and output it has its dials and paper-tape printouts. Acting as a control unit and as an arithmetic unit, it has the mass of levers, switches, and gears that are seen when one opens its case. In our modern world we can find many other machines with the functional parts we have considered. They take many forms, although we are not used to calling the toast from the automatic toaster or the dinner we cook on the electric range an output. If we examine these household items from the point of view of computers, the toast and dinner are indeed outputs.

A major question is still unanswered. Can we make a machine that reads and writes? If we look at modern electric typewriters, we have the answer to half of our question; they can at least write. In one sense, they can also read, for such a machine reads information in the form of depressed keys, transforms it into electrical impulses for storage, and then uses the impulse to output the desired letter. However, this does not answer the question of whether or not we can build a machine that understands and acts on words the way a man can. Note again something pointed out in Chap. 1, that symbols may be assigned any meaning we wish and, conversely, that we may use almost any symbol we wish for an object. Our numerals comprise the simplest organized set of symbols, or language.

If we wanted to build a machine that understood a language, numerals might be the easiest language to use. We already have machines that understand numerals as a language, provided that the correct "grammar" or "sentence structure" is used. For example, 17-35-41 might mean "unlock" to the lock on a safe. And 483-9112 might instruct a telephone "connect me with John Jones at 415 South Market Street." In these examples the machine relates a combination of numerals to action. In the case of the safe, there are only two actions: lock and unlock. However, the telephone provides us with an example of a machine that understands millions of "words." The step to a machine where 42 means "add" or 67 "subtract" is really a small one logically, if not physically.

In practice the machines we are referring to are computers. Many of them can be "spoken" to in our everyday Hindu-Arabic numeral system. Few, if any, of them record and work with these base-10 numerals. Rather, they convert and store them in the form of binary, or base-2, numerals. We have seen that arithmetic operations are simple in base-2. To store a numeral mechanically or electrically in base-10 requires a device with 10 distinct "states," one to stand for each of the 10 possible digits. To store base-2 numerals requires a device with only two states, such as an electrical switch that in state 1 is on and in state 2 is off. An on-off switch corresponds, then, to the binary numerals, 0 and 1. When a switch is on it could correspond to a 1, and when it is off it could mean 0. Because vacuum tubes and transistors can be used and, indeed, are used as a form of electrical switch, machines can be designed and built with internal storage and operation that is electrical and thus extremely fast. To such a machine 100011_2 is as reasonable a symbol for 35 as any other, for these machines do not have prior experience that is limited to base-10 numerals.

Although we have considered what functional parts our machine should have, we might ask what our machine should be able to do with them. Clearly, its memory is central to its performance. The machine must have some method of identifying and retrieving the data it has stored in the memory unit. The control unit must be able to follow a prescribed set of directions. If these directions require arithmetic operations, the control must be able to use the arithmetic unit and store the result of the operation in the memory for later use. The control should be able to change the order of the operations based on simple tests, such as when a value has become zero or when the sum of two numbers is positive; that is, the control must be able to make simple decisions based on intermediate results and then modify its sequence of actions, based on these decisions, to meet the needs of the problem.

3-4 EXERCISES

1. Identify the input, storage, and control functions of a television set?

2. Repeat Exercise 1 for an automatic washer.

3. Besides the example given, list other machines that "understand" numerals as words and instructions?

4. Besides language and our numerals, name other sets of organized symbols for objects?
5. Identify the steps of the program in Sec. 3-3 that make use of:

   (*a*) An input device
   (*b*) An output device
   (*c*) A memory unit
   (*d*) An arithmetic unit
   (*e*) A control unit

6. What is the name for a set of symbols that are kept secret?
7. The computer memorizes numbers in base-2. This is not a problem for it, because it electrically records 11 in its memory when it is presented with the number 3, and similarly with any other base-2 number, and it is not hampered by prior knowledge of base-10. In much the same way you could learn an alternate for your alphabet and could, with practice, read it very rapidly. For example, decipher the following simple code.

   5-1-3-8 9-19 1 16-1-18-20 15-6 20-8-5 13-1-9-14

REVIEW EXERCISES

1. What is an algorithm?
2. Give two examples of problems normally solved or objectives reached by an algorithmic procedure.
3. A "little league" determines the league champion on the basis of the highest won-lost percentage. Assume there are six teams in the league, and a listing of the number of games won and lost by each team is available. Write a set of directions, or a program, that could be used to determine the league winner.
4. In Sec. 3-2 a program was flow-diagrammed to find the largest number in a column. Consider the 50 numbers listed here and, following the steps in the flow diagram, find the largest number in the set of numbers:

367	593	252	921	263	277	725	408	631	295
459	62	967	582	567	427	364	269	773	591
214	841	600	725	955	914	962	115	935	392
638	372	742	11	503	60	12	491	624	213
124	815	366	976	157	72	664	405	919	418

5. Name some advantage in following a program in Exercise 4 over finding the largest number by inspection.
6. What jobs should the parts of a computer be able to perform?
7. A standard and simple problem for computers is the preparation of pay warrants. Assume that you are the head of a company employing a number of men. Make a flow diagram for the preparation of their checks, assuming all are paid on an hourly basis.
8. It has been pointed out that the control unit of a computer should be able to change the sequence of instructions. Discuss how this need arises in the program presented in Sec. 3-3.

Computer display unit through which program changes can be made by use of a light pen.
COURTESY IBM CORPORATION.

chapter four

COMPUTERS AND PROGRAMMING

From Chap. 3 it is clear that computers are not simply an extension of the adding machine. For they are able to perform a series of operations in a predetermined and changeable sequence. This ability to modify the sequence of execution, both during and between problems, is perhaps their most important aspect. In a modern computer the list of operations, or program, is stored electronically, in the form of binary digits, in the computer itself. This concept, of the internally stored program, was a major breakthrough in computer technology, because it put all the processes a computer uses on the same basis. It was proposed by the American mathematician John von Neuman in the late 1940s. In the present chapter we examine how one talks to and uses these machines, which "think" in binary digits.

4-1 TALKING TO COMPUTERS

In early computers, programs were loaded into the machine in the form of punched paper cards, like the modern IBM card. The machine accepted a card from a hopper and performed the operation indicated by the holes punched in the card. Then it accepted the next card and performed the operation it indicated, and so on. The instructions on the cards referred to numbers stored in the computer. An early loom, invented in France about 1728, used a similar procedure to control the pattern of the weave. Every time a thread was drawn through the loom, the color of the thread was selected by needles passing through a wooden "card" that had holes punched in it. At the next pass a new card was placed in front of the needles, and in this manner the pattern was controlled. The idea of an internally stored program is analogous to a loom card whose holes are changed, thus making it unnecessary to change cards.

An adding machine stores numbers with its dials. Each dial stores a single base-10 digit. This means that each storage "cell" (dial) must have ten distinct "states" or conditions, one corresponding to each digit in our numeral system. From Chaps. 1 and 2 it is clear that a numeral storage system need not store ten but only two kinds of digits. This is important in designing a storage device, because only two and not ten states are needed. One of the simplest two-state devices is the electrical switch. A switch is either on, permitting current to flow, or off, breaking the flow of current. A set of switches can easily be used to store a number. First the number is written in a base-2 representation. Then the binary digits of the representation are stored as ons and offs of the set of switches. Almost all modern computers follow these steps exactly. They accept input data and convert them and store them in the form of binary digits as ons and offs of extremely small electronic switches.

If a machine can be built to store numbers, it is relatively simple to build one that can understand them as directions to be followed. Two machines of this kind were mentioned in Chap. 3, namely, the combination lock and the dial telephone. Imagine a machine that contains two hundred blocks of switches, each block capable of storing an eight-digit base-10 numeral. In addition each block is identified by a three-digit base-10 numeral. In other words, the machine contains 200 cells, each having a numeral associated with it, and being numbered from 001 to 200. The numeral identifying each cell is known as its "address." Now suppose that this set of cells is connected to a second machine, one not unlike those found in a telephone switching office. Instead of switching telephones and making connections based on the numerals dialed, this machine can "reach" into the first machine electronically and copy eight-digit numerals from a preselected memory cell. Then it "dials" the numeral it has copied.

However, instead of connecting two callers, as on a telephone, this automatic dialing connects circuits that cause numbers in the storage cells of the first machine to be added, subtracted, multiplied, or divided. For example, the switching circuits can be set up so that, if 0 and 1 are the first two digits dialed, addition takes place. Suppose the eight-digit numeral dialed is in the following form:

01 $d_1d_2d_3$ $d_4d_5d_6$

$d_1d_2d_3$ represents a three-digit numeral between 001 and 200, and $d_4d_5d_6$ another three-digit numeral also between 001 and 200. The circuits can be set up so that dialing 01 causes the number stored in the memory cell identified by $d_1d_2d_3$ to be added to the number stored in the memory cell identified by $d_4d_5d_6$, the answer being stored in memory cell $d_1d_2d_3$. This is usually stated in terms of addresses. The number at address $d_1d_2d_3$ is added to that at address $d_4d_5d_6$, the result being stored at address $d_1d_2d_3$. For example, initially the following conditions existed:

Stored in memory cell 103: 14,792
Stored in memory cell 152: 118

the result of dialing 01103152 would be

Stored in memory cell 103: 14,910
Stored in memory cell 152: 118

Note that 14,910 is the sum of 14,792 and 118.

Similarly the control unit could interpret 02 as an order to subtract. If initially the following conditions existed:

Stored in memory cell 103: 14,792
Stored in memory cell 152: 118

the result of dialing 02103152 would be

Stored in memory cell 103: 14,674
Stored in memory cell 152: 118

It could be that 03 meant multiply and 04 divide. If the initial conditions were the same as in the two previous examples, the result of dialing 03103152 would be

Stored in memory cell 103: 1,745,456
Stored in memory cell 152: 118

and of dialing 04103152

Stored in memory cell 103: 125
Stored in memory cell 152: 118

The result of the example of division is simply the nearest whole-number quotient. The remainder of 42, which occurs when 14,792 is divided by 118, was lost.

Actual computers function in the same manner. Their control unit is more complex than that of the imaginary machine described above, and they can dial much more complex directions. A sequence of directions is usually loaded into consecutive memory cells, and then the control unit is told, by turning switches and pushing buttons, to copy the numeral from the first cell and follow its direction. The control unit is set up so that, after it has finished following a direction, it automatically goes to the next cell in order and dials its direction. This process continues until all the desired directions have been dialed or until the control unit dials a direction that causes it to jump to a new memory cell rather than the next one in sequence or is instructed in some other way, such as external control, to modify the sequence of directions.

Not all computers are built in the same way or follow the same pattern of numerical codes. Nor do they all use the same set of operations to perform various functions. In general the simpler the computer, the fewer the instructions it can follow. In any computer the numerical codes corresponding to addition, subtraction, and so on, together with the format in which these codes must be used, are known collectively as the computer's "machine language" or "absolute language." In the early days of programming, programmers had to learn these codes so that they could talk to computers.

When writing a program in a modern context the programmer is freed from the task of learning the machine's language and instead uses an intermediate or "programming language," which is somewhere between the English of the program and the absolute language of the machine. The concept of a programming language is simple, but its creation is a complex and difficult task.

Consider the basic problem in using a computer. Anyone wishing to use a computer thinks and speaks in one of the many natural languages—in our case, English. However, a computer works in an artificial numerical language. The basic problem, then, is to translate between the two languages. The usual way translation between two natural languages is achieved is for one of the two parties involved to learn the language of the other. This is the method used in the early days of computers; the programmer had to learn the language of the computer, because the language of man was far too complex to "teach" to the machine.

In translating natural languages, it is not always necessary for one of the parties to master the language of the other completely. Effective communication can often be carried on in a mixture of the two languages. This approach has provided an excellent solution to the problem of talking to computers. Over a period of years several intermediate languages have been developed for writing programs. They are not purely numerical or purely verbal or alphabetic. Rather they are close enough to the natural language of the programmer so that he learns them easily and close enough to the numerical language of the computer so that it can be given the task of completing the translation. As mentioned above, there are a number of programming languages. Whichever a programmer uses depends on the problem to be solved. Among these languages there is Cobol, short for common oriented business language; Cogo, short for coordinate geometry, which is designed for civil engineers; and Pl-I and Algol and a host of others, each for programmers in specific areas. One of the most widely used programming languages is Fortran, short for formula translation. Fortran, in its many variations, is used in engineering and science. Because several versions of Fortran are simple, it is instructive to study one and attempt a few programs written in it.

Any program consists of a list of statements, whose directions are to be carried out in order unless otherwise indicated. This is why the set of steps in the sample problem in Sec. 3-3 was considered a program. In the list of statements in a Fortran program only a limited number of possible statements are available. To identify each statement, or at least those which require identification, a numeral can be assigned to it. These "statement numbers" are not used to order the statements of the program but only to identify them. For this reason the statements in a Fortran program are not usually in the order of their statement numbers; that is, statement 2 may not appear in the program after statement 1, or the only numbered statement in the whole program may be 36 or 876. Any identifying numeral from 1 to 9999 may be assigned to a program statement. The only requirement is that a statement number may be used only once in a program; that is, each statement identified must be done so uniquely.

In Fortran programs there are four kinds of statements that can be used: arithmetic, control, input-output, and specification. "Arithmetic" statements are used to indicate calculations. This kind of statement is easy to recognize, because it always contains an = sign. On the left side of the = sign there can be only a single "name" of a quantity. The Fortran name of a quantity is a letter or combination of letters. It stands for one of the quantities in the problem. On the right side of the = sign there can be any combination of numbers and names for numbers connected with the + sign to indicate addition, the − sign to indicate subtraction, the * to indicate multiplication, and the / to indicate division. For example, the statement

```
Z = 3.2 * A + 6. * B - C/D
```

means that 3.2 times the value assigned to A is to be added to 6 times the value assigned to B, and from this the quotient of the numbers represented by C divided by D is to be subtracted. The resulting value is to be assigned to Z. If Z had a value assigned to it before the execution of this statement, this value is lost when a new value is assigned to Z. By assigning values and performing appropriate calculations, most arithmetic processes can be carried out.

"Control" statements modify and control the sequence in which the instructions are executed. Consider the following example of the simplest kind of control statement—the GO TO statement. Suppose somewhere in a program the following three statements appear:

```
Z = X + 1.

GO TO 7

Y = T + 2. * X
```

Because the computer executes instructions in order, it first finds the value of Z by adding 1 to the value of X, given to X in a previous instruction. To execute the control statement

```
GO TO 7
```

the computer "jumps" to program statement 7, which appears somewhere else in the program, and executes this instruction rather than the arithmetic statement

```
Y = T + 2. * X
```

Another kind of control statement, which is used in the examples that follow, is known as the IF statement. Like the GO TO statement the IF statement instructs the computer to jump to a new program statement, instead of simply proceeding to the next one. Unlike the GO TO statement, which offers no choice and specifies only one jump location, the IF statement makes a selection from three possible jump locations. The IF statement selects the destination of the jump based on the algebraic sign of a specified quantity. A typical IF statement is

IF (X) 7,5,9

The IF statement is executed in two steps. First the computer calculates the value of the quantity indicated in the ( ) and looks at its sign. Here this is simply the quantity indicated by X. If this quantity is negative, the computer takes its next instruction from the program statement whose number is given in the first of the three positions created by the use of the two commas. This would be statement 7. If the quantity has a value of zero, the computer jumps to the statement indicated in the middle of the three positions. This would be statement 5. If the quantity is positive, the computer takes its next instruction from the program statement whose number is given in the last of the three positions. This would be statement 9. Summarizing: if X is negative, the computer jumps to statement 7; if zero, to statement 5; if positive, to statement 9. The IF statement gives the computer a choice, so that it can perform simple decision-making tasks.

The other two kinds of Fortran statements are the "input-output" statement and the "specification" statement. The job of the input or output statement is obvious—to get data into the memory of the computer, and to get results out of the computer in a form usable to the programmer. Although there are many ways these functions can be accomplished, only two methods are examined here—the IBM card for input and the computer-controlled electric typewriter for output.

Data are "written" on IBM cards by punching holes with a machine called a "key punch." This system was developed by Herman Hollerith in 1889 and 1890. Hollerith was an American engineer who worked for the Bureau of the Census during the late nineteenth century. The whole concept was developed to help reduce the data collected in the 1890 census. The punched card was a major breakthrough, because it permitted a machine to both "read" and "write" data for the first time. Hollerith was also one of the founders of the IBM Corporation.

A modern punched card has space for 80 columns of 12 rows each. One character of "alphanumeric" data is punched per column. As in Figs. 4-1 and 4-2, 10 of the rows are numbered, so that there is a row of zeros, a row of 1s, a row of 2s, and so on, to the row of 9s at the bottom of the card. Above the row of zeros there is space for two more rows. The

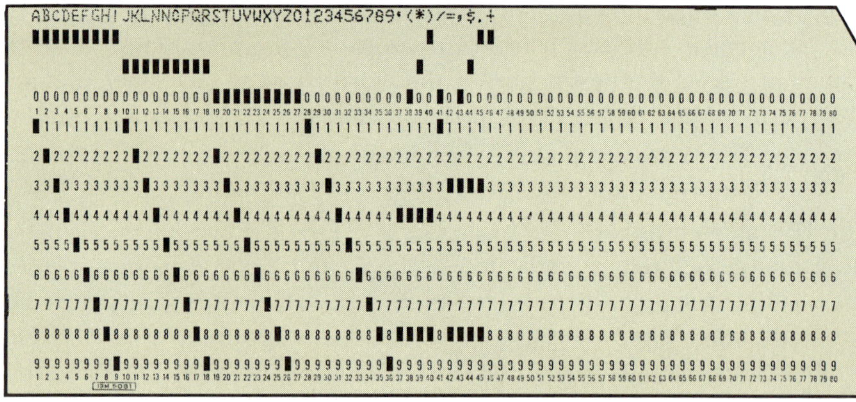

FIG. 4-1 PUNCHED CARD SHOWING THE ALPHANUMERIC-PUNCH RELATIONSHIP.

top space is called the 12 row, and the second the 11 row. The printing on the card is only to help determine what is on the card. The characters are coded in the following manner. If a column is to store a zero, a hole is punched in the zero row. If it is to store a 1, a hole is punched in the 1 row. Other digits are punched in the same manner. To code a letter, two holes are used. The letter A is coded as a 12 punch and a 1 punch in the same column. A B is coded as a 12 punch and a 2 punch. A C is coded as 12 punch and a 3 punch, and so on, to I, which is coded as a 12 punch and a 9 punch. J is then coded as an 11 punch and a 1 punch. The rest of the letters are coded in a similar manner. Special charac-

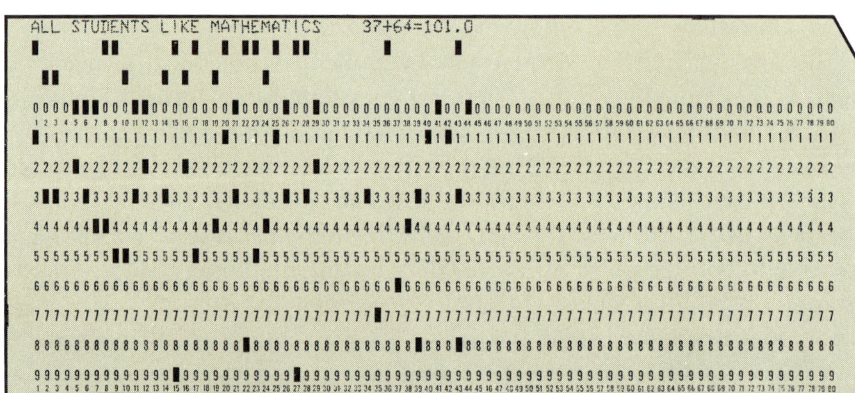

FIG. 4-2 SIMPLE MESSAGE ON A PUNCHED CARD.

ters, such as periods and commas, are coded as three punches in a single column. Figure 4-1 shows a typical character set and the corresponding punches. Figure 4-2 shows a message punched on a card. The punches corresponding to a particular character are in the column headed by that character.

The Fortran instruction for reading a numerical value from a punched card and assigning it to a selected symbol has four parts:

8 READ 37, XRAY
I II III IV

Part I is the program statement number, and as with any Fortran statement number it identifies the statement for future reference. Here the identifying number is 8. The word READ always makes up part II. This tells the computer that the form of data input is a punched card. If another form of input is to be used, another code word appears in the place of READ. In part III there is always a reference to another statement, in the form of its number, in this case, 37. The statement referred to here is always a specification, which tells the translator program the manner in which the data are punched on the card. This means that specification statements are used during the translation of programs from programming language to machine language rather than being steps of the problem-solving process. Part IV of the input statement "names" the quantity being read in, or in the case of output statements, which have the same general form, typed out. The following is an example of an output statement:

8 TYPE 37, XRAY
I II III IV

Parts I and II of this example have the same function as in the example of a READ statement given above; that is, part I is the program statement number, and again it is 8. Part III, again 37, tells the translator program the location of the format it is to use in typing out the value. In this example the value is that assigned to XRAY, which is identified in part IV of the output statement. Part II is the word TYPE, which indicates that the value is to be typed on the electric typewriter of the computer. Not all computers include this relatively slow form of data output, so that this command could be invalid on larger machines.

Only one kind of specification statement is presented in the sample programs that follow. It is the FORMAT statement and is related to the input-output function. Assume that the sample FORMAT statement below is that referred to in part III of the sample input-output statements above. The FORMAT statement has three parts:

```
37 FORMAT (F15.5)
I     II      III
```

Part I of it is the program statement number. Here it is 37. FORMAT statements are always identified, because they are always related to and referred to in an input or output statement. The word FORMAT identifies the purpose of the statement, namely, to specify the format in which the data are to be read into the computer or that which the computer is to use in typing them out. Part III, which is always enclosed in parentheses, is the specification itself in a coded form. The code used here is simple, based on the fact that two kinds of quantities are used in Fortran programs: one is restricted to whole numbers, and the other uses decimals. The translator program distinguishes between the two on the basis of the symbol chosen to represent the quantity. Symbolic quantities whose name begins with the letter I, J, K, L, M, or N are whole numbers. If the symbolic name of a quantity begins with any other letter, it requires a decimal.

If part III of a FORMAT statement contains the letter F, for floating point, the translator knows that it is dealing with a quantity that requires a decimal. If the letter I is in the parentheses, a fixed point or integer is being used. In the example the specification is F15.5. The 15.5 tells the translator that the quantity contains 15 characters, with the decimal between the fifth and sixth digits from the right. Generally in an F specification the number following the F and on the left of the decimal point indicates the number of characters or digits, and that on the right the number of digits on the right of the decimal in the quantity itself.

If a whole number, like NCA or M, had been used in the input-output examples above, the letter I would have appeared in the parentheses. For example,

```
   READ 62 NC
62 FORMAT (I6)
```

The READ statement instructs the computer to read a value from a punched card and assign it to the symbol NC. It also instructs the translator to take the specification for the quantity from program statement 62, which contains the code letters I6. The I indicates a whole number, and the 6 informs the translator that six digits are to be used.

The last Fortran statement to be discussed is a special FORMAT statement. It is called the H Format, in honor of Hollerith, and it makes it possible to print alphabetic information along with numerical data:

100 FORMAT (11HTHE AREA IS)

In this FORMAT statement the 11H is a code telling the computer to print the characters in the 11 spaces following the H. Note that THE AREA IS requires 11 spaces if the space between the words is counted. This FORMAT causes the computer to print THE AREA IS when executing a TYPE 100 statement. Another example follows:

105 FORMAT (27HTHE HIGHEST TEMPERATURE WAS)

Note that THE HIGHEST TEMPERATURE WAS requires 27 spaces if the space between the words is counted. This FORMAT statement causes the computer to print THE HIGHEST TEMPERATURE WAS when executing a TYPE 105 statement. To use this type of FORMAT statement one has only to include in the program a statement of the form TYPE xx, where the xx indicates the FORMAT statement containing the desired message.

4-1 EXERCISES

1. Identify three everyday devices that have two distinct states and thus can be used as binary storage devices.
2. How many distinct states does a device need to store a single base-5 digit? A base-7 digit? A base-8 digit?
3. How many base-2 storage devices are needed to store a base-8 digit?

Exercises 4 to 7 refer to the machine language described in the present section. With the following additional information:
The number 3 is stored in the memory cell #52
The number 5 is stored in the memory cell #77
The number 16 is stored in memory cell #113
The number 20 is stored in memory cell #154
The number 2 is stored in memory cell #160

4. Write a machine-language statement that finds the sum of 5 and 16, with the result being stored in memory cell 77.
5. Write a machine-language statement that, by using division, results in the number 10 being stored in memory cell 154.

For Exercises 6 and 7 suppose that the following additional instruction can be used:

$05d_1d_2d_3d_4d_5d_6$

When the machine executes this instruction, the number stored in memory cell $d_1d_2d_3$ is also placed in memory cell $d_4d_5d_6$. This is often known as a "store" instruction.

6. Write a machine-language program that has the following results. The number 25 is stored in cell 77, and the number 0 in cell 53.
7. Write a machine-language program that results in the number 3 being stored in cell 52, the number 9 in cell 77, the number 27 in cell 113, and the number $81 = 3 \times 3 \times 3 \times 3$ in cell 154.
8. Identify the four basic kinds of statements which can appear in a Fortran program.
9. What is the difference between an executable and a nonexecutable Fortran statement?
10. Write a Fortran statement that finds the sum of three numbers L, M, and N and calls it I.
11. Write a Fortran statement that can be used to type out a five-digit number the computer is storing as NUM.
12. Write a Fortran program that generates and prints odd numbers.
13. Write a Fortran program that reads in a four-digit whole number and prints it out if it is less than 5000.
14. Identify the message punched on the following card, using the key card in Fig. 4-1.

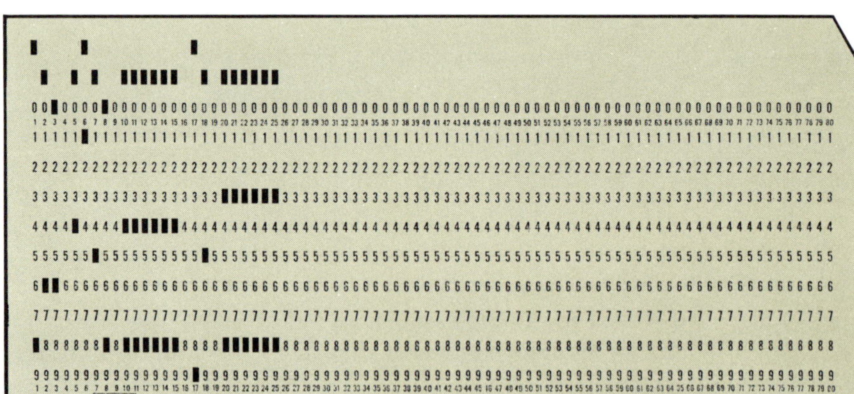

FIGURE FOR EXERCISE 4-1-14

15. If a computer can be set up to distinguish among one, two, or three holes punched in a column of an IBM card, what is the largest number of different characteristics it can be made to distinguish; that is, in a column where there are 12 possible punch positions, how many characters can be represented by holes punched in a column if one, two, or three holes can be used to represent a character?

4-2 THE SEARCH PROBLEM

To see how the few simple instructions described in Sec. 4-1 can be put together to solve meaningful problems, several complete examples are now considered.

Suppose that you have a stack of punched cards, such as those described in Sec. 4-1. On each card a single number is punched and typed. Assume that the numbers are located in the first 15 columns of the card and that they are all positive. The problem is to find the largest number.

If the number of cards is small, the problem is not hard. One has only to scan the cards and pull out the one with the largest value. However, if the number of cards is large, a more systematic procedure should be used, so that no value is omitted and the largest one is found. If a computer is used, the job is to find a method that takes advantage of a computer's great speed and simple-minded devotion to duty. One of the usual ways of finding such a method is to determine a usable "hand" method and attempt to extend it in a form suitable to the computer.

A hand method for this problem comes readily to mind. First one picks up the top card of the deck. Then he picks up the next card and compares the numbers on the two cards. He keeps the card with the larger value and discards the other card. Then he picks up the next card and compares its value to that on the card being held. Again the card with the larger value is kept, and the other card is discarded. This process is repeated until all the cards have been examined. The remaining card has the largest value. To make a flow diagram (Fig. 4-3) for this procedure, suppose that the card being kept is held in the left hand and the right hand is used to pick up and discard cards.

There are two decisions to be made in finding the largest value: the larger of two values must be chosen, and the last card must be indicated. There was nothing in Sec. 4-1 to indicate that the computer could understand instructions like "choose the larger value" or "stop at the last card." There are a number of ways to set the computer to make these decisions. The problem of informing the computer that all the cards have been ex-

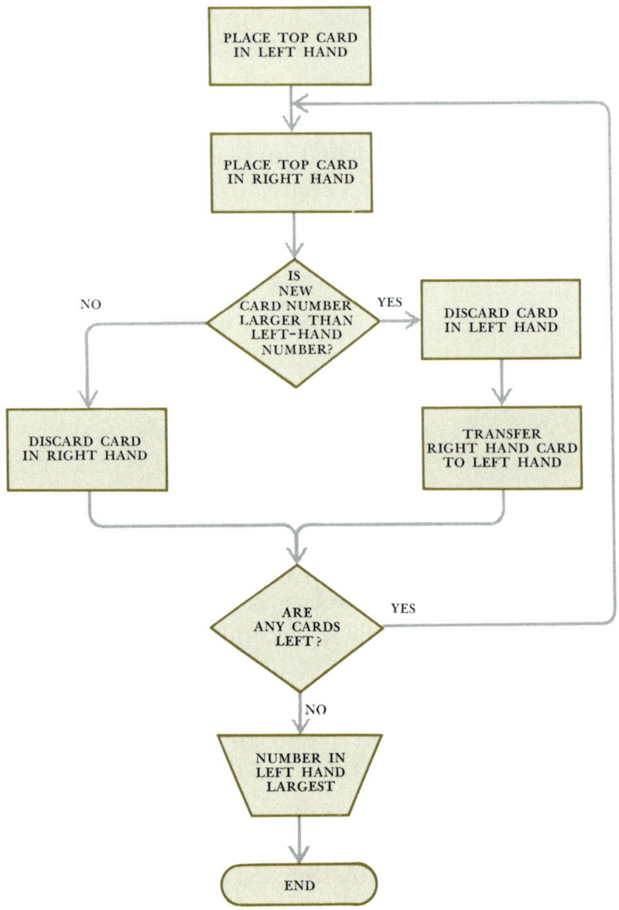

FIG. 4-3 FLOW DIAGRAM OF THE HAND SOLUTION OF THE SEARCH PROBLEM.

amined takes advantage of the fact that none of the data cards has the value of zero. As the value of each card is read into the computer, it can be checked to see if it is zero. Then by adding a card with a zero to the bottom of the stack, the presence of zero informs the computer that all the cards have been read in. The IF statement reacts to an input value of zero, ending the reading of values when zero is read in.

A similar procedure can be used to compare the value on the cards. Suppose that the value on the first card read in to the computer is assigned the symbol A, and the value on the next card the symbol B. The program could then instruct the computer to calculate a third quantity Z = A − B. If Z is positive, A is larger than B; if Z is negative, B is larger than A. Using this approach, a flow diagram (Fig. 4-4 on page 94) for the computer solution to this problem can be devised.

Figure 4-4 can be translated into a Fortran program:

Fortran statement	*Step number*
READ 101,A	1
4 READ 101,B	2
101 FORMAT (F15.5)	3
IF (B) 2,2,3	4
3 Z = A − B	5
IF (Z) 5,4,4	6
5 A − B	7
GO TO 4	8
2 TYPE 101,A	9
END	10

In this listing a step number has been assigned to each statement to identify it in the discussion. As far as the computer is concerned, only statements with program statement numbers are identified. These are the instructions to be loaded into the computer to find the largest value, and they can be analyzed to see how the computer would interpret them and carry them out. The program is in a form that could be punched on cards and would work in almost all computers capable of handling Fortran as a programming language.

A computer operates on this program in two phases: first translation and then execution. The following analysis considers the execution cycle only. In execution the computer begins with the first statement in the list and then follows each step in order unless instructed otherwise. The exception to this occurs with statements like FORMAT and END, which are involved in the translation from Fortran to machine language. Therefore execution begins with step 1.

Step 1 READ 101,A This instruction tells the computer to read a value from a card in the card reader and assign it to the symbol A.

Step 2 4 READ 101,B This instruction tells the computer to read a value from a card in the card reader and assign it to the symbol B.

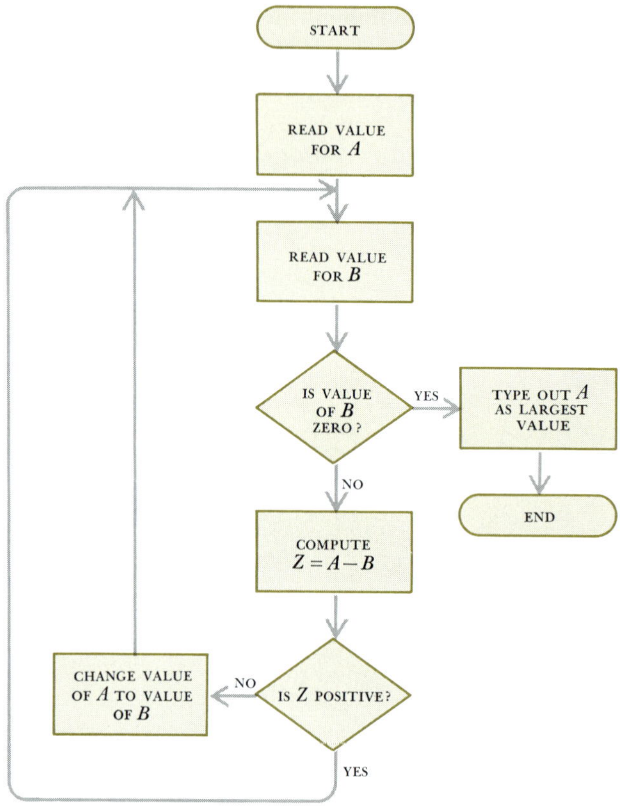

FIG. 4-4 FLOW DIAGRAM OF THE COMPUTER SOLUTION OF THE SEARCH PROBLEM.

Step 3 101 FORMAT (F15.5) This is not an executable statement but a specification to the Fortran translator program. It specifies that numbers read in or typed out under its control are to be considered 15 characters long, with the decimal to be placed between the fifth and sixth characters from the right side of the number.

Step 4 IF (B) 2,2,3 This instruction tells the computer to examine the algebraic sign of B. If it is negative, the computer is to take its next instruction from program statement 2 (step 9). If the value of B is zero, the next instruction is also statement 2. If the value of B is positive, the next instruction is to be taken from program statement 3 (step 5).

Step 5 3 Z = A − B This instruction tells the computer to calculate the value of A − B and assign it to Z.

Step 6 IF (Z) 5,4,4 This instruction tells the computer to examine the value assigned to Z. If it is negative, then the computer is to take its next instruction from program statement 5 (step 7). If the value is zero or positive, the computer is to take its next instruction from program statement 4 (step 2).

Step 7 5 A = B This instruction tells the computer to assign the value of B to A. The old value of A is lost, but B retains its value.

Step 8 GO TO 4 This instruction tells the computer to take its next order from program 4 (step 2).

Step 9 2 TYPE 101,A This instruction tells the computer to type out the value of A. The form to be used will be found in program statement 101, which must be a FORMAT statement.

Step 10 END This is a message to the translator program telling it that the last instruction of the program has been read in and that it can complete the translation into machine language and set up the computer to begin the execution.

The key steps of this program are 4 and 6. They are the decision-making steps. In step 4 the input value is checked to see if it is zero. If it is, then all the data cards have been read in and the largest value has been found and should be typed out. This is done by telling the computer to jump to the instruction for typing out the value of A. If there are more data cards to be read, this instruction tells the computer to continue. Step 6 looks for the largest value by keeping the value of A if it is larger, that is, if A − B is positive, and replacing it if it is smaller, that is, if A − B is negative.

To use this program each statement is punched on a card. These cards, in order, together with certain key cards telling the translator program about the general nature of the program, are then loaded into the computer. The computer, under the control of the translator program, then translates the program from Fortran to machine language. Finally it goes to the beginning of the resulting machine-language program and carries it out. In the present program this means that the computer turns on its card reader and begins looking for the largest value.

Modern computing systems have card readers capable of reading 1500 or more punched cards per min. In many systems even this rate of data input is too slow, so that the cards are not read directly. Instead they are read by a second machine, which transfers the data from the cards to magnetic tape, much like the tape used in a tape recorder. The main machine then reads the information from the tape.

The present program indicates that the output is to be typed, in this case, by an electric typewriter controlled by the computer. More complex machines are not restricted by the speed of 2 characters per sec of an electric typewriter but use a "supertypewriter" known as "high-speed printer." These devices can produce 1100 or 1200 lines of 130 to 150 typewritten characters per min. Some models, which project a picture on special paper, can print out more than 30,000 lines of information per min. Although the speed of the card reader seems fast by human standards, we should remember that the speed of calculation of a computer is still faster.

If a man's output were related to his speed of mental calculation in the same way a computer's is, he could write only one alphanumeric character every 20 hr. Devices such as the high-speed printer and card reader are only 10,000 or so times faster than their corresponding human activities. Of course, the input and output of computers is not limited to these two devices. Input takes many forms, such as direct temperature readings from a thermostat or weight readings from a scale. It also takes the form of typing on an electric typewriter or electronic signals from a radar or radio. Pictures sent by space probes often take the form of coded numbers. The picture is taken in the form of hundreds of dots of gray, as in a newspaper photograph. Each dot is assigned a numerical value, depending on how light or dark it is. These numerical values are then transmitted directly back to a computer on earth that translates the numbers back into a picture. The radio receiver may be an integral part of the computer itself. Output also takes many forms. It takes the form of sound when electrical patterns are converted into sound over a loudspeaker. It also takes the form of a rush of water from a dam as a computer remotely opens a valve. And it takes the form of jet thrusts as a computer controls the speed and direction of a spaceship reentering the earth's atmosphere. Some of the most useful input-output devices are built around television-like screens. A computer translates mathematical equations into pictures and displays them on the screen, and a man using the computer can communicate directly by "writing" on the screen, using electronic "light pens" to modify, adjust, and interpret the electronic pictures he sees. This form of data input is known as "stylus" input. As more and more computers become equipped with these "picture-tube" input-output devices, many problems of communication with the machine are being solved.

4-2 EXERCISES

1. What aspect of the program in the present section prevents the use of zero and negative data values?
2. Modify the program so that the input data can be expanded to allow negative if not zero values.
3. A simple change in program statement 3 can be used to modify the program so that it looks for the smallest rather than the largest value. What is this modification, and why does it produce such a change?
4. A variation of the problem calls for typing out values every time a number larger than any previous one is found in the data stack. Modify the program so that, every time a value of B larger than any previous one is read in, it is typed out.

Fortran can be used to read in more than one quantity from a single card. When this is to be done, the names of the quantities to be put in are listed in a single READ statement. Naturally the specification used in the FORMAT statement must also be modified. For example, suppose that one wished to read in two quantities: one to be called A and the other N. The value of A is typed in the first 15 columns of the input card with the decimal five places from the right, and the value of N is punched in the next seven card columns. READ and FORMAT statements that could be used to put in these values would look as follows:

```
   READ 56,A,N
56 FORMAT (F15.5,I7)
```

One can see that the first quantity identified in the list of quantities of the READ statement is associated with the first specification in the FORMAT statement and that the second quantity in the READ statement is associated with the second specification in the FORMAT statement. If more quantities were included in the single input statement, additional specifications would have to be added for each new quantity.

Reading in two quantities at once could be very useful in problems such as the one considered here, for example, if the data were modified so that the zero or negative quantities could appear. If data cards with zero are to be allowed, the method used to check for the last data card has to be modified, because it depends on the only zero data card being the last, an artificially introduced card. To make this modification, one might modify statement 4 of the program to read in two quantities: B as before and a new quantity C. And C could be the number punched in the second set of 15 columns on the data cards. Because there are no punches in these columns, the value of C would be zero for all the regular data cards. A special data card with the number 1 punched in the second set of 15 columns could then be placed at the end of the data. With the value 1 assigned to C, the program could then be modified to use 1 to key the switch to the output portion of the program.

5. Modify the flow diagram in Fig. 4-4 to allow zero and negative data input values in the manner described above.

6. Modify the Fortran program given in the present section to allow zero and negative data input values.

7. Assume that there were 48,677 data cards to be examined. Further assume that a clerk could be hired to examine them for the largest value. The clerk can examine, on an average, and including moving boxes of cards, and so on, one card per 5 sec. The clerk costs the company a basic salary of $400 for a 22-day work month (8 hr per day) with an additional cost of $125 overhead per month. How much would it cost to find the largest of the 48,677 values under these conditions?

8. Assume that a basic computer model rents for $44,250 per month and that a work month for the machine is 30 days each 22 hr long. The $44,250 includes the cost of the space and the personnel needed to run the machine. If the computer requires $\frac{1}{1000}$ sec to read a value from a data card, and $\frac{1}{100,000}$ sec to check the value it has read in, and $\frac{1}{100}$ sec to print out the final answer, what is the cost of finding the largest value if there are 48,677 data cards?

9. If one wished to change the name of the variables in the program in the present section from A and B to M and N, what other changes would have to be made?

10. Suppose that step 1 of the program was changed to read A = 0. Why or why not would the program still find the largest value?

11. Five of the statements in the program are identified by program statement numbers. Why is this necessary?

12. When students are asked to write a program of this kind, a common error is to write program statement 5 as B = A rather than A = B. How would this error affect the answer printed out by the computer?

13. Modify the program so that it finds and types out both the largest and the smallest value.

4-3 THE AVERAGE PROBLEM

A problem that is slightly more complex but similar to the one in Sec. 4-2 illustrates more of the thinking used in writing computer programs. Suppose that the same data used in the search problem are used again. The problem involves a deck of cards, with one 15-digit number punched on each card. The number of cards in the deck is unknown, and there is no convenient way to count them, short of using a computer. Also, as in the search problem, all the numbers on the cards are positive. The problem is to find the average value of the numbers.

Because the average value of a set of numbers is defined as their total value divided by their number, the problem is twofold. First the total value of all the numbers has to be found, and then the number of numbers. The average is found by division, the total being divided by the number of numbers. How would one go about doing this problem "by hand," that is, with a desk calculator rather than a computer. In all probability a plan

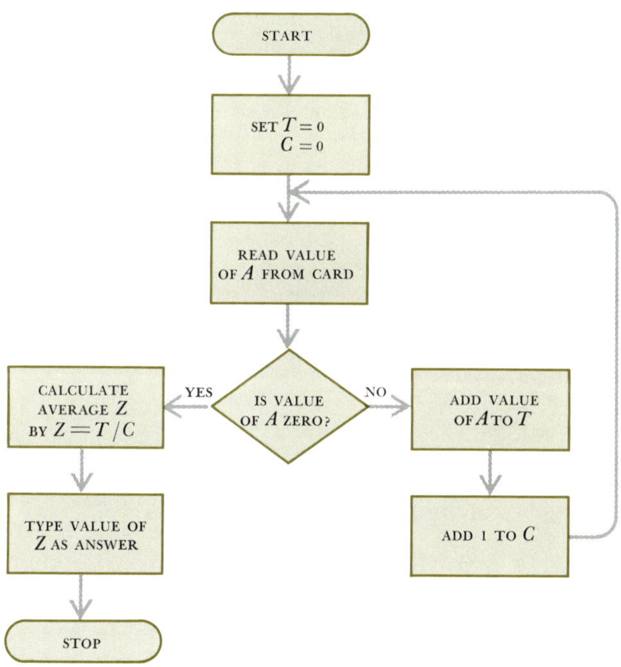

FIG. 4-5 FLOW DIAGRAM OF THE COMPUTER SOLUTION OF THE AVERAGE PROBLEM.

along the following lines would be used. First the calculator would be cleared. Then the value from the first card would be entered. Because the number of cards must be determined, a tally indicating that a value had been entered would be made on a separate piece of paper. As each value was read in, it could be added to the total simply by not clearing the machine, and the tally would be increased by one. Once all the values had been read in, the tally could be totaled and divided into the total of values to get the average.

To translate this method into a Fortran program, symbolic names have to be assigned to the quantities. Let A stand for the value read from an individual card. Let C stand for the number of cards that have been added to the total at any point. Let T stand for the total value of the cards that have been read by the computer. Every time a value is read in and assigned to A, it should be added to a running value of T and the value of C increased by 1. Figure 4-5 is a flow diagram using this approach.

As in the search problem, the solution uses a special zero data card placed at the end of the deck. The Fortran program is as follows:

Fortran statement	*Step number*
T = 0	1
C = 0	2
62 READ 307,A	3
307 FORMAT (F15.5)	4
IF (A) 9,10,9	5
9 T = T + A	6
C = C + 1.	7
GO TO 62	8
10 Z = T/C	9
TYPE 307,Z	10
STOP	11
END	12

A detailed analysis of these steps follows.

Step 1 T = 0 This instruction tells the computer initially to assign the value of zero to T. It performs the same function as the initial clearing of the desk calculator.

Step 2 C = 0 This instruction tells the computer initially to assign the value of zero to C.

Step 3 62 READ 307,A This instruction tells the computer to read a value from a card and assign it to A. The numeral 62 identifies the statement for future reference. 307 informs the translator program that the form in which the data are to be punched on the card will be found in program statement 307, which must be a FORMAT statement.

Step 4 307 FORMAT (F15.5) This is an information statement for the translator program. It tells it that 15 characters will be involved in input or output statements that use statement 307 as a FORMAT reference. The decimal is to be placed five characters from the right.

Step 5 IF (A) 9,10,9 This instruction tells the computer to examine the sign of A. If it is positive or negative, the computer is to take its next instruction program statement 9 (step 6). If the value of A is zero, the computer is to take its next instruction from program statement 10 (step 9).

Step 6 9 T = T + A At first glance this instruction does not seem to make much sense. However, the computer is set up to perform the operations on the right of an equal sign first. This means that, in executing this instruction, the computer first calculates the value of T + A. It then assigns this value to T, destroying the old value of T. This is precisely the desired result here. It has the effect of assigning to T a running total of all the A values read in. Because T is initially zero, when all the cards have been read in, it will have a value equal to the total of all the card values.

Step 7 C = C + 1. This is similar to step 6. Every time the computer executes this step, the value of C is increased by 1. The computer first calculates the value of C + 1., and then assigns this value to C. Because the value of C is initially zero and the computer executes this instruction once every time a card is read, when all the cards have been read, the value of C will be equal to the total number of cards.

Step 8 GO TO 62 This instruction tells the computer to take its next order from program statement 62 (step 3).

Step 9 10 Z = T/C This instruction tells the computer to calculate the value of T/C, or T divided by C. This value is to be assigned to Z. Z is the desired average, because this instruction is not executed until all the nonzero cards have been read.

Step 10 TYPE 307,Z This instruction tells the computer to type out the value of Z. It is to use the format described in program statement 307, which is a FORMAT statement.

Step 11 STOP This is a control statement of a kind not yet discussed. It tells the computer to stop. When the start key is pressed, the computer automatically goes on to the next program and begins its translation and execution.

Step 12 END This step tells the translator program that the entire program has been read in and that it should begin the conversion to machine language.

There are variations of this simple kind of calculation. In problems a number of quantities can be punched on each data card. This kind of program can easily be modified to read in and calculate many of a number of quantities needed in statistical analysis. Several of these quantities, together with the method of calculation are considered in Chaps. 9 and 10. It is also possible to modify the input instructions so that all the "raw" data can be read in and electronically stored in the computer memory. Thus the data can be searched or examined without going through the long process of reading in the individual cards again.

Almost all modern computer systems use two kinds of memory. One is characterized by the high speeds with which data are read in and read out. The standard "hardware" or equipment used in this kind of memory is a small (about $\frac{1}{50}$ in. in diameter) ferroceramic "doughnut," known as a "core." Each of these little circles, made of iron and ceramic materials, is magnetic. Like any magnetic material each core has polarity; that is, it has a North Pole and a South Pole. Depending on how the polarity in a core is orientated, the core stores a zero or a 1. A large computer may have as many as 10 million cores in its high-speed memory. Newer computers have even more exotic memories, based on advanced technology. Although these memory units may bear strange names, such as "thin film memory" and "integrated superconducting memory," they are still based on the relatively simple concept of storing a zero or a 1 magnetically. Experimental memory units have been produced that can store 250,000 or more such zeros or ones on a single 5- by 5-in. card. In many of these newer kinds of memory the time required to retrieve information is measured in nanoseconds (one-billionth of a second).

To translate the size of these memories into more meaningful terms, recall that we could write all the numerals from 1 to 31 in only five 0,1 binary numerals. In computing jargon a binary numeral of memory is called a "bit," and five bits store numbers from zero to 31. Remember $31 = 11111_2$. Because there are 26 letters in English, five bits could be used to store any letter with a simple $1 = A$, $2 = B$, and so on, code. For example, the word *computer* could be stored in 8 five-bit groups. The first group would store 00011_2 or 3 for C, the third letter of the alphabet. The second group would store 01111_2 or 15 for the fifteenth letter of the alphabet, O. The third group would store 01101_2 for M, the fourth 01000_2 for P, and so on until every letter was stored in numerical form within the computer memory. With this kind of system a 10 million core memory could be thought of as storing or retaining about 2 million English letters.

The second kind of memory found in almost all computer systems is characterized by its ability to store large amounts of data. This kind of memory could be called "high-volume" memory. High-volume memories are required to hold the information and data required of many problems. High-volume memories are considered slow, because the time required to store data and retrieve them is longer than a similar operation in high-speed memories. The time it takes for such information retrieval is fast when compared with that required for a man to write the corresponding symbols or even to focus his eyes on what he has written. The majority of high-volume memories in use are based on the same idea used in an everyday sound tape recorder; that is, they store information by magnetizing a thin film of magnetic material coated or deposited on a surface. In an ordinary tape recorder and in many computer applications this coating is placed on plastic tape and in others on disks like phonograph records rotated at high speed under "read-write heads" for data input and output. Such devices may hold up to 800 million bits, and in many applications several such machines may be linked together. Other high-volume devices, based on electronic circuitry and photographic technology, have been designed to hold up to 1 trillion bits of data. With such devices, the total records of a major company or government subdivision could, in a matter of seconds, be electronically stored with any part of the information available.

There are advantages to reading and storing large amounts of data and working with them internally. For example, in the average problem, once the average had been found, it might have been useful to go back through the whole mass of data and search for values near the average. This is a simple matter if all the data have been electronically stored in the computer itself, for the relatively long process of reading the cards does not have to be repeated. If all the data are in the computers, they can be ordered, systematically modified, or adjusted at electronic speeds rather than by tedious changes. The logical extension of this idea is the complete and permanent storage of records in electronic form. Magnetically stored data are, in many ways, as stable a form of information storage as written records. And they have a number of advantages. They are more compact, and any piece of data can be retrieved in a fraction of the time it would take to search comparable written records.

4-3 EXERCISES

1. Name two machines that do whatever they do in the same way you would do it by hand.

2. Name two machines that do whatever they do radically differently from the way it would be done by hand.

3. Steps 1 and 6 are very closely related in this program. In view of the computer's method of following the instruction given in step 6, why would the execution of the program stop at step 6 if step 1 were omitted?

4. A student makes an error in typing up the cards to be used to put the program into the computer. As a result step 5 is typed as

    IF (A) 10,10,9

    In the problem as described, why or why not would this error make a difference in the average found by the computer?

5. Suppose the following changes were made in the program: step 1 changed from T = 0 to 6 T = 0 and step 11 changed from STOP to GO TO 6. What would such a change accomplish?

6. Modify the program in the present section so that the computer types out a list of the data values as well as finding the average value.

7. Calculate the largest number of base-10 numerals that can be stored in a high-speed memory of 10 million cores.

8. Calculate the largest number of English letters that can be stored in a memory of 10 million cores.

9. Suppose a high-speed printer could type 1200 lines of 150 characters per min. How long would it take it to "dump" or print out every bit in a memory of 800 million bits (1 bit = 1 character of type in this case)?

10. Assume that the average student at a university has 14 letters in his name. If there are 30,000 students and each is assigned an eight-digit identifying number, how many bits of memory are required to store a listing of students and their numbers?

11. Modify the program in the present section so that it finds the largest as well as the average input value. (*Hint:* Make use of the program in the preceding section.)

12. Modify the program so that it types out the average only if it is larger than 10.

4-4 THE PRIME-NUMBER PROBLEM

In many kinds of problems computers lead to completely new methods of solution, or if not new methods, they make older methods acceptable that formerly were not. A relatively simple problem from the theory of numbers illustrates this.

The theory of numbers is the branch of mathematics dealing with some of the properties of whole numbers, numbers such as 0, 1, 2, 3, ... One of the more interesting areas of this field is prime numbers. Whole numbers can be divided into two classes: those like 6, 15, and 60 which can be broken down into the product of other whole numbers, and those like 3, 7, and 31 which cannot be written as the product of whole numbers other than themselves and 1; that is, $6 = 3 \times 2$ and $15 = 3 \times 5$ and $60 = 5 \times 12$ or $60 = 5 \times 4 \times 3$ or $60 = 2 \times 2 \times 3 \times 5$, but the only way to write 3 as a product is 3×1 or 31 as a product is 31×1. Numbers that do not have whole-number factors, that is, cannot be written as the product of other whole numbers except themselves and 1, are known as "prime" numbers. All other whole numbers are nonprimes. A list of primes and nonprimes less than 100 follows:

Prime numbers	*Nonprimes*
1, 2, 3, 5, 7, 11, 13, 17, 19	4, 6, 8, 9, 10, 12, 14, 15, 16, 18, 20
23, 29, 31, 37, 41, 43, 47, 53	21, 22, 24, 25, 26, 27, 28, 30, 32
59, 61, 67, 71, 73, 79, 83, 89	33, 34, 35, 36, 38, 39, 40, 42, 44, 45
97	46, 48, 49, 50, 51, 52, 54, 55, 56, 57
	58, 60, 62, 63, 64, 65, 66, 68, 69, 70
	72, 74, 75, 76, 77, 78, 80, 81, 82, 84
	85, 86, 87, 88, 90, 91, 92, 93, 94, 95
	96, 98, 99

It is easy to see that, with the exception of the number 2, prime numbers are odd. It is also easy to see that not all odd numbers are prime. How does one know if either or both of the numbers 1991 and 1993 are prime?

This is all background to the basic problem considered in the present section, namely, testing numbers to see if they are prime. There is one sure and simple method of testing a number: divide it by all numbers small than itself (actually one has only to try all the odd prime numbers smaller than the square root of the number). If one wishes only to test for "primeness," as soon as a number divides evenly into that being tested, it is identified as nonprime. Only after all the divisors have been tried is the number being tested verified as prime. The square root of both 1991 and 1993 is near 44.5; that is, $44 \times 44 = 1936$ and $45 \times 45 = 2025$. If two whole numbers are to be multiplied together and the result is to be 1991 or 1993, one must be less than 45. If both were greater than 45, the product of the two would have to be greater than 2025. From the list of primes above a second list can be made of "trial divisors" necessary to test 1991 and 1993 for primeness. The list would contain 3, 5, 7, 11, 13, 17, 23, 29, 31, 37, 41, and 43. A "by hand" test follows:

$1991 \div 3 = 663$	remainder 2
$1991 \div 5 = 398$	remainder 1
$1991 \div 7 = 284$	remainder 3
$1991 \div 11 = 181$	remainder 0

Thus 1991 is a nonprime, because it can be written as $1991 = 11 \times 181$. Testing 1993 in the same manner,

$1993 \div 3 = 664$	remainder 1
$1993 \div 5 = 398$	remainder 3
$1993 \div 7 = 284$	remainder 5
$1993 \div 11 = 181$	remainder 2
$1993 \div 13 = 153$	remainder 4
$1993 \div 17 = 117$	remainder 4
$1993 \div 19 = 104$	remainder 17
$1993 \div 23 = 86$	remainder 15
$1993 \div 29 = 68$	remainder 21
$1993 \div 31 = 64$	remainder 9
$1993 \div 37 = 53$	remainder 32
$1993 \div 41 = 48$	remainder 25
$1993 \div 43 = 46$	remainder 15

Because every trial divisor leaves a remainder, 1993 must be a prime number. The most systematic method would be to divide by every odd number less than the square root of the number being tested, because it is easy to list all such numbers. This would be harder but one would not need any knowledge of which odd trial divisors were prime and which were not. Let N represent the number being tested and M a trial divisor, and begin with $M = 3$. N must be odd. 2 is the only even prime. The method of testing is outlined in Fig. 4-6. To convert the flow diagram in Fig. 4-6 into program it is necessary to use another aspect of Fortran. If in a Fortran program I, J, K, L, M, and N are used as symbols for quantities, then the quantities are not carried with a decimal point but to the nearest

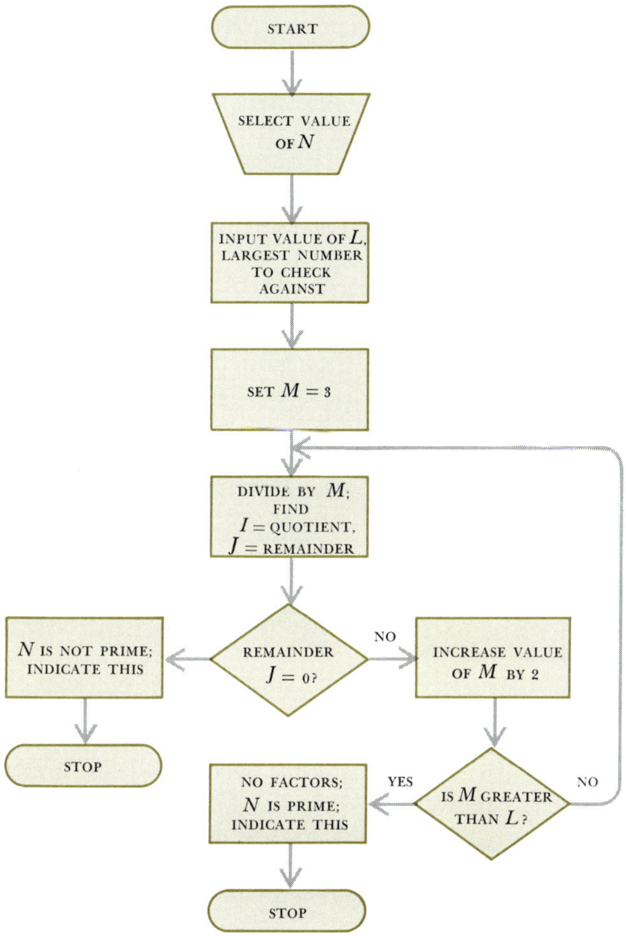

FIG. 4-6 FLOW DIAGRAM OF CHECKING A NUMBER AS PRIME.

whole number. Thus if N = 28 and M = 3, then N ÷ M = 9 and the remainder of 1 is lost. This principle can be used to find remainders. The number being tested N can be divided by the trial divisor M, the result being called I. I is then the largest whole number of times M divides N. If I is multiplied by M, the result is not necessarily N but may be the largest whole-number product of M less than N. If I * M is subtracted from N, the result is the remainder obtained by dividing M into N. The Fortran program is as follows:

Fortran statement	*Step number*
READ 462,N	1
READ 462,L	2
462 FORMAT (I10)	3
M = 3	4
7 I = N/M	5
J = N − I * M	6
IF (J) 3,4,3	7
3 M = M + 2	8
K = M − L	9
IF (K) 7,7,8	10
8 TYPE 1	11
1 FORMAT (5HPRIME)	12
STOP	13
4 TYPE 2	14
2 FORMAT (8HNONPRIME)	15
STOP	16
END	17

Step 1 READ 462,N

Step 2 READ 462,L

Step 3 462 FORMAT (I10) These statements are to read in values for the number being tested N and the nearest whole number to the square root of N, which is L and has been calculated previously. The indicator in the FORMAT statement has been changed from that of the previous programs, because only whole numbers are used here.

Step 4 M = 3 This statement sets the initial test divisor at 3.

Step 5 7 I = N/M

Step 6 J = N − I * M These two statements calculate the quotient of N ÷ M to the nearest whole number and also the remainder. Because whole-number division loses the remainder, step 6 is necessary to recover it. For example, suppose N = 34 and M = 5. Then I = N/M is $I = \frac{34}{5} = 6$, and the remainder J can be calculated thus:

$$J = 34 - 6 \times 5$$
$$J = 34 - 30$$
$$J = 4$$

Thus J = 4 is the amount "lost" because of the kind of division used.

Step 7 IF (J) 3,4,3 The job of this statement is to check the remainder.

If it is zero, the number is not prime and the computer "jumps" to statement 4 to indicate this. If the remainder is not zero, the next instruction to be followed is statement 3 (step 8).

Step 8 3 M = M + 2

Step 9 K = M − L

Step 10 IF (K) 7,7,8 Step 8 increases the value of the trial divisor by two units. Step 9 calculates the difference between this new trial divisor and the test value entered in step 2. If K is negative or zero, meaning M is still not larger than L, then the program instructs the computer to return to statement 7 (step 5) for its next direction. If K is positive, M, the trial divisor, is greater than L, the test number. In this case the program goes to statement 8 (step 11) and indicates that no factors can be found.

Step 11 8 TYPE 1

Step 12 1 FORMAT (5HPRIME)

Step 13 STOP These directions instruct the computer to type the information in statement 1 (step 12), which is simply the word *prime*, and then stop.

Step 14 4 TYPE 2

Step 15 2 FORMAT (8HNONPRIME)

Step 16 STOP

Step 17 END These statements are to type out the word *nonprime* if this is so. END informs the computer that the end of the program has been reached, and translation into absolute language can begin.

In many respects this simple program, to test numbers to see if they are prime, illustrates what may be the most important aspect, that is, computer speed of calculation, which has steadily and dramatically increased since their commercial introduction in the early 1950s. Early computers used vacuum tubes, which act as an electronic switch and are thus suitable for storing binary information. In these computers thousands of operations could be performed in a single second. Transistors had been developed in 1948, and an attempt was made to apply them to computers. However, it was not until 1958 that a transistor was developed with sufficient speed to be practical for computers. Almost all the computers in use by the middle 1960s were of the transistor or solid-state kind. They were called second-generation computers.

More research has led to a third generation of computers. In these machines, 150 transistors wired in a complex circuit can be replaced by a single chip of silicon $\frac{6}{100}$ in. square that is to transistor technology what transistors were to the vacuum tubes they replaced. In 1 hr a vacuum-tube computer performed a "man-year" or more of calculations. These same calculations take only $\frac{1}{2}$ sec on a third-generation computer. In other words 7.2 million operations or more per sec are possible with third-generation and later computers.

To illustrate the significance of this, suppose that a method of solving a problem involved 120,000 separate calculations. If a trained operator could perform 1 calculation per min and assuming that he did not make any errors and did not go mad, he could solve the problem in 1 year (including a 2-week vacation). For the sake of argument, assume that the operator cost his employer \$6000 per year. Then the solution to the problem cost \$6000 and a year's delay. If the solution were nonsequential, that is, the calculations at one point did not depend on those at others, the problem could be divided among several operators. This would not reduce the cost but it would reduce the delay. It should be mentioned, however, that almost all problems are sequential and cannot be divided. If a third-generation computer were set up to do the same calculations, the total time of calculation would be reduced to a fraction of a second. In reality 2 or 3 sec would have to be added because of the delay in the input and output functions, which are relatively slower than calculation. Using a time of 3 sec, let us estimate the cost of the solution. Assume that the computer system costs \$300,000 per month. However, it runs 24 hr per day, 7 days per week. Assume that a month is 30 days and the machine can be used only 22 hr per day, the other 2 hr being spent on maintenance. In a "work month" there would then be $30 \times 22 \times 60$ min, or 39,600 min. Thus the computer would cost

$$\$300{,}000 \div 39{,}600 = \$7.58 \text{ per min}$$

Because the solution takes 3 sec, or $\frac{1}{20}$ min, it costs $\$7.58 \times \frac{1}{20} = 38$ cents as opposed to \$6000. Actually there is more time involved than 3 sec, unless the program for solving the problem is already available. The important point is that methods of solving a problem, whose major objection to their use was time and expense, are becoming the fastest and cheapest way, thanks to computers. In the sample problem for finding prime numbers, this means that older methods are now acceptable and even the cheapest in many cases. The routine and dull work can be left to the machine.

Few of us understand how fantastic the speed of calculation by computers is. They have not been used long enough. It is something like what would have happened between 1870 and 1890 if transportation speeds had gone from the 14-day coast-to-coast speed of the train to the 6-hr speed of the modern subsonic jet.

4-4 EXERCISES

1. Use the method outlined in Fig. 4-6 to test 2009 for primeness.

2. Test 2011 to see if it is prime.

3. Test 2013 to see if it is prime.

4. Modify the program so that it types out a factor if it divides the number being tested evenly.

5. Modify the program so that it types out all the numbers that divide the number being tested evenly, that is, with a remainder of zero.

6. Many modern industries depend on the speed of modern transportation or communication. Give two examples of each kind.

7. Assume that a certain computer can calculate the product of two 16-digit numbers in $\frac{1}{300,000}$ sec and that it can type the 32-digit answer at the rate of 1 digit per $\frac{1}{180,000}$ min. Assume that the ratio of a man's rate of calculation to his rate of outputing data, that is, writing down the answer, is the same as the computer's. If it takes him 3 sec to multiply 24 times 5 in his head, how long would it take him to write down the answer?

8. There are a number of modern industries that could not exist without modern computers. One example is the large, multibranch bank, which may have to process the transactions of 1000 or more branches every day. Suppose that such a banking system would have to "post" or process 10 million transactions between the close of business on one day and the opening on the next. If one clerk, using modern billing machines, can process a transaction in 30 sec, how many clerks, each working an 8-hr shift, would be needed to process 10 million transactions?

9. Among the many other Fortran statements one can use is one that permits the computer operator to modify processing as the computer executes the program. A simple example of such a command, or problem statement, is the IF SENSE SWITCH statement, which has the following form:

```
IF (SENSE SWITCH 1) 3,7
```

When the computer executes this command, it examines the state of a switch located on the control console. If the switch is on, the computer jumps to the program statement listed first, in this case, program statement 3. If the switch is off, the computer jumps to the second program statement identified in the IF SENSE SWITCH instruction, in this case, program statement 7. Incorporate an IF SENSE SWITCH statement and any other necessary statements into the program so that, if the switch is on, the computer types out only the words *prime* or *nonprime*, and if the switch is off, the computer types out factors of the number being tested, which proves that it is not a prime, or the word *prime* if this is so.

10. Using the method of finding remainders and quotients outlined here, write a Fortran program that reads in two integers, the one the number to be divided and the other the divisor, and then types out the quotient and remainder of the division.

4-5 THE CONVERSION PROBLEM; CONCLUSIONS AND COMMENTS

It would be unfair to leave our discussion of computers and computing without outlining a program to carry out the change-of-base problem in Sec. 3-3. For a flow diagram we can use the one given there. The Fortran program can make use of the ideas examined in the previous sections of the present chapter:

Fortran statement	*Step number*
READ 9,NC	1
9 FORMAT (I10)	2
NS = NC	3
NA = 1	4
NB = 0	5
4 NQ = NC/7	6
NR = NC − NQ * 7	7
NB = NB + NA * NR	8
NA = NA * 10	9
IF (NQ) 5,6,5	10
5 NC = NQ	11
GO TO 4	12
6 TYPE 9,NS	13
TYPE 10	14
10 FORMAT (20HWRITTEN IN BASE 7 IS)	15
TYPE 9,NB	16
STOP	17
END	18

In this program, all the quantity names are prefixed with N so that the computer will treat them as whole-number rather than decimal quantities.

Step 3 NS = NC This step is to "save" the original value of NC for use later in typing the final result. (Remember that the value of C was changed two or three times in the problem in Chap. 3.) In the original outline for this problem the fourth instruction read as follows: divide the value of C by 7 to the nearest whole number and find a quotient and remainder. In the Fortran program, this instruction becomes:

Step 6 4 NQ = NC/7

Step 7 NR = NC − NQ * 7 These instructions follow the pattern developed in Sec. 4-4 to recover a remainder. The next two steps in Sec. 3-3 read as follows: multiply the remainder of step 4 by the value of A; change the value of B by adding the result of step 5 to the present value of B. These two instructions are compressed into one:

Step 8 NB = NB + NA * NR Since the calculations on the right are done first, NA is multiplied by NR and the result is added to NB. The function of NA is the same as before, to correctly position the digit determined by the division. Its value is changed every time with:

Step 9 NA = NA * 10 The remaining steps are for checking to see if the calculation is finished and for typing out the answer. These follow the same scheme as similar steps in the previous sections.

To nonprogrammers and nonoperators the most important thing about computers is that they are making us think again about the nature of problem solving. Their great speed makes such methods possible as that used for the prime-number problem in Sec. 4-4. Modern weather forecasting uses numerical procedures that can be carried out in a few minutes with computers. However, if computers could not be used, data taken today for the forecast of tomorrow's weather would require weeks to reduce to usable form and would be useless. Many other things would be impossible without computers. A large banking firm might have to process 6 to 10 million transactions a day, clearly impossible without machines, as illustrated in the exercises in Sec. 4-4. And without computers, almost every woman in the country would be needed as a telephone operator. Computers can also help to individualize studies of large groups of people. In the past any attempt to organize and deal with large numbers of individuals required the creation of broad classes into which they were placed, all members of a class being treated identically. The schools are a classic example. Unless a child made trouble and got special attention for it, he went through school more or less in lock step with all the other students. His individual needs could not be met, not because of a lack of desire on the part of educators, but because of the number of children and the economic impossibility of supplying each with the optimum individual instruction. Computers are permitting much more individual treatment, by aiding in more careful record keeping, by assisting school officials in flexible programming of instruction, and by assisting in the preparation of materials. In some cases computers are becoming the instructor in those phases of education where they are appropriate. The student uses computers as he would a book or a teacher asking questions. In many areas of the country there are no longer students in the first, second, or third grades; students are studying the primary-grade subjects at their own pace and advancing when they are ready. This approach to education requires modern computer-oriented data-handling techniques.

The second and perhaps more important area where computers are making startling changes is in the way man thinks. To design machines that can solve problems, man has had to examine basic questions relative to logic, problem solving, language, and communication. In devising machine and program languages and in their translation, he has had to study many areas of communication. Logic and the logical design of circuits to perform mathematical functions help to tie together many far-reaching concepts many of which are discussed in later chapters.

Computers will certainly not solve all the ills of mankind, but they are causing great changes. We are not only in the atomic age; we are also in a computer, electronic, and information revolution.

4-5 EXERCISES

1. Modify the program to convert numbers to base 8 rather than base 7.

2. Modify the program so that it begins by reading two values. The first number it reads in is to be the base it is to convert into and the second number the base-10 number to be converted.

3. Does the computer used in this program have to know anything about number bases, or does it have to do any arithmetic in bases other than 10? Explain!

4. The number being converted in the program outlined in Sec. 3-3 was "saved," as step 3 of the program in this section saved the number. How was this number saved in Sec. 3-3?

5. How many base-7 digits does the largest number, 9,999,999,999, this program converts have?
6. Why is 9,999,999,999 the largest number the program converts?

REVIEW EXERCISES

1. What single aspect of a modern computing system permits it to be used as a problem-solving device?
2. Why is there no such thing as a computer but computing systems?
3. One of the most useful aspects of modern computers is their speed. What is it about a computer that seems to be most responsible for this speed?
4. What aspects besides their size make computer systems different from a desk calculator?
5. What is Fortran short for?
6. What are the four kinds of statements in a Fortran program?
7. What is the difference between an executable and a nonexecutable Fortran statement?
8. How many items can appear on the left side of a Fortran arithmetic statement? On the right side?
9. Name two programming languages other than Fortran. Why is there more than one programming language in use?
10. What is the function of a programming language?
11. Is a programming language, such as Fortran, necessary in order to use a computer? What are the alternatives?
12. Programming languages are known as computer "software," and the computer itself as "hardware." Why have these names come into use?

A typical modern computer logic circuit.

FROM A FULL-COLOR PHOTOGRAPH COURTESY OF MOTOROLA SEMICONDUCTOR PRODUCTS, INC.

chapter five

SYMBOLIC LOGIC

In Chaps. 3 and 4 we indicated that computers can make logical decisions. But how can this be? How can we design a machine that uses logic? What is there about the structure and nature of logic that lends itself to a mechanical process? To answer these questions, we must clarify the nature of the principles of logic. We all believe that we think logically, although we admit that we do not do so all the time. Almost everyone claims that he can follow a logical argument and reason logically. Few of us have examined what actually occurs when we make logical decisions.

5-1 THE PLACE OF LOGIC IN MATHEMATICS

Since the beginning of recorded history man has wondered about the nature of reasoning. One of the most famous writers on the subject was Aristotle, who attempted to analyze reasoning in his own language, as did other logicians before 1800. Two nineteenth-century English logicians, Augustus De Morgan and George Boole, thought that one of the major problems in studying logic was that their only tool was everyday language. They thought that the processes of thought and inference could not be adequately examined using this method, and they set out to create a new and special language in which they could make a model of the logical operations of English. Just as an aircraft manufacturer might make a model of an aircraft to study its characteristics, or as an engineer might make a model of a new machine to see how well it worked, De Morgan and Boole made a model of logical reasoning. This model took the form of an algebra, which is a series of formulas and rules for working with a set of symbols. When one studies algebra in high school, he is studying the algebra of numbers. Mathematical logicians created an algebra of statements; that is, instead of studying how numbers add or multiply, or how to interpret a formula like $F = \frac{9}{5}C + 32$, they studied how statements are linked together to make logical patterns. Instead of working with symbols for numbers, they worked with symbols for statements. In Chap. 7 we examine still another algebra, that of sets. For the present, we consider how mathematicians have formalized and studied the logic they require. We also consider some of the results and ideas De Morgan and Boole encountered. As with many of the ideas we have discussed, these also have implications outside of mathematics. Two of the most important involve the uses of logic in argument and the idea of symbolic models. The power of model building has also become evident as it has been applied to areas other than mathematics. Mathematics itself has been developed, for the most part, as a model to study other subjects, notably the physical nature of the universe. Many difficult areas in the nature of thinking and model building have been encountered and worked on as mathematical problems. The nature of logical reasoning and proof are only two of these. To begin our study of symbolic logic, we must examine the basic "unit" of reasoning —the statement.

5-2 LOGICAL STATEMENTS AND CONNECTIVES

Statements of symbolic logic do not encompass all the statements of English. Rather "logical statements" are restricted to those which can be classified as either true or false, but not both at the same time. The references we make to "statements," then, refer to "logical statements." Such expressions as "Drive carefully" and "Are you feeling better?" cannot be considered either true or false and are outside of our discussion. By excluding phrases that are simultaneously true and false, we eliminate certain semantic paradoxes. Consider the statement "I always lie." If we assume the statement to be true, that is, that I am an inveterate liar, then the fact that I say it makes it false. Hence, this statement is, at once, true and false. Some sentences that are simple logical statements are:

Today is Friday.
All dogs smell.
Calculus is fun.
Interplanetary travel will be possible by 1978.

Much of our study of symbolic logic will involve combining simple statements into compound or more complex ones. We should, then, define "simple statement." However, we are hard pressed to do so. As is often the case in mathematics and logic, the basic terms are difficult to define. We said that a statement of logic is a sentence that can be classified as either true or false. We may say that a simple statement of logic contains only one idea. Now we are faced with the prospect of defining "idea." If you looked up the word *point* in an old dictionary, it would be defined as the intersection of two lines, and a line would be defined as a set of points. Such "circular" definitions are usually satisfactory for words used in everyday language. In mathematics, it is recognized that not every word can be defined in terms simpler than itself. We must have a starting point. The "starting-point" words are called "undefined terms." Some examples of these from geometry are point, line, and plane. Such concepts, or terms, are also sometimes called "primitive concepts." The idea of a simple statement will be a primitive concept for our study of logic, so that we shall not define it. It is enough to say that "I am sick" is a simple statement of logic and "I am sick and tired" is a compound statement of logic, and compound statements are combinations of two or more simple statements.

In logic, we are concerned with the compounding of simple statements in general, but not with specific simple statements; that is, we study general formulas somewhat like $A = LW$. In this equation for the area of a rectangle, A, L, and W are used in place of numbers. They may be replaced by various numbers at various times. We shall use letters to replace simple statements. For example, the letter p could be used to represent the statement "I am a student"; q could represent "I am a good dancer"; t could represent "Today is a holiday." As we make more complicated statements, we shall also use symbols for the words that link simple statements together. The truth or falsity of a compound statement depends on the truth of its simple components and how they are connected. For the compound statement "I am a student and I am a good dancer" we have defined

"I am a student" as p and "I am a good dancer" as q. To represent this statement as a formula, we could write "p and q." However, we wish to eliminate the English language from our logic, and so we choose a symbol for the conjunction *and*. This symbol is $\wedge$, and the above phrase can be written $p \wedge q$. In the same manner $\vee$ will be substituted for *or* and $\sim$ for *not*. The symbols $\sim$, $\wedge$, and $\vee$ are called "connectives." We know that $\sim$ does not really connect statements, but we shall group it, somewhat arbitrarily, with symbols that do. Later we shall introduce other connectives. For the present, some examples will familiarize you with these.

Let p stand for the statement "I am going water skiing" and q for "I am going to play golf." Then, if we wished to abbreviate the compound statement "I am going water skiing and I am going to play golf," we should write $p \wedge q$. For a less active schedule we might say "I am going water skiing or I am going to play golf." This is symbolized by $p \vee q$. If we hate water, we might say: "I am not going water skiing." In symbolic form: $\sim p$. For more examples, let r be the sentence "Me Tarzan" and s be "You Jane":

$r \wedge s$ is "Me Tarzan and you Jane."
$\sim r \wedge s$ is "Me not Tarzan and you Jane."
$r \vee \sim s$ is "Me Tarzan or you not Jane."

When a statement is preceded by $\sim$, as in $\sim p$, it is read "not p." However, when p has meaning, as above, the word *not* is inserted to make the sentence grammatically correct. So $\sim p$ is read "I am not going water skiing."

Our problem will be to decide on the "truth value" of compound statements, given the truth or falsity of the component simple statements. If a statement is true, we shall say it has truth value T or is true; if it is false, we shall say it has truth value F or is false. For example, suppose r stands for "Today is Saturday." Further, suppose r is true; that is, it is, in fact, Saturday. Then, what is the truth value of $\sim r$? Clearly, $\sim r$ is false.

5-2 EXERCISES

Which of the following are statements of logic?

1. Why did the Yankees lose?
2. The last time I saw Paris was early in the fall.
3. He went down in a blaze of glory.
4. Do not pass "go."

5. You should behave yourself.
6. This statement is false.

Write each of the following compound sentences in symbolic form. Use p and q for simple statements and use the connectives $\sim$, $\vee$, and $\wedge$.

7. Today is either Monday or Tuesday.
8. Today is Monday and not Tuesday.
9. It is neither Monday nor Tuesday.
10. It is not Monday, but it is Tuesday.
11. It is impossible that it is Monday and Tuesday.

Let p be "It is raining" and q be "The streets are slippery." Translate the following symbols into English.

12. $p \vee q$
13. $p \wedge q$
14. $p \vee \sim q$
15. $\sim p \vee \sim q$
16. $\sim (p \vee q)$
17. $\sim (p \wedge q)$
18. $\sim p \wedge \sim q$
19. $\sim (\sim p \wedge \sim q)$
20. $\sim [\sim (\sim p \wedge q) \vee (p \vee \sim q)]$

Let r be "John is smart" and s be "Bill is handsome." Assume that both r and s are true. Which of the following compound statements are true?

21. John is smart and Bill is handsome.
22. John is smart and Bill is ugly.
23. Bill is handsome but John is stupid.
24. John's intellect leaves little to be desired; however, Bill's appearance is totally unpleasing.
25. It is absurd to assume that John is stupid or Bill is ugly.
26. It is not the case that Bill is handsome or John is stupid.

5-3 TRUTH TABLES FOR *NOT*, *AND*, AND *OR*

The truth or falsity of a compound statement depends on the truth or falsity of its simple components. We wish to know how the truth of its components affects the truth of a compound statement. For example, how does the truth value of p affect the truth value of $\sim p$? It is clear that, if p is true, then $\sim p$ is false, and if p is false, then $\sim p$ is true. A convenient way of illustrating this is a truth table:

p	$\sim p$
T	F
F	T

The left column contains the two possible truth values of p, and the right column the corresponding truth value of $\sim p$. The first row shows that, when p is true, $\sim p$ is false; the second row that, when p is false, $\sim p$ is true. The "negation" of p is written as $\sim p$, the negation of $p \wedge q$ as $\sim(p \wedge q)$.

The truth table for $\wedge$ is more complicated because it connects two simple statements. We must examine all four possible truth values of p and q to make the $p \wedge q$ truth table. p and q can both be true; they can both be false; p could be true and q false; or p could be false and q true. These four combinations are listed in the table below:

p	q	$p \wedge q$
T	T	
T	F	
F	T	
F	F	

Now we must decide on the truth value of $\wedge$ for each of the four pairs of truth values of p and q. It seems reasonable to assume that, if p is true and q is true, then a statement $p \wedge q$ is true. Similarly if both are false, then $p \wedge q$ is false. For the case of one component true and the other false, consider the following example. If an acquaintance of yours claimed that he could run a mile in a half hour and high-jump twelve feet, would you say that he lied? Although the first part is probably true, we consider the entire claim to be false, because of the exaggeration in the last part. On such a basis, the truth table for $\wedge$ is

p	q	$p \wedge q$
T	T	T
T	F	F
F	T	F
F	F	F

For simplicity we use $\wedge$ to mean *and*. If you recall the exercises in Sec. 5-2, $\wedge$ can also be interpreted as *but* or *however*. To take care of this ambiguity, we shall refer to $\wedge$ as the "conjunction" of two statements rather than as a particular word. Then, the truth table above shows that the conjunction of any but two true statements is false.

The statement $p \lor q$ is called the "disjunction" of p and q. The truth table for the disjunction is somewhat more difficult to make than that of the conjunction. If we proceed in the same manner, first writing all the truth combinations for p and q, the table is

p	q	$p \lor q$
T	T	
T	F	
F	T	
F	F	

It seems reasonable to assign the value *false* to the disjunction when both components are false. The nature of our intuitive definition of $\lor$ as *or* causes us to assign the value *true* to the disjunction when one statement is true and the other is false. After all, only one component of the statement "Our next child will be a boy or a girl" can be true. Thus we have filled in the truth table for disjunction except for the case in which both components are true:

p	q	$p \lor q$
T	T	?
T	F	T
F	T	T
F	F	F

The difficulty with the last part of the truth table is that, in English, *or* is used in two different ways. For example consider the statements "I shall buy a color television or I shall buy a stereo" and "In the fall, I shall join the Marines or I shall join the Air Force." In the first case, it is possible that this person may become extravagant and buy both appliances; that is, he may do one or the other or both. In the second case, it is inconceivable that he can do both; that is, this disjunction means one or the other but not both. For our model of English to be accurate, we must distinguish between these two uses of *or*. We cannot ignore one of them, because both are correct. Therefore, we shall call $p \vee q$ the "inclusive disjunction." It will mean p or q or both and will have the truth table

p q	$p \vee q$
T T	T
T F	T
F T	T
F F	F

$p \veebar q$ will be called the "exclusive disjunction." It will mean p or q but not both, and its truth table is

p q	$p \veebar q$
T T	F
T F	T
F T	T
F F	F

"I shall buy a color television or I shall buy a stereo" is of the form $p \vee q$, and "In the fall, I shall join the Marines or I shall join the Air Force" is of the form $p \veebar q$.

5-4 TRUTH TABLES FOR COMPLEX STATEMENTS

Negations, conjunctions, and disjunctions can link statements together to form very complicated statements, and parentheses are used to mean "do this first." For example, to make the truth table for $\sim(p \wedge q)$, begin by writing the four possible pairs of truth values for p and q:

p q	$\sim(p \wedge q)$
T T	
T F	
F T	
F F	

Next write the truth values for $p \wedge q$, momentarily ignoring the negation:

p	q	$\sim (p \wedge q)$
T	T	T
T	F	F
F	T	F
F	F	F

The table is completed when the correct truth values are filled in under the $\sim$. To do this, examine the truth values under $(p \wedge q)$. If a T appears, its negation is F, and vice versa. Thus the final truth table is

p	q	$\sim$	$(p \wedge q)$
T	T	F	T
T	F	T	F
F	T	T	F
F	F	T	F

The column under the $\sim$ gives the truth values for the entire statement $\sim(p \wedge q)$. If p and q are both true, the statement is false; otherwise, it is true. For a more complicated example let us make the truth table for $(p \vee q) \wedge \sim q$. First fill in the truth values for the disjunction, in parentheses:

p	q	$(p \vee q)$	$\wedge$	$\sim q$
T	T	T		
T	F	T		
F	T	T		
F	F	F		

Next fill in the appropriate truth values for the negation of q:

p	q	$(p \vee q)$	$\wedge$	$\sim q$
T	T	T		F
T	F	T		T
F	T	T		F
F	F	F		T

The final column is located under the ∧ and will contain the truth values of the entire compound statement for the values of p and q. To determine the first truth value under the conjunction, note that the disjunction on the left is true but that the negation of q on the right is false. Thus in the first row we have the conjunction of a true statement and a false statement. From our definition of conjunction, the entire statement is false. In the second row we have the conjunction of two true statements; thus the conjunction is true. Proceeding in the same manner, we fill the final two positions with F's. The table appears thus:

p	q	$(p \vee q)$	$\wedge$	$\sim q$
T	T	T	F	F
T	F	T	T	T
F	T	T	F	F
F	F	F	F	T

Note that the statement is true only when p is true and q is false. In this example we had to rewrite the truth table three times. In practice, it is written only once, and the columns of truth values are written in turn. The truth table of $\sim[(p \vee q) \wedge \sim(p \wedge q)]$, presented below, has been completed in steps as you will do in the exercises. The words below the columns indicate the order of completion. In problems that have parentheses inside brackets, work from the inside out. Thus the last answer will be under the ~, that is, outside the brackets:

p q	$\sim$	$[(p \veebar q)$	$\wedge$	$\sim$	$(p \wedge q)]$
T T	T	F	F	F	T
T F	F	T	T	T	F
F T	F	T	T	T	F
F F	T	F	F	T	F
first	last	second	fifth	fourth	third

5-4 EXERCISES

Construct truth tables for the following.

1. $\sim(p \vee q)$

2. $(p \vee \sim q) \wedge \sim p$

3. $\sim(p \vee q) \vee \sim(q \vee p)$
4. $p \vee \sim(p \wedge q)$
5. $\sim[(p \vee q) \wedge (\sim p \vee q)]$
6. $\sim[\sim(\sim p \wedge \sim q) \wedge \sim(p \wedge \sim q)]$
7. Let p be the statement "It is raining," and let q be "The streets are slippery." Write the following in symbolic form and construct its truth table: "It is neither raining nor are the streets slippery."
8. Repeat Exercise 7 for "It is not the case that it is both raining and the streets are slippery."

In an advanced logic course it is shown that all the connectives of logic can be replaced by a single symbol that is called the "alternate denial" or simply "slash" and is written $p|q$. It is defined as "p and q are not both true," and its truth table is

p	q	$p\|q$
T	T	F
T	F	T
F	T	T
F	F	T

Note that the truth table is a faithful representation of the definition; namely, $p|q$ is true whenever p and q are not both true.

9. Write a compound statement that has the same truth table as $p|q$ but uses the connectives $\sim$ and $\wedge$ exclusively.
10. Construct a truth table for $\sim(p|q)$.
11. Construct a truth table for $p|p$.
12. Construct a truth table for $(p|q)|(p|q)$.
13. Where have you seen the truth table in Exercise 11 before?
14. Where have you seen the truth table in Exercise 12 before?

5-5 THE CONDITIONAL AND BICONDITIONAL

Some of the most important English sentences are constructed in the form "If . . . , then . . . ." In our logic, we define the symbol $\rightarrow$ to show this construction; $p \rightarrow q$ is read, "If p then q." Now we may translate into symbols such phrases as "If I were a millionaire, then I shouldn't be in this course" and "If a woman is elected President, I shall move to Canada."

In the statement $p \rightarrow q$, p is called the "antecedent" and q the "consequent." In the second example above, the antecedent is "a woman is elected President," and the consequent is "I shall move to Canada." The entire statement $p \to q$ is called a "conditional." There are some variations as to how $p \to q$ is read. We may read it "If p then q" or "p conditions q" or "p implies q." The word *implies* is sometimes reserved for a special kind of conditional. Although we make no such distinction here, we shall have more to say about this later.

In order to continue our analysis of reasoning, we must decide on the truth value of the conditional. Consider the conditional "If the weather is fine tomorrow, I shall take you water-skiing," and suppose that a young man has made this promise to a young woman. Let us examine the conditions under which we should consider him truthful and those under which he is not. It seems reasonable to assume that he is truthful if the weather is fine the next day and he takes her water-skiing; that is, the conditional is true when both the antecedent and consequent are true. This is shown in the first row of the truth table below. He is not truthful if the weather is fine and he does not take her water-skiing. So the second row of the truth table shows the conditional to be false when the antecedent is true and the consequent false. We have now the following:

p	q	$p \to q$
T	T	T
T	F	F
F	T	?
F	F	?

For the third line, suppose that the weather is bad but the young man decides to take the young woman water-skiing anyhow. Could we say that his original offer was false? Probably not. The date depends on fine weather. If the weather is bad, he is not obligated to take her water-skiing but he may still do so if he wishes. Thus we shall agree that the conditional is true if the antecedent is false and the consequent is true. The last line of the truth table is true. If the weather is bad, he is not obligated to take her water-skiing. Thus the complete truth table for the conditional is

p	q	$p \to q$
T	T	T
T	F	F
F	T	T
F	F	T

In everyday language, statements are connected according to certain "implied" rules. A statement such as "If you will attend the rally, then ice cream is pink" would be thought nonsensical, because there is no relationship between the antecedent and consequent. However, these rules for linking only certain statements are different for various societies and even for age groups within one society. Therefore, a connective will be entirely defined by its truth table. To a mathematician the statement "If the moon is made of green cheese, then the liberals and conservatives will unite" is obviously true, and this is his main concern. A semanticist or a psychologist would examine this statement in another light. We are concerned here with truth or falsity and shall allow any combination of statements and connectives as long as the simple statements can be ajudged true or false.

The last connective we consider is called the "biconditional," which is a stronger version of the conditional and is described by "if and only if." The biconditional $p \leftrightarrow q$ is read "p if and only if q." To construct the truth tables we use the sentence "A triangle is isosceles if and only if it has two equal sides." This sentence asserts that, if the triangle is isosceles, it has two equal sides, and, furthermore, if it is not isosceles, it does not have two equal sides; that is, the biconditional states that both statements are true or both are false and no other possibility exists. Thus its truth table is

p	q	$p \leftrightarrow q$
T	T	T
T	F	F
F	T	F
F	F	T

In the rest of the present chapter we work with the six connectives: negation, conjunction, disjunction (inclusive and exclusive), conditional, and biconditional; their truth tables are repeated below and should be memorized to facilitate understanding of later discussions:

p	q	$\sim q$	$p \wedge q$	$p \vee q$	$p \veebar q$	$p \rightarrow q$	$p \leftrightarrow q$
T	T	F	T	T	F	T	T
T	F	T	F	T	T	F	F
F	T		F	T	T	T	F
F	F		F	F	F	T	T

We may now make complex statements using all or any of the six connectives. For example, the truth table for $(p \rightarrow q) \wedge \sim p$ is

p	q	$(p \rightarrow q)$	$\wedge$	$\sim p$
T	T	T	F	F
T	F	F	F	F
F	T	T	T	T
F	F	T	T	T
		first	last	second

So far our truth tables have contained no more than two simple statements. Statements of logic do not have to be so confined, and we may connect any number of simple statements. For example, let us construct the truth table for $(p \veebar q) \wedge \sim r$. The first step is to write every combination of truth values for p, q, and r:

p	q	r	$(p \veebar q) \wedge \sim r$
T	T	T	
T	T	F	
T	F	T	
T	F	F	
F	T	T	
F	T	F	
F	F	T	
F	F	F	

With the introduction of a new statement r, the truth table must be extended to eight lines. Note the pattern of T's and F's under p, q, and r. Under p, there are four T's and then four F's. Under q, there are two T's, then two F's, then two more T's, and then two more F's. Under r, the T's and F's alternate. It is helpful to remember this pattern when constructing a truth table for three or more statements, to ensure that the table contains all combinations of the truth values of the simple statements. To continue with our truth table, we next fill in the truth values under $\veebar$ for all eight rows:

p	q	r	$(p$	$\veebar$	$q)$	$\wedge$	$\sim r$
T	T	T		F			
T	T	F		F			
T	F	T		T			
T	F	F		T			
F	T	T		T			
F	T	F		T			
F	F	T		F			
F	F	F		F			

Then the same is done for $\sim r$:

p	q	r	$(p$	$\veebar$	$q)$	$\wedge$	$\sim r$
T	T	T		F			F
T	T	F		F			T
T	F	T		T			F
T	F	F		T			T
F	T	T		T			F
F	T	F		T			T
F	F	T		F			F
F	F	F		F			T

And, finally, we use the truth values under $\veebar$ and $\sim r$ to fill in the values under $\wedge$, thus completing the table:

p	q	r	$(p \veebar q)$	$\wedge$	$\sim r$
T	T	T	F	F	F
T	T	F	F	F	T
T	F	T	T	F	F
T	F	F	T	T	T
F	T	T	T	F	F
F	T	F	T	T	T
F	F	T	F	F	F
F	F	F	F	F	T
			first	last	second

5-5 EXERCISES

Construct truth tables for the following.

1. $p \wedge \sim p$
2. $\sim(p \wedge \sim p)$
3. $\sim p \rightarrow \sim q$
4. $(p \rightarrow q) \leftrightarrow (\sim p \rightarrow \sim q)$
5. $p \rightarrow (q \vee r)$
6. $(q \rightarrow q) \leftrightarrow (q \rightarrow \sim q)$
7. $(p \vee \sim q) \wedge \sim r$
8. $[(p \rightarrow q) \wedge (q \rightarrow r)] \rightarrow (p \rightarrow r)$

Let a, b, c, and d be the following statements for Exercises 9 to 15:

(a) The Frankenstein monster is lovable.

(b) $2 + 2 = 75$.

(c) Guadalajara is in Mexico.

(d) The gravitational attraction between two bodies is inversely proportional to the square of the distance between them.

Translate the following into English and, referring to the statements above, give the truth value of each.

9. $a \wedge \sim b$
10. $a \rightarrow \sim b$
11. $\sim b \wedge d$
12. $c \leftrightarrow d$
13. $\sim c \rightarrow \sim d$
14. $(a \vee d) \rightarrow b$
15. $(c \vee d) \wedge (a \vee d)$
16. Define $p|q$ as in the exercises in Sec. 5-4. Write a truth table for $(p|p)|(q|q)$.
17. Write a truth table for $[(p|p)|(p|p)]|(q|q)$.
18. Where have you seen the truth table in Exercise 16 before?
19. Where have you seen the truth table in Exercise 17 before?

Another symbol that may be used to replace all the other connectives is called the "joint denial" or "arrow slash." It is written $p \downarrow q$ and is defined as "both p and q are false" or "not p and not q." The truth table for arrow slash is

p q	$p \downarrow q$
T T	F
T F	F
F T	F
F F	T

20. Write a truth table for $p \downarrow p$. What other compound statement has this truth table?

21. Write a truth table for $(p \downarrow q) \downarrow (p \downarrow q)$. What other compound statement has this truth table?

22. Write a truth table for $(p \downarrow p) \downarrow (q \downarrow q)$. What other compound statement has this truth table?

23. Write a truth table for $[(p \downarrow p) \downarrow q] \downarrow [(p \downarrow p) \downarrow q]$. What other compound has this truth table?

5-6 LOGICALLY EQUIVALENT STATEMENTS

It is possible in English to say the same thing in many different ways. To verify this, one has only to listen to an electioneering politician. In logic, one must know how to recognize two statements that have exactly the same meaning, or how to "verify" that two statements have the same meaning. Toward this end, consider the following statements: "I am not hungry and I am not tired" and "It is not true that I am hungry or that I am tired." Both statements mean the same thing, namely, that I have had enough food and enough sleep. If we write them in symbolic form, the first is $\sim p \wedge \sim q$ and the second $\sim(p \vee q)$. The truth tables for these statements are

p q	$\sim$	$(p \vee q)$	$\sim p$	$\wedge$	$\sim q$
T T	F	T	F	F	F
T F	F	T	F	F	T
F T	F	T	T	F	F
F F	T	F	T	T	T

Notice that both statements have the same truth table. This is to be expected. If two statements are logically equivalent (have the same meaning), then their truth values must be the same. Hence we define "logically equivalent statements" to be those which have identical truth tables. In Chap. 6 we replace statements with their logical equivalents to simplify expressions.

Let the symbol $\equiv$ mean "is logically equivalent to." Then we have shown above that $\sim(p \vee q) \equiv \sim p \wedge \sim q$. Some logical equivalents are used often and have become "classics" of symbolic logic. The equivalence above is known as De Morgan's law. We introduce others in the exercises and in Chap. 6.

5-6 EXERCISES

1. Prove that $\sim(p \wedge q)$ is equivalent to $\sim p \vee \sim q$. This is another example of De Morgan's law.

2. Which of the following are equivalent?

(*a*) $p \rightarrow q$	(*b*) $\sim p \wedge \sim q$	(*c*) $q \rightarrow p$	(*d*) $\sim(p \wedge \sim q)$
(*e*) $\sim q \rightarrow \sim p$	(*f*) $q \vee \sim p$	(*g*) $q \vee p$	

3. Four of the statements in Exercise 2 are equivalent. Let p be "You are lovely" and q be "You will have many dates," and write these four statements in English.

4. Using only $\sim$ and $\vee$, construct a compound statement equivalent to $p \wedge q$.

5. Using only $\sim$ and $\wedge$, construct a compound statement equivalent to $p \vee q$.

6. What is the relationship between $p \leftrightarrow q$ and $(p \rightarrow q) \wedge (q \rightarrow p)$?

7. Prove that $p \rightarrow [p \wedge \sim(q \vee r)]$ is equivalent to $\sim p \vee (\sim q \wedge \sim r)$.

5-7 VARIATIONS OF THE CONDITIONAL

The conditional is the most confusing of the connectives. This is as true in everyday conversation as in symbolic logic. One of the most common fallacies associated with the conditional is that given that a conditional is true, assume that the conditional formed by interchanging the antecedent and consequent of the original is also true; that is, we should agree that the following conditional is true: "If it is raining, then the streets are wet." Now, interchanging the antecedent and the consequent, we have "If the streets are wet, then it is raining." Must the second conditional also be true? The answer is no. Wet streets do not necessarily mean it is raining; they may have just been washed.

Conditionals formed by interchanging the antecedent and the consequent are called "converses." Thus the second conditional above is the converse of the first, and the converse of a conditional is not logically equivalent to that conditional. How can the conditional be modified, and are any of the modifications equivalent to the original? Below are truth tables for the modifications of the conditional. These new conditionals are famous in the study of logic, and each has been given a name:

p q	*Conditional* $p \to q$	*Converse of conditional* $q \to p$	*Inverse of conditional* $\sim p \to \sim q$	*Contrapositive of conditional* $\sim q \to \sim p$
T T	T	T	T	T
T F	F	T	T	F
F T	T	F	F	T
F F	T	T	T	T

By examining the truth tables, we see that, as we suggested, the conditional is not equivalent to its converse. However, the conditional is equivalent to its contrapositive (the contrapositive is the basis for some "indirect proofs," a concept we have more to say about in Chap. 6). Also, note that the converse is equivalent to the inverse. The inverse is sometimes called the "contrapositive of the converse," a rather clumsy title but a very good description. Note that the antecedent and the consequent of the converse are interchanged and negated in the inverse. The same description applies between the conditional and its contrapositive. For example, the contrapositive of "If it is raining, then the streets are wet" is created by interchanging the antecedent and the consequent and negating both. It becomes "If the streets are not wet, then it is not raining." To summarize, we list the converse, inverse, and contrapositive of "If it is raining, then the streets are wet":

Converse: "If the streets are wet, then it is raining."
Inverse: "If it is not raining, then the streets are not wet."
Contrapositive: "If the streets are not wet, then it is not raining."

The meaning of a conditional statement can further be complicated by using words other than *if* and *then* to connect simple statements or by transposing *if* and *then*. For example, "I shall run for President if I get the Democratic nomination" means "If I get the Democratic nomination, then I shall run for President." In the first case, the antecedent is placed after the consequent and *then* is not written. Thus we may say "q if p," which is equivalent to $p \rightarrow q$. Some other "atypical" conditionals are:

"I shall go swimming only if the temperature of the water is above 60°."
"Calling me names is sufficient reason for me to punch you in the nose."
"To be a successful scuba diver, it is necessary to have an underwater spear gun."

All three examples are conditionals. The logician must determine how they relate to $p \rightarrow q$. Let us examine them in the order in which they appear, and let s be "I go swimming" and a be "The temperature of the water is above 60°." Then the first example is of the form *s only if a*. However, this example could be translated to "If the temperature of the water is not above 60°, then I shall not go swimming," the meaning being the same in both cases. Thus the statement *s only if a* states $\sim a \rightarrow \sim s$, but $\sim a \rightarrow \sim s$ is the contrapositive of $s \rightarrow a$; and because the conditional and its contrapositive are equivalent, *s only if a* means $s \rightarrow a$. In the second example, it is clear that the speaker is saying "If you call me names, then I shall punch you in the nose." Thus sentences of the form *r is sufficient condition for s* mean $r \rightarrow s$. It is sometimes helpful to remember that the *sufficient* part of the sentence is the antecedent of a conditional and the rest is the consequent. In the third example, does the fact that one owns a spear gun automatically make him a successful scuba diver? Of course not. It takes much practice and experience, among other things. Thus we cannot say "Ifyou have an underwater spear gun, then you are a successful scuba diver," and the example must mean "If you are a successful scuba diver, then you own an underwater spear gun." The *necessary* part is the consequent of a conditional; the remaining part is the antecedent. Thus a statement of the form *r is a necessary condition for q* means *if q then r*. The table below gives the basic statement *if* p *then* q on the left and all its equivalent forms on the right:

Conditional	Equivalent statement
$p \to q$ (if p then q)	1. q if p 2. p only if q 3. p is a sufficient condition for q 4. q is a necessary condition for p

Finally, we consider the statement p *is a necessary and sufficient condition of* q. This could be rewritten as p is a necessary condition for q and p is also a sufficient condition for q. Referring to the table above, we see that this translates to $(q \to p) \wedge (p \to q)$. Note that this translation is the conjunction of a conditional and its converse. Its truth table is

p	q	$(q \to p)$	$\wedge$	$(p \to q)$
T	T	T	T	T
T	F	T	F	F
F	T	F	F	T
F	F	T	T	T

Thus p *is a necessary and sufficient condition for* q is equivalent to p *if and only if* q or $p \leftrightarrow q$.

5-7 EXERCISES

Let p represent "Freud was a genius" and q represent "Psychoanalysis is a useful therapeutic technique." Translate the following into symbolic form.

1. Freud was a genius only if psychoanalysis is a useful therapeutic technique.
2. Psychoanalysis being a useful therapeutic technique is a sufficient condition to believe that Freud was a genius.
3. It is necessary for Freud to be a genius for psychoanalysis to be a useful therapeutic technique.

4. Psychoanalysis is a useful therapeutic technique if Freud was a genius.
5. Freud was not a genius unless psychoanalysis is a useful therapeutic technique.
6. Freud's being a genius is a necessary and sufficient condition for psychoanalysis to be a useful therapeutic technique.
7. Psychoanalysis is a useful therapeutic technique if and only if Freud was a genius.

Use the conditional "If you like Wagner, then you like the Valkyries," to answer the following (Exercises 8 to 10).

8. Form the converse of the above.
9. Write the contrapositive.
10. Write its inverse.
11. To what is the conjunction of the inverse and the contrapositive equivalent? Prove your contention.
12. What is the contrapositive of the contrapositive?

Write the following, using the connectives $\rightarrow$ and $\leftrightarrow$ only.

13. A necessary condition for r is s.
14. For p, q is sufficient.
15. Only if q, will p happen.
16. A necessary and sufficient condition for t is q.

REVIEW EXERCISES

Let p stand for "I like to play pinochle" and q stand for "I should not learn to play bridge." Put the following into symbolic form.

1. I should necessarily learn to play bridge, because I like to play pinochle.
2. Only if I like pinochle should I learn to play bridge.
3. The fact that I do not like to play pinochle is a necessary and sufficient reason that I should not learn to play bridge.
4. Hating to play pinochle is a sufficient reason that I should not learn to play bridge.

Use the statement "If the United States is not larger than Canada, then France is smaller than Texas" to do Exercises 5 and 6.

5. Write, in words, the contrapositive of the given statement.
6. Write, in words, the contrapositive of the converse of the given statement.

7. Construct a truth table for $(p \wedge q) \rightarrow \sim[\sim p \vee (r \leftrightarrow q)]$.

SELECTED REFERENCES

Dinkines, Flora: "Introduction to Mathematical Logic," Appleton-Century-Crofts, Inc., New York, 1964.

This paperback contains numerous examples and exercises related to the topics in the present chapter.

Kemeny, John G.: "Introduction to Finite Mathematics," Prentice-Hall, Inc., Englewood Cliffs, N.J., 1966.

The explanation of the topics in the present chapter are good if somewhat brief.

Stoll, Robert R.: "Set Theory and Logic," W. H. Freeman and Company, San Francisco, 1963.

AXIOMATICS
PRECLUDES
CHAOS!
USE
LOGIC
NOW!

chapter six

ARGUMENT AND AXIOMATICS

To almost everyone logic means common sense, the ability to reason accurately. With the material from Chap. 5, we now have the background to analyze the process of reasoning.

6-1 ARGUMENT

Reasoning takes the form of drawing a conclusion from one or more facts. For example, one might reason that a candidate would make a good governor, because he wanted to be President and, to have a chance at the Presidency, he must make a superior record in the governor's office. However, the dissenter would argue that, because the candidate's eyes are on the Presidency, he would not see the problems of the state as significant issues and would make a bad governor.

An argument, then, is a series of statements, the last of which is called the "conclusion." The conclusion is said to follow (hopefully) from the previous statements, the "premises" of the argument. An important task of the logician is to establish the validity or fallacy of the argument. This does not mean that a good logician could decide whether a candidate would be a good governor or not. The truth or falsity of a statement and the validity or invalidity of an argument are completely different. It is possible for arguments to be valid and their conclusions to be false, for example:

I am free to burn buildings, or the United States is a dictatorship.
I am not free to burn buildings.

Therefore, the United States is a dictatorship.

The difficulty with this is the first premise. If false "facts" are allowed in an argument, almost anything can be proved.

Perhaps the most dramatic examples of argument take place in the courtroom. The primary assumption of a court of law is that the testimony is true. According to this assumption, an attorney attempts to lead the testimony of a witness to some conclusion. The opposing attorney must be on guard, because it is also possible for the conclusion to be true although the argument is invalid:

If John Wilkes Booth committed murder, then he had a warped personality.
He had a warped personality.

Therefore, he committed murder.

This argument is not valid, because the truth of the conclusion is not established by the premises. The validity of an argument depends on the form in which the premises are stated and not at all on the truth or falsity of the conclusion.

It is possible to argue anything if false premises are allowed. Thus we define an argument to be valid if the conclusion is true whenever all the premises are true. Otherwise an argument is invalid or fallacious. An argument in symbolic form follows. The statements above the line are the premises, and that below the line the conclusion. The $\therefore$ is an abbreviation of *therefore*:

$$\begin{array}{l} p \rightarrow q \\ \underline{\sim q} \\ \therefore \sim p \end{array}$$

A truth table can be used to determine whether this argument is valid or not. By our definition, all we have to do is make a truth table for each premise and the conclusion and see if the conclusion is true every time both premises are. The truth tables are

p q	$p \to q$	$\sim q$	$\sim p$
T T	T	F	F
T F	F	T	F
F T	T	F	T
F F	T	T	T

This argument is valid. Note that both premises are true only in the last line of the tables, and here the conclusion is also true. Therefore this argument conforms to our definition of *valid.* For another example, we test the validity of the following argument:

$$\begin{array}{l} p \to q \\ \sim q \to \sim r \\ \hline \therefore r \to p \end{array}$$

Its truth tables are

p q r	$p \to q$	$\sim q \to \sim r$	$r \to p$
T T T	T	T	T
T T F	T	T	T
T F T	F	F	T
T F F	F	T	T
F T T	T	T	F
F T F	T	T	T
F F T	T	F	F
F F F	T	T	T

This argument is invalid. In line 5, note that both premises are true but the conclusion is false. This is sufficient reason for saying that the argument is invalid. For if it is possible to use true premises and reach a false conclusion in only one case, the argument is useless. The argument is invalid although in lines 1, 2, and 8 the premises and conclusion are all true. Our definition of a valid argument requires that the conclusion is true in every case in which the premises are true; one failure and the whole argument is invalid.

In everyday reasoning we use certain memorized valid forms again and again in our discourse. The truth-table method of establishing validity has the advantage of being purely mechanical. Since all statements are either true or false, a computer could be used to check the validity of an argument by assigning the on position of switches to mean true and the off position to mean false. In fact, computer experts have devised a program language to aid mathematicians and logicians in the study of logically valid patterns.

6-1 EXERCISES

Test the validity of the following arguments.

1. $$\begin{array}{l} p \vee \sim q \\ \underline{q} \\ \therefore p \end{array}$$

2. $$\begin{array}{l} p \leftrightarrow q \\ \underline{\sim q} \\ \therefore p \end{array}$$

3. $$\begin{array}{l} \sim q \rightarrow p \\ \underline{p \wedge q} \\ \therefore \sim p \end{array}$$

4. $$\begin{array}{l} \underline{\sim p} \\ \therefore \sim (p \wedge \sim q) \end{array}$$

5. $$\begin{array}{l} \underline{\sim (\sim p \wedge \sim q)} \\ \therefore q \end{array}$$

6. $$\begin{array}{l} p \rightarrow q \\ \underline{q \rightarrow r} \\ \therefore p \rightarrow r \end{array}$$

7. $$\begin{array}{l} t \vee p \\ t \rightarrow s \\ \underline{p \rightarrow q} \\ s \vee q \end{array}$$

8. $$\begin{array}{l} p \veebar q \\ \sim r \rightarrow q \\ \underline{\sim (p \wedge r)} \\ \therefore \sim p \end{array}$$

Translate the following into symbols and test for validity.

9. One is necessarily tense when working too hard.
   I am tense.
   ___
   Therefore I am working too hard.

10. Montana is a southern state or New Yorkers drink beer, only if Mother is overweight.
    ___
    Therefore New Yorkers' drinking beer is sufficient reason for Mother to be overweight.

11. Either I do not study hard or I get good grades.
If I do not get good grades, then my parents are displeased.
Therefore if my parents are displeased, then I did not study hard.

12. I am not a surgeon or I am intelligent.
I receive a passing grade in basket weaving or I am stupid.
I am not lazy or I shall not receive a passing grade in basket weaving.
Therefore I am not lazy, if I am a surgeon.

6-2 AN INDIRECT METHOD OF CHECKING VALIDITY

The truth-table method of checking the validity of an argument can be tedious under the best conditions. If an argument has many premises, the truth table becomes so large as to be impossible. To avoid making truth tables for every argument, we have to remember two facts only: first, an argument is either valid or invalid, and, second, an argument is invalid if we can find even one instance in which all the premises are true and at the same time the conclusion is false. The indirect method of checking validity involves trying to make an argument invalid by finding one case in which the premises are true and the conclusion false. If we cannot find such a case, the argument is valid. For example,

$p \to q$
q
$\therefore p$

First, assume that the argument is invalid; that is, assume that the premises are true and the conclusion false, and indicate this with the letters T and F:

	Assume
$p \to q$	**T**
q	**T**
$\therefore p$	**F**

The second premise and the conclusion are single statements, and we have assumed their truth values to be true and false respectively. The only "complicated" statement is the first premise, which we have assumed to be true. What remains to be done is to check the first premise to see if its assumed truth is compatible with the assumed truth values of p and q. Note that p is assumed to be false, and q to be true:

F **T**	Assume
$p \to q$	T
q	T
$\therefore p$	F

Thus the conditional $p \to q$, with p false and q true, is true, as we assumed. We have a case here in which the premises of the argument are true and the conclusion is false. The argument is invalid.

Another example follows:

	Assume
$p \to q$	**T**
$\sim q$	**T**
$\therefore \sim p$	**F**

We begin by assuming the argument to be invalid and write the appropriate truth values of the premises and the conclusion to the right. Because $\sim p$ is false, p is true. This is noted by the letter T above the p in the first premise:

T	Assume
$p \to q$	T
$\sim q$	T
$\therefore \sim p$	F

Next, q must be false, because $\sim q$ is assumed to be true. This is noted by the letter F above the q in the first premise:

T **F**	Assume
$p \to q$	T
$\sim q$	T
$\therefore \sim p$	F

The diagram now shows the truth value of each simple statement and those of the compound statements. Have we successfully found values of p and q that cause the argument to be invalid? Are the premises true and the conclusion false? No, they are not. In the example above, p is true and q is false. Thus $p \to q$ is false, not true, as we assumed. It is not possible to make any adjustment in the truth values of $\sim p$ and $\sim q$ to cause the argument to be invalid, as we assumed it was. And because we cannot make the argument invalid, it must be valid.

One way of describing this method is to say that it is backward. We begin with the conclusion and work up. In the example below we again assume the argument to be invalid by inserting the appropriate truth values to the right of the statements:

	Assume
$p \to q$	**T**
$\sim r \to \sim q$	**T**
$\therefore p \to r$	**F**

Next consider the conclusion $p \to r$. The only way a conditional can be false is when the antecedent is true and the consequent is false. So p is true and r is false:

	Assume
$p \to q$	T
$\sim r \to \sim q$	T
$\therefore p \to r$	F
T **F**	

Write the truth values of p and $\sim r$ in the premises:

	Assume
T	
$p \to q$	T
T	
$\sim r \to \sim q$	T
$\therefore p \to r$	F
T F	

In the second premise $\sim r \to \sim q$, $\sim q$ must be true; if it were not, the premise would be false, but we have assumed it to be true:

	Assume
T	
$p \to q$	T
T **T**	
$\sim r \to \sim q$	T
$\therefore p \to r$	F
T F	

Finally, the fact that $\sim q$ is true forces the q in the first premise to be false:

T F	Assume
$p \rightarrow q$	T
T T	
$\sim r \rightarrow \sim q$	T
$\therefore p \rightarrow r$	F
T F	

This, then, completes the assigning of truth values. Note that in the first premise there is a contradiction. We assumed $p \rightarrow q$ to be true, but its antecedent is true and its consequent is false, and this means that it is false. Thus our assumption of an invalid argument leads to a contradiction, and the argument must be valid.

Another example follows:

	Assume
$p \rightarrow q$	**T**
$\sim q \vee s$	**T**
$s \vee t$	**T**
$\therefore \sim p \vee t$	**F**

In the conclusion, $\sim p$ and t must both be false, because a disjunction is false only when both statements are false:

	Assume
$p \rightarrow q$	T
$\sim q \vee s$	T
$s \vee t$	T
$\therefore \sim p \vee t$	F
F **F**	

Fill in other truth values of p and t:

T	Assume
$p \rightarrow q$	T
$\sim q \vee s$	T
F	
$s \vee t$	T
$\therefore \sim p \vee t$	F
F F	

In the third premise, s must be true for the disjunction to be true, and in the first premise, q must be true for the conditional to be true:

$$\begin{array}{llll}
 & \text{T} \quad \textbf{T} & & \text{Assume} \\
 & p \rightarrow q & & \text{T} \\
 & \qquad \textbf{T} & & \\
 & {\sim}q \vee s & & \text{T} \\
 & \textbf{T} \quad \text{F} & & \\
 & \underline{s \vee t} & & \text{T} \\
\therefore & {\sim}p \vee t & & \\
 & \text{F} \quad \text{F} & & \text{F}
\end{array}$$

Lastly, in the second premise, ${\sim}q$ is false:

$$\begin{array}{lll}
\text{T} \quad \text{T} & & \text{Assume} \\
p \rightarrow q & & \text{T} \\
\textbf{F} \quad \text{T} & & \\
{\sim}q \vee s & & \text{T} \\
\text{T} \quad \text{F} & & \\
\underline{s \vee t} & & \text{T} \\
{\sim}p \vee t & & \text{F} \\
\text{F} \quad \text{F} & &
\end{array}$$

There are no contradictions. All the premises are true, and the conclusion is false. Our assumption of an invalid conclusion is justified, and the argument is invalid.

6-2 EXERCISES

1 to 12. Check the validity of the exercises in the previous section by the indirect method.

Translate the following into symbolic form and test their validity.

13. If George dates whom he pleases, it is necessary that he be popular. He is not in many activities only if he is not popular. For him to be in many activities is sufficient for his grades to suffer.

Therefore if his grades do not suffer, then he may not date whom he pleases.

14. If a check is on that file, then I am anxious about it. Checks are not marked by a cross only if they are payable to the bearer. They are

returned to me only if they are not honored by the bank. If they are marked with a cross, they are necessarily for amounts of more than \$50. Not being on the file is sufficient reason for them to be marked "Not negotiable."
The bank has not failed to honor all your checks.
If a check is returned to me, this is sufficient reason for me to be anxious about it.
Checks marked "Not negotiable" are for amounts of less than \$50.

Therefore if a check is yours, then it is not payable to the bearer.

6-3 ARGUMENT FORMS

When we reason during conversation, we do not make mental truth tables or check our arguments by the indirect method in Sec. 6-2. Rather, we have memorized, perhaps subconsciously, a few valid argument forms, and we select one of them whenever we need it. If the statements of our discussion do not fit our argument framework, we rephrase them so that they do; that is, we substitute sentences that have equivalent meaning. The most frequently used argument forms have special names.

Two of them, *modus ponens* and *modus tollens*, are derived from the Latin *modus* meaning manner, *ponere* meaning to affirm, and *tollere* meaning to deny. Thus the argument form *modus ponens* is a manner of affirming the parts of a conditional and *modus tollens* a manner of denying its parts.

The most common argument forms are

Modus ponens	*Modus tollens*	*Hypothetical syllogism*	*Disjunctive syllogism*
$p \to q$	$p \to q$	$p \to q$	$p \vee q$
p	$\sim q$	$q \to r$	$\sim p$
$\therefore q$	$\therefore \sim p$	$\therefore p \to r$	$\therefore q$

They have all been proved valid in previous exercises or in examples, and we shall not prove them again.

An argument in the *modus ponens* form is:

If a basketball player averages 40 points a game, then he will be an all-American.
Joe averages 40 points a game.

Therefore Joe will be an all-American.

Modus tollens:

If a basketball player averages 40 points a game, then he will be an all-American.
Joe will not be an all-American.

Therefore Joe does not average 40 points a game.

Hypothetical syllogism:

If it rains, I shall get my feet wet.
If I get my feet wet, then I shall get sick.

Therefore if it rains, I shall get sick.

Disjunctive syllogism:

You are a Republican or a Democrat.
You are not a Democrat.

Therefore you are a Republican.

The hypothetical syllogism may be extended to an argument that contains more than two conditionals as premises. For example,

$$\begin{array}{r} p \rightarrow q \\ q \rightarrow r \\ r \rightarrow s \\ s \rightarrow t \\ \hline \therefore p \rightarrow t \end{array}$$

The argument

It is not the case that I am neither Republican nor Democrat.
I am not a Democrat.

Therefore I am a Republican.

does not seem to fit any of the four valid argument forms given. However, if we replace the first premise with its logical equivalent:

I am a Republican or I am a Democrat.
I am not a Democrat.

Therefore I am a Republican.

the argument is seen to be a disjunctive syllogism and therefore valid. Thus if we have a list of valid argument forms and a list of logically equivalent statements, we can check validity by changing arguments to the equivalent valid form. Under these conditions we can also supply a conclusion to a list of premises by choosing a conclusion that completes a valid argument form. The logically equivalent statements that were proved in Chap. 5 are

$\sim p \wedge \sim q \equiv \sim(p \vee q)$	De Morgan's law
$\sim p \vee \sim q \equiv \sim(p \wedge q)$	De Morgan's law
$p \rightarrow q \equiv \sim q \rightarrow \sim p$	Contrapositive
$\sim p \vee q \equiv p \rightarrow q$	The last law
$p \vee q \equiv q \vee p$	Disjunction is commutative
$p \wedge q \equiv q \wedge p$	Conjunction is commutative
$p \equiv \sim \sim p$	Double negation

These equivalences are patterns that can be altered slightly for different situations. For example, $\sim t \vee s \equiv s \vee \sim t$ by the commutative property of disjunction. We can simplify $\sim(\sim p \wedge \sim q)$ by De Morgan to $\sim \sim p \vee \sim \sim q$ and finally to $p \vee q$. Thus $\sim(\sim p \wedge \sim q) \equiv p \vee q$. And again $\sim p \rightarrow \sim q \equiv \sim \sim p \vee \sim q$ by the last law, and by double negation $\sim p \rightarrow \sim q \equiv p \vee \sim q$.

The following argument can be proved valid by changing it to a valid argument in one of the standard forms:

$\sim(p \wedge \sim q)$
p

$\therefore q$

transforms to

$p \rightarrow q$ *modus ponens*
p

$\therefore q$

The first premise $\sim(p \wedge \sim q) \equiv \sim p \vee \sim\sim q$ by De Morgan, and by double negation $\sim(p \wedge \sim q) \equiv \sim p \vee q$, which gives $\sim(p \wedge \sim q) \equiv p \rightarrow q$. Thus the argument is *modus ponens.*

What is the conclusion to the following premises?

$$\begin{array}{l} t \rightarrow p \\ \sim p \vee q \\ \underline{\sim(s \wedge \sim t)} \\ \therefore ? \end{array}$$

Change the second and third premises to

$$\begin{aligned} \sim p \vee q &\equiv p \rightarrow q \\ \sim(s \wedge \sim t) &\equiv \sim s \vee \sim\sim t \\ &\equiv \sim s \vee t \\ &\equiv s \rightarrow t \end{aligned}$$

Thus

$$\begin{array}{l} t \rightarrow p \\ p \rightarrow q \\ \underline{s \rightarrow t} \\ \therefore \end{array}$$

At first, this seems to fit no argument form. However, if we make $s \rightarrow t$ the first premise:

$$\begin{array}{l} s \rightarrow t \\ t \rightarrow p \\ \underline{p \rightarrow q} \\ \therefore \end{array}$$

then the premises are those of the modified hypothetical syllogism. Thus the conclusion is $s \rightarrow q$:

$$\begin{array}{l} s \rightarrow t \\ t \rightarrow p \\ \underline{p \rightarrow q} \\ \therefore s \rightarrow q \end{array}$$

What is the conclusion to

$$\begin{array}{l} (s \veebar t) \rightarrow (a \vee b) \\ \underline{\sim(a \vee b)} \\ \therefore ? \end{array}$$

It is $\sim(s \veebar t)$ by *modus tollens.* Quantities inside parentheses are treated as if they were single statements:

$$\begin{array}{l} (s \veebar t) \rightarrow (a \vee b) \\ \underline{\sim(a \vee b)\qquad\qquad} \\ \therefore \sim(s \veebar t) \end{array}$$

is of the form

$$\begin{array}{l} p \rightarrow q \\ \underline{\sim q\quad} \\ \therefore \sim p \end{array}$$

where p stands for $(s \veebar t)$ and q stands for $(a \vee b)$.

6-3 EXERCISES

Supply a conclusion to each of the following by translating to one of the four valid argument forms. In each exercise state the argument form used.

1. $\begin{array}{l} p \rightarrow q \\ \underline{\sim r \rightarrow \sim q} \\ \therefore \end{array}$

2. $\begin{array}{l} \sim p \vee q \\ \underline{\sim q\qquad} \\ \therefore \end{array}$

3. $\begin{array}{l} \sim q \rightarrow \sim p \\ \underline{\sim \sim p\qquad} \\ \therefore \end{array}$

4. $\begin{array}{l} \sim(p \wedge \sim q) \\ \underline{\sim q\qquad\quad} \\ \therefore \end{array}$

5. $\begin{array}{l} (p|p) \vee q \\ \underline{\sim q\qquad\quad} \\ \therefore \end{array}$

6. $\begin{array}{l} (s \vee r) \rightarrow q \\ \sim t \rightarrow \sim q \\ \underline{\sim t \vee (s \wedge t)} \\ \therefore \end{array}$

7. If motherhood is good, then Christmas does not come on the 25th.
   Christmas does come on the 25th.

   Therefore

8. Animals are always mortally offended if I fail to notice them.
   If an animal belongs to me, then it is in that field.
   For an animal to guess a riddle it is necessary that it has been properly trained in a boarding school.
   An animal is in that field only if it is not a badger.
   If an animal is mortally offended, it always rushes about wildly and howls.

My noticing an animal is sufficient condition for it to belong to me.
If an animal has been properly trained in a boarding school, then it does not rush wildly about and howl.

Therefore

9. If a husband gives his wife a new dress, then he is not cross.
For a husband to come home for tea, it is sufficient that he is methodical.
He hangs his hat on a gas jet only if his wife does not keep him under control.
If he is a good husband, then he gives his wife a new dress.
If he is not methodical, then he will hang his hat on a gas jet.
When his wife does not keep him under control, he is necessarily cross.

Therefore

6-4 PROOFS

In ordinary conversation, a proof is a collection of statements designed to establish a truth, or to convince someone that a certain conclusion does come from certain information. For a scientist or mathematician, a more formal definition of a proof is needed, because he must be able to establish his results, not simply to convince others, but to use the results to explore further. With the ideas of logically valid arguments and logical equivalence that have been developed, we can build a more formal definition of a proof.

A formal proof is a chain or sequence of statements linking a set of known or assumed facts with a conclusion to be proved. The conclusion must be the last statement in the chain; the "given" facts may appear anywhere in the chain. Each statement in the chain must be one of the given facts, or be a logically valid conclusion derived from statements preceding it in the chain, or be logically equivalent to a statement preceding it in the chain. In other words, in a formal proof we demonstrate that we can proceed from the known facts or assumptions to the conclusion by drawing intermediate conclusions using valid argument forms and restating items where necessary. Consider, for example, the following facts:

Statement 1 John or Jim will drive a boat in the hydroplane race.

Statement 2 If Jim is to drive in the race, then he must be a qualified swimmer.

Statement 3 Jim does not know how to swim.
Let us now "prove" that the driver of the boat will be John:

Statement 4 If Jim is to drive in the race, then he must be a qualified swimmer.
This is given statement 2.

Statement 5 Jim is not a qualified swimmer.
This is a restatement of 3.

Statement 6 Jim cannot drive in the race.
This is concluded from statements 4 and 5 by *modus tollens.*

Statement 7 John will drive the boat in the hydroplane race.

The proof is complete, because statement 7 follows logically from statements 1 and 6 by disjunctive syllogism. This proof is, of course, a verbal version similar to the symbolic step-by-step examples we examined in Sec. 6-3. Instead of a sequence of symbolic statement forms, we have a sequence of sentence statements building a chain from the facts to the conclusion. If we were given the following symbolic facts:

Statement 1 $A \vee B$

Statement 2 $B \rightarrow D$

Statement 3 $\sim D$
and wished to conclude
$\therefore A$
we could do so by using the following intermediate steps. First, replace statements 2 and 3 by $\sim B$, which follows logically from them, because they form the premises of a *modus tollens* argument to which $\sim B$ is the conclusion. Second, conclude A, which follows logically from statement 1 and $\sim B$, because they form the premise of a disjunctive syllogism whose conclusion is A. We could say that

$$\begin{array}{l} A \vee B \\ B \rightarrow D \\ \underline{\sim D \qquad} \\ \therefore A \end{array}$$

leads to

$A \vee B$
$\underline{\sim B}$
$\therefore A$

by *modus tollens,* and this proves A by disjunctive syllogism. If statement A above is: John drives the boat in the hydroplane race, and B is: Jim drives the boat in the hydroplane race, and D is: Jim is a qualified swimmer; then the proof given in symbolic form above is identical to the verbalized proof that John will be the driver.

This "sequence-of-statements" proof is the kind we usually encounter when someone tries to "prove something" to us; that is, if someone wants to show us that a certain conclusion "follows" from certain information and if a direct jump from the facts to the conclusion is too great for us, he builds a chain of intermediate conclusions leading from the facts to the final conclusion. He may also find it necessary to restate certain ideas and results so that we understand them more easily. He may wish to say something in a different way. Thus a proof is a valid argument. The method of establishing validity is to reduce the number of premises by replacing groups (usually pairs) of premises by a single premise that is the conclusion to a valid argument form and by substituting for some premises their logical equivalents. For example, prove s follows from $\sim(\sim p \wedge \sim q)$, $\sim p$, $q \rightarrow s$. In order to do so, we show that the following is valid:

$\sim(\sim p \wedge \sim q)$
$\sim p$
$\underline{q \rightarrow s}$
$\therefore s$

Thus

$\sim(\sim p \wedge \sim q)$
$\sim p$
$\underline{q \rightarrow s}$
$\therefore$

yields

$p \vee q$
$\sim p$
$\underline{q \rightarrow s}$
$\therefore$

by De Morgan, which leads to

$$\begin{array}{l} q \\ \underline{q \to s} \\ \therefore s \end{array}$$

by disjunctive syllogism, and this is *modus ponens.* First we replaced $\sim(\sim p \wedge \sim q)$ by its equivalent $p \vee q$ by De Morgan's law. Then we replaced the two premises $p \vee q$ and $\sim p$ by q, which follows from them by disjunctive syllogism. The final argument is *modus ponens,* and s is proved.

Thus our formal "proof" is nothing more or less than a valid argument form; that is, in theory, if not in practice, we could check the soundness of a conversational "proof" by checking the validity of the argument in symbolic form. In practice, of course, this is not possible, because to reduce even the most elementary discussion to symbolic form, or simply to identify all the elementary statements involved, would be too complex. And if the symbolic form were identified, the checking of each case, or simply searching for a linking set of statements, would consume more time than it would be worth. Instead of a formal analysis, complete in every detail, we omit many steps and usually get only an impression of the soundness of the proof. In mathematics and science the usual procedure is to indicate, in an informal way, our intermediate conclusions that relate our beginning ideas to our conclusion.

What is implied in our discussion to this point is that the soundness of a proof has nothing to do with its subject. A weakness of beginning students, in any subject, is that, when attempting to make a careful evaluation of a result, they lose sight of the importance of the structure of the argument, being diverted by the subject itself.

There are important differences between a proof built on a chain of intermediate statements and one that is demonstrated by the truth-table method. A chain of statements can usually be checked to see if each statement in the chain is sound, but if we cannot find such a chain, we have gained nothing; that is, once a chain is worked out and checked, we can be sure that we have "proved" the result, but not being able to find such a chain indicates only that we have failed to establish that it exists, not that the conclusion is invalid. A truth-table analysis would definitely establish the validity or invalidity of the conclusion. However, both methods of "proof" are sterile, for they do not tell us what conclusions to draw from the known facts. They permit us only to check the "reasonableness" of our conclusions after we have arrived at them by other methods. For example, the ancient Greek scholar Archimedes developed a technique to determine volumes and areas of such things as spheres and cylinders. He used something he called a "method" to determine his results. His method was a forerunner of modern calculus, but because he refused to discuss it with any of his contemporaries, it was not discovered that he knew of this method until 1906. Archimedes used his method to discover results, which he proved valid by a rigorous procedure known as "the method of exhaustion." This involved proving that all other possible answers were incorrect, leaving the desired result as the only available answer. Archimedes refused to reveal his method, because he could not establish formally that it always gave the correct answer. Using modern concepts of calculus, scholars, during the eighteenth, nineteenth, and twentieth centuries, have been able to prove that Archimedes' method, or at least its modern counterpart, always gives correct results. Once this fact had been established, it was no longer necessary to show individual results "valid," because they were being found by a logically sound method. We illustrate here the difference between proving the general validity of a method and proving specific results. We might "prove" the following about the numbers 3, 4, and 5:

Equation 1 $3(4 + 5) = 3 \times 4 + 3 \times 5$
We know that $4 + 5 = 9$. Therefore $3(4 + 5) = 3 \times 9$. But $3 \times 9 = 27$; therefore the left side of equation 1 is equal to 27. We also know that $3 \times 4 = 12$ and $3 \times 5 = 15$. Therefore $3 \times 4 + 3 \times 5 = 12 + 15$. But $12 + 15$ is also equal to 27, so that the right side of equation 1 has the same value as the left side. We have shown equation 1 to be true. We could then repeat this pattern to prove that

Equation 2 $5(7 + 4) = 5 \times 7 + 5 \times 4$
or that

Equation 3 $6(13 + 2) = 6 \times 13 + 6 \times 2$
or any number of such formulas.

A better procedure would be to prove a more general result, namely, that if a, b, and c are symbols standing for any numbers, then

General principle 1 $a(b + c) = a \times b + a \times c$
Once the truth of this formula is established, separate proofs of equations 1, 2, or 3 or any other such specific statements are not necessary. Because they all follow this pattern, their validity is established in the proof

of the general principle. This particular general principle is known as the "distributative law for real numbers."

However, no matter how many specific examples of a principle we show to be true, we have not established the general principle. But if we establish a general principle, we have proved the specific examples. Suppose that someone believes that the expression

$x^4 - 13x^3 + 53x^2 - 83x + 42$

is equal to zero when any numerical value is substituted for the symbol x. We could try a few values for x, remembering that x^4 means x times x times x times x. Let us try $x = 1$:

$1^4 - 13 \times 1^3 + 53 \times 1^2 - 83 \times 1 + 42$
$1 - 13 + 53 - 83 + 42$
$96 - 96$
0

Let us try $x = 2$:

$2^4 - 13 \times 2^3 + 53 \times 2^2 - 83 \times 2 + 42$
$16 - 13 \times 8 + 53 \times 4 - 83 \times 2 + 42$
$16 - 104 + 212 - 166 + 42$
$270 - 270$
0

Perhaps we should try a larger number, say $x = 7$:

$7^4 - 13 \times 7^3 + 53 \times 7^2 - 83 \times 7 + 42$
$2401 - 13 \times 343 + 53 \times 49 - 83 \times 7 + 42$
$2401 - 4459 + 2597 - 581 + 42$
$5040 - 5040$
0

Can we then conclude that this assumption is true? After all, it worked in all cases we tried. Let us try one more example, say $x = 10$:

$10^4 - 13 \times 10^3 + 53 \times 10^2 - 83 \times 10 + 42$
$10{,}000 - 13 \times 1000 + 53 \times 100 - 83 \times 10 + 42$
$10{,}000 - 13{,}000 + 5300 - 830 + 42$
$15{,}342 - 13{,}830$
1512

which is hardly equal to zero.

If we had stopped at the end of three trials, we should have thought that the assumption was true. In fact, we could also have tried $x = 3$, and the expression would have again been equal to zero.

Detailed methods of proof vary, depending on the subject and the degree of formality required. However, underlying any method of actual proof are the concepts we have examined.

6-4 EXERCISES

1. What is the relationship between a formal proof and a logically valid argument?

2. What are the two kinds of statements that link the given facts with the conclusion of a formal proof?

3. A man is running for the office of mayor of a city. In one of his speeches he claims to "prove" that he is the best man for the job with the following statements. The statements are given in symbolic form only:

$$\begin{array}{l} A \rightarrow B \\ \underline{\sim A} \\ \therefore \sim B \end{array}$$

You know that all three statements are true. What would you conclude about the man's reasoning ability?

4. Substitute logically equivalent statements and conclusions from valid argument forms to show that the following premises yield R:

$$\begin{array}{l} Q \\ Q \rightarrow \sim P \\ \underline{P \vee R} \\ \therefore R \end{array}$$

5. By substituting logically equivalent statements and by reaching intermediate conclusions, using the valid argument forms in Sec. 6-3, show that $\sim P \vee Q$, $\sim(Q \wedge R)$, P, $S \vee R$ yields S.

6. Consider the following set of statements:

If today is Tuesday, then there is a baseball game today.

There are never both a baseball game and a basketball game on the same day.

There is a basketball game today.

Today is not Tuesday.

Using the following symbolic replacements (*a*) develop the symbolic patterns of the above statements, and (*b*) prove the last one:

T: Today is Tuesday

B: There is a baseball game today.

C: There is a basketball game today.

7. Suppose that you were unable to find a formal proof for the last statement in Exercise 6. What could you say about its truth value?
8. "Prove" a false statement by showing it to be true in a number of cases.

6-5 AXIOMATICS

Our discussion of logic, argument, and proof has been concerned with one phase of logic only: "deduction." Deduction applies general principles to a set of facts to reach a specific goal. For Sherlock Holmes, for example, the general principles were those of police science, and the specific goal the apprehension of the guilty party. Another example of deduction is the process you go through in trying to determine what your final grade will be in a course. You might reason thus: "In mid-terms I received two B's and two C's. The B's were not very high B's, but I handed in all work and responded well in class discussion. Therefore I shall receive a B in the course." In this example, the general principles are those of grading policy. They are used to reach the specific conclusion, "the grade will be a B."

Deduction, then, is reasoning from the general to the specific. In many cases, the general principles used in deductive reasoning cannot be established deductively. We cannot prove deductively that the sun will come up every day, although we know that it will. This suggests another kind of reasoning called "induction." Induction has to do with observation, guess, and intuition and is reasoning from specific instances to general principles. We say that the sun comes up every morning because we have watched it do so in the past. Suppose that Holmes has identified a criminal, and it is now necessary to find and arrest him. The apprehension could involve inductive reasoning in that the criminal has been observed to eat lunch in the same restaurant on the last seven Mondays. He concludes, inductively, that the criminal eats there every Monday. Hence the police wait for him there. Induction is used to establish general principles, and deduction to apply them to special cases. If the principles that are established inductively are false, then deductions based on them might also be false, although the deductions were logically valid. We cannot be sure that the deductions are false, however, simply because we discover that the principles from which they were drawn were in some way false. In effect, we are reminding ourselves here that, if $p \rightarrow q$ is true and p is false, we have no knowledge as to whether q is true or false.

These notions, together with related discoveries in theoretical physics that took place toward the end of the last century, caused mathematicians, physicists, and logicians to examine the nature of the assumptions they required in the study of mathematical and physical systems. In physics these examinations led to the modern theories of the atom and Einstein's theories of the universe. In mathematics and logic, they led to a "new" approach to the study of mathematical and logical systems, in which assumptions and proofs were analyzed separately. The result of this separation was to establish a new area of study known as "axiomatic systems."

An axiomatic system is a group of definitions and "laws" describing something, usually some physical object or property. You have been working in such a system in the present chapter. Symbolic logic is an axiomatic system that is a model of reasoning. The geometry taught in high school is an axiomatic system that is supposed to describe the universe. We say "supposed to describe," because there is some question as to how good a job it does. Any axiomatic system contains the following elements:

1. Undefined or "primitive" terms, such as *simple statement of logic, point,* and *line.* These terms are to be understood intuitively.
2. "Technical" terms, involving the use of the primitive terms. The word *triangle* is a technical term, involving the primitive terms *line* and *point.* The term *compound statement* is a technical definition from logic.
3. A list of properties assigned to the primitive terms. These properties are called "axioms" and are usually established inductively. In logic, some axioms include the basic truth tables for *and* and *or.* In geometry, an axiom states that a line can be extended indefinitely. Axioms are really assumed truths.
4. Rules of logic, such as *modus ponens,* that can be used to deduce more facts from the given axioms.

An axiomatic system is a chain that begins with primitive terms, technical terms, and axioms and proceeds to the establishment of new facts, which are called "theorems" and which can be used to derive more theorems. In our study of logic we began with simple statements, then defined compound statements, and proceeded to prove some theorems. One such theorem, although we did not call it so, was that a conditional is logically equivalent to its contrapositive.

To understand better how an axiomatic system works, we define a small system and attempt to prove one or two theorems in it. In this example, the game of ticktacktoe, we do not give all the technical terms or all the assumptions (axioms):

Undefined terms Squares, zeros, and crosses.

Technical term "Winning combinations" are three crosses or three zeros in a line.

Axiom It is possible to make a grid of nine squares on which crosses or zeros can be drawn:

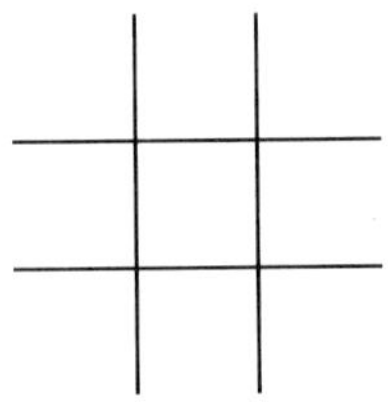

We now deduce theorems:

THEOREM 1 There are eight winning combinations: three vertical, three horizontal, and two diagonal.

THEOREM 2 There are nine possible first moves.

THEOREM 3 There are eight possible second moves.

We could keep on deducing theorems, perhaps hundreds more. They would then become the things we "know" about the game of ticktacktoe. The things we "know" about the universe and about mathematics have come to us in much the same way. The mathematical facts that are used every day, calculations with numbers, for example, can be deduced from a set of axioms created by the Italian mathematician Peano. The facts we know about our universe come from a set of axioms.

In ticktacktoe, the axiom is completely obvious. This is usually so with axioms; they are obvious "truths." One of Euclid's axioms states that, if two lines intersect a third so that the sum of the interior angles on one side of the third line are less than 180°, then the two lines meet on this side of the third. An obvious fact? In attempting to prove this axiom, the Russian mathematician Nicolai Ivanovich Lobachevsky created a new geometry called "noneuclidean." There is some evidence that noneuclidean geometry is a better model of the universe than is Euclid's.

Axiomatic systems can be evaluated, relative to the "real" world, as useful or not useful. They are not true or false. The measure of their usefulness might be how well deduced results agree with observed occurrences. There is a danger in this, because sometimes the axiomatic system is assumed to be the real world, and an observation at odds with the system is discredited. One of the goals of the axiomatic approach is to help overcome this problem and make the creation of theories about the nature of the universe more flexible. There is ample historical evidence of the trouble caused when men do not know the difference between assumption and fact. For example, men like Copernicus, who tried to point out the nonuseful nature of the established aristotelian theory as to the makeup of the system of stars and planets. Galileo was the victim of a similar conflict with the ideas of Aristotle. When his observations contradicted established theorems, he was forced to abjure his beliefs.

Another example of an axiomatic system is known as "finite geometry." This geometry uses the words *point* and *line*, but the axioms of the system restrict these terms so that they are not the "usual" point and line. The entire system has a quadrilateral and its diagonals as its model:

Undefined terms Point, line.

Technical terms Two lines are "parallel" when they have no point in common.

Axioms

1. Any two points determine one line.
2. Any line contains exactly two points.
3. For any point P and any line l, there is exactly one line through P and parallel to l.
4. There are at least three points.

THEOREM 1 There are at least four points.

Proof

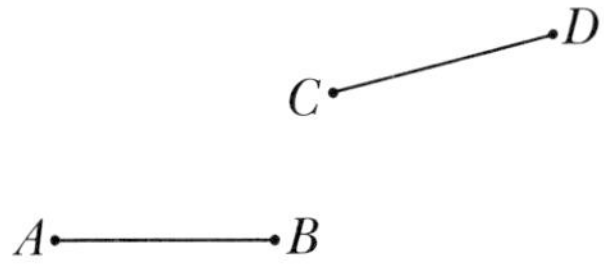

1. Draw three points and label them A, B, and C (axiom 4).
2. Draw a line from point A to B and call this line AB (axiom 1).
3. There is one line through C and parallel to AB (axiom 3).
4. The line through C and parallel to AB contains a second point; call it D (axiom 2).
5. There are now four points: A, B, C, and D. This proves theorem 1.

THEOREM 2 There are at least six lines.

Proof

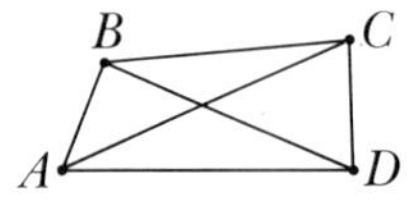

1. There are at least four points: A, B, C and D (theorem 1).
2. The following pairs of points form lines: AB, AC, AD, BC, BD, and CD (axiom 1).
3. By step 2 above, there are six lines, and this proves theorem 2.

THEOREM 3 There cannot be more than four points.

Proof

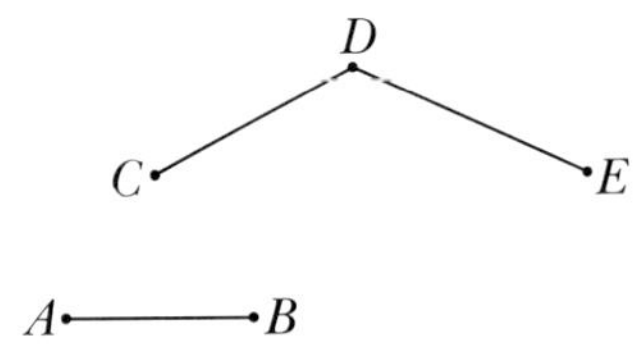

1. Suppose that there is a fifth point called E.
2. Then lines CD and DE are parallel to AB by the definition of *parallel.*
3. It is not possible to have two lines through one point D and parallel to a third line AB (axiom 3).
4. Thus there cannot be a fifth point.

It is possible to prove more theorems in this system. However, the number of theorems is limited, because we have defined only one technical term.

The selection of a set of axioms is the foundation of any axiomatic system. There are several characteristics that such a set of axioms must have. The most important is that they do not give contradictory results. For example, our number system would not be very useful if it could be shown that $2 + 2 = 5$. The choice of axioms is, of course, constantly studied in mathematics and in physics. However, axioms are studied even more closely in philosophy and theology, for both disciplines involve axiomatic systems. Different religions are established because of different axiom sets (different basic beliefs). Thus a mathematician that concentrates his studies in the field of axiomatics is a philosopher, and conversely many of the greatest philosophers and theologians were mathematicians.

Axiomatic systems have been a major factor in preventing rigid points of view in mathematics and science. From the assumption of such a system, the conclusions "follow logically." The usefulness of the conclusions is measured, not by their logical nature, but by their reliability in forecasting useful results. This approach has been a powerful one for mathematicians and scientists in general and is finding application in other areas of man's interest.

6-5 EXERCISES

1. Give two examples of inductive reasoning.
2. Give two examples of deductive reasoning.
3. Deduction is nonproductive. Explain.
4. What is an axiom?
5. In the game of golf, state some primitive terms, technical terms, and a possible theorem.

The following defines an axiomatic system:

Primitive terms Point, line.

Technical term A "squiggle" is any figure made up of six points and three lines.

Axioms

(*a*) There are at least three points not on the same line.
(*b*) Each line contains exactly three points.
(*c*) Any two points determine a line.

6. Show that there are at least six points in this system.
7. Show that the system must contain at least two squiggles.
8. Is the primitive term *line* in this system synonymous with your definition? Explain.

Given the following axiomatic system:

Undefined terms Point, line.

Technical term Two lines are "parallel" if they have no point in common.

Axioms

(*a*) Any line contains exactly two points.
(*b*) Each point is contained in exactly two distinct lines.
(*c*) Any line has exactly five lines parallel to it.
(*d*) There exist at least two distinct points, and these determine a line.

9. Show that there are at least eight points in this system.
10. Name two possible geometric models for this system.

REVIEW EXERCISES

Check the validity of the following arguments.

1. $(w \vee l) \wedge \sim p$
$w \rightarrow s$
$s \rightarrow p$
$\therefore l$

2. $\sim t \rightarrow \sim r$
$\sim s$
$t \rightarrow w$
$r \vee s$
$\therefore s$

3. $p \rightarrow (r \rightarrow s)$
$\sim r \rightarrow \sim p$
p
$\therefore s$

4. Supply a valid conclusion to the following argument. If John dates whom he pleases, he must be popular. His not being in many activities is a sufficient condition for him not to be popular. He is not in many activities, or his grades suffer. Therefore . . .
5. Why is the following argument valid?

$(C \wedge D) \rightarrow (\sim A \vee \sim B)$
$A \wedge B$
$\therefore C \wedge D$

6. Define *primitive term.*

SELECTED REFERENCES

Fujii, John N.: "An Introduction to the Elements of Mathematics," John Wiley & Sons, Inc., New York, 1961.
The sections on formal proof and axiomatic systems are particularly good.

Kemeny, John G.: "Introduction to Finite Mathematics," Prentice-Hall, Inc., Englewood Cliffs, N.J., 1966.

Good discussion of valid argument forms, particularly the indirect method of establishing validity.

Stoll, Robert R.: "Sets, Logic and Axiomatic Theories," W. H. Freeman and Company, San Francisco, 1961.

A formal presentation of the topics in the present chapter.

Is it true, as the Venn diagram indicates,
that the intersection of the set of men
and the set of horses
is the set of men on horseback?

chapter seven

SETS

As mathematicians began to explore and develop the patterns of logic and the fundamental nature of their discipline, they found that every phase of their subject had a common factor. Everywhere they turned, they found that they had to deal with specific collections of objects or with problems involving the relationships among two or more collections. It was natural that they eventually began to study the properties of collections in general. The term "set" came into use as the name for such a collection. We should make it clear that in Sec. 7-1 we are not referring to a particular collection but to the properties of collections themselves.

7-1 INTRODUCTION AND DEFINITIONS

The formal study of sets began with the German mathematician Georg Cantor (1845–1918). Although Cantor developed set theory as a specific tool for working with advanced topics, by 1890 the theory had grown to a point where it was studied for its own sake. Cantor's genius soon became more evident as mathematicians found set theory to be one of the most important concepts ever introduced into the study of mathematics. In later sections we examine some of the ways this theory can be used to study problems in many other branches of mathematics and even outside mathematics.

To begin our study, we identify what we mean by sets in a more formal manner. For our purposes, a set is any well-defined collection of objects. We identify the words *object* and *collection* as primitive terms, as in Chap. 6. In other words, we assume that their meaning is understood and do not attempt to explain them in simpler terms. We do this, of course, to avoid the circularity of definition. However, we expand the meaning of the term *well defined.* A collection is well defined if, given any object, we can decide whether it belongs in the collection or not. For example, the collection of persons 6 ft tall or taller is well defined. We can check any object to see if it is a person and if this person is 6 ft tall or taller. On the other hand, the collection of tall people is not well defined, because the word *tall* is not specific enough to check an object's classification. Is a man of 5 ft $11\frac{3}{4}$ in. tall or not? The collection of logically valid argument forms is a set, because it is well defined, but that of great symphonies is not, because it is not well defined, on account of the word *great.* We restrict our attention to well-defined collections.

To study sets, it is necessary to introduce one or two other primitive terms. One is the idea of an object "belonging" to a set or being a "member" of the set. These objects, or "elements," as they are called, are not necessarily material; sets can be made of ideas. For example, a collection of economic theories is a set, containing such objects as capitalism and socialism. The concept of a set "containing" elements is also a primitive term in our discussion. In general, we do not specify the nature of the elements of the sets we are considering. We deal with specific sets to illustrate our discussion, but the reader should remember that we are studying how sets are related in general. The properties of specific sets may not apply to all sets.

It is also necessary to have special symbols for sets, their elements, and the relationships among sets. This is necessary so that we can specify properties of sets without making specific references. The usual notation involves the use of lowercase letters as symbols for elements. For example, we might say that a is an element of set A or, more simply, that a is an element of A. In this context we are using A as a symbol for a set and a as a symbol for an object that we claim is an element of the set. The phrases *is an element of* and *belongs to* and *is a member of* are used often enough so that we replace them with the single symbol $\in$. This notation results in statements like $a \in A$ and $b \in A$ and $b \notin B$. These are read a is a member of set A, and b is an element (or member of, belongs to) set A, and b is not a member of set B. In the last expression we follow the standard mathematical convention of drawing a slash through a special symbol to mean that the relationship of the symbol does not hold. (For example, = means equal, and $\neq$ means not equal.)

One problem arises in using this uppercase-lowercase distinction between sets and elements. Sets themselves are objects and can be elements of other sets. Let A stand for the set of students who take freshman English at 10 A.M. at a university. If d stands for a student in this class, we write $d \in A$. On the other hand, let C stand for the set of classes taught at the university at 10 A.M. Then we write $A \in C$; that is, A, the 10 A.M. freshman English class, is an element of C, the set of 10 A.M. classes. Note that $d \notin C$, because C is a set of classes and d stands for one student. Later, we discuss sets made of sets.

There are two basic ways of specifying a particular set. One is to list the members of the set. The other is to describe that characteristic which makes them members of the set. For example, consider the set of even numbers between 1 and 9. Our sentence specifies this set by describing its members. If we specified this set by listing its members, we would write: "The set of numbers 2, 4, 6, 8." For the sake of neatness and clarity, we usually enclose such a list in braces, as $\{2,4,6,8\}$. This results in statements like

The set of vowels = $\{a,e,i,o,u\}$

or

The set of A students = {John, Jim, Joan, Janet}

or even such strange statements as $x \in \{a,e,i,o,u\}$. This statement makes us remember that the letter x is used as a symbol, in this case to stand for each of the five letters a, e, i, o, and u.

Describing sets as collections of well-defined objects in no way implies order; that is, suppose $H = \{1,2,a,b\}$. The H could be written

$H = \{1,2,b,a\}$

or

$H = \{a,1,2,b\}$

or

$H = \{b,2,a,1\}$

The reordering of the elements of a set does not change it any more than changing the seating arrangement changes your mathematics class.

Two sets arise quite often in our discussion and deserve special mention and special symbols. One is the set of 53-ft-high pink poodles in Texas, which is the same as the set of astronauts who have landed on Mars or the set of former Presidents of the United States who have been elected five times to that office. These are all the same set, that which contains no elements. Although these examples may seem silly, it is to our advantage to have this "empty" set, or set without any elements. The empty set, or "null" set, deserves a special symbol. We assign it two: $\emptyset$ and $\{\ \}$. The following is an example of its use:

The set of elements shared by the set of even numbers and the set of odd numbers is $\emptyset$.

The other set worthy of special mention is that composed of all elements we are considering in a specific problem. This is usually known as the "universal" set, and we use U as its symbol. When working with portions of a set, we often refer to the set as the universal set with reference to our discussion.

Let us now turn our attention to simple relationships among sets. Consider the following sets:

$A = \{1,2,3,4,5,6,7,8,9\}$
$B = \{2,4,6,8\}$
$C = \{1,2,3,4,5\}$
$D = \{1,3,5,7,9\}$

The elements of sets B, C, and D are special in that they are all also elements of the set A. When this occurs, when all the elements of one set are also elements of another, we say that one set is a "subset" of the other. Thus B is a subset of A, because B contains elements of A only. Similarly C is a subset of A, and so is D. We use the symbol $\subseteq$ to replace the phrase *is a subset of*. Then we write $B \subseteq A$, $C \subseteq A$, and $D \subseteq A$. If A and B are unspecified sets, a formal definition for a subset is:

If for every choice for $x \in B$ it is also true that $x \in A$, then $B \subseteq A$.

That is, $B \subseteq A$ if every element of set B is also an element of set A. There is nothing about the definition of subset to assure us that if $B \subseteq A$, then $B \neq A$; that is, if all we know about set B is that B is a subset of set A, we cannot be sure that B is not simply another symbol for set A. By implication, we write $A = B$ if and only if A and B are to be two distinct symbols for the same set.

If we know that $A \subseteq B$ and that $A \neq B$, then we call A a "proper" subset of B. The notation for proper subset is $\subset$. For example, when we write $A \subset B$, we mean that all the elements of A are elements of B, but A and B are not the same set. The only way this can happen is if there is at least one element in B that is not in A; that is, set B contains an element that is not a member of set A. Using special symbols, we write:

$A \subset B$ if, for every $a \in A$, then $a \in B$, and there is at least one element $b \in B$ such that $b \notin A$.

For a few examples of this notation, consider the following sets:

$Q = \{a,e,i,o,u\}$
$R = \{e,a,i,o,u\}$
$S = \{a,b,c,d,e,i,o,u,v,w,x,y,z\}$

and for the universal set let

$U = \{a,b,c,d,e,f,g,h,i,j,k,l,m,n,o,p,q,r,s,t,u,v,w,x,y,z\}$

or using a set description, U is the set of English letters. The following symbolic statements would then be true (remember / written through a symbol means *not*):

$Q \subseteq U$		
$R \subseteq U$		
$R \subseteq Q$	but	$R \not\subset Q$
$R \subset S$	and	$R \subseteq S$
$\emptyset \subseteq Q$	and	$\emptyset \subset Q$

A comment about notation. Since $A = B$ is reserved to mean that A and B are two different symbols for the same set, one may say that $A = B$ if and only if $A \subseteq B$ and $B \subseteq A$.

7-1 EXERCISES

1. Give some examples of sets; also give some examples of collections that are not sets, and explain how they fail to meet the requirements of being a set.
2. Convert the following set descriptions into listings of set elements.

   (*a*) The set of states of the United States whose name begins with the letter C.
   (*b*) The set of odd numbers that are greater than 3 but less than 20.
   (*c*) The set of letters in the name of the first President of the United States.
   (*d*) The set of letters in the name of the sixteenth President of the United States.
   (*e*) The set of even numbers that are greater than 9 and less than 7.

3. Write a description of the following sets.

   (*a*) $\{2,4,6,8\}$
   (*b*) $\{a,b,c\}$
   (*c*) {T. Roosevelt, W. H. Taft, W. Wilson, W. G. Harding, C. Coolidge, H. Hoover, F. D. Roosevelt, H. Truman}
   (*d*) {Hawaii}
   (*e*) $\{\emptyset\}$

4. Why do we say that *containing* is a primitive term of set theory?
5. Translate the following from symbols to words.

   (*a*) $z \in B$
   (*b*) $x \in P$ and $x \notin Q$
   (*c*) $A \subseteq B$ and $B \subset C$
   (*d*) $A \subset B$ and $B \in C$
   (*e*) $\emptyset \subseteq A$
   (*f*) $A \subseteq B$ but $B \not\subset A$
   (*g*) $A \subseteq B$ and $B \subseteq A$
   (*h*) $A \subset B$ and $B \subset A$

6. Assume that A, B, C, P, Q, Z, and X are the same as in Exercise 5.

   (*a*) If statement 5*c* is true, is it also true that $A \subseteq C$?
   (*b*) If statement 5*c* is true, is it also true that $A \subset C$?
   (*c*) Can statement 5*h* be true?

(*d*) Without any knowledge about set A, what can you say about the truth statement 5*e*?

(*e*) If statement 5*b* is true, which of the following can be true: $P \subseteq Q$, $P \subset Q$, $Q \subseteq P$, $Q \subset P$?

7. The collection of even numbers is clearly a set according to our meaning of the concept. We have used the method of description to specify it. Could we make a listing of this set? Explain.
8. Two methods of specifying sets, listing and description, were given. Give some reasons why both methods are necessary.

7-2 OPERATIONS ON SETS

With the concepts and notation we defined in Sec. 7-1, we are now ready to examine ways of combining and working with sets. In Sec. 2-6 we referred to a "binary operation" of numbers. Our first operation on sets are binary compositions of sets; that is, we consider methods of combining two sets to produce a third. In addition to binary compositions we examine at least one operation on sets that generates a new set from a single set.

Suppose that we have two sets A and B. We define their "intersection" as that set of elements which belongs to them both; that is, the intersection of sets A and B is the set of elements they share. For example, consider the sets $\{1,2,3,4,5\}$ and $\{2,4,6,8\}$. Their intersection is the set $\{2,4\}$. More formally we use the symbol $\cap$ to replace the word *intersection*. Thus:

$A \cap B =$ the set of all objects x for which it is true that $x \in A$ and $x \in B$.

$A \cap B$ is the symbol for this new set, as $3 + 5$ is a symbol for the number we usually identify with the numeral 8. A few more examples follow. Suppose that

$P = \{a,b,c,d\}$
$Q = \{c,d,e,f\}$
$R = \{e,f,g,h\}$
$S = \{a,d,f\}$

Then we have these results:

$P \cap Q = \{c,d\}$
$P \cap R = \emptyset$

$Q \cap P = \{c,d\}$
$Q \cap R = \{e,f\}$
$S \cap P = \{a,d\}$
$S \cap Q = \{d,f\}$

We use these same sets to illustrate our second operation on sets. The set $\{a,b,c,d,e,f\}$ has a special relationship to the sets P and Q. If we examine it closely, we notice that it is the set of those elements which appear as either elements of P or elements of Q or both. The set $\{a,d,c,e,f\}$ bears a similar relationship to sets Q and S. We make the following more formal definition. The "union" of two sets is the set formed from all the objects that appear as elements of either of the original sets. We use the symbol $\cup$ for this operation and write:

$A \cup B =$ the set of all objects x for which it is true that $x \in A$ or $x \in B$.

Here the *or* is the $\vee$ kind of *or* rather than the $\underline{\vee}$ kind.

Using P, Q, R, and S as given above, we have the following results of forming unions:

$P \cup Q = \{a,b,c,d,e,f\}$
$Q \cup R = \{c,d,e,f,g,h\}$
$P \cup S = \{a,b,c,d,f\}$
$P \cup R = \{a,b,c,d,e,f,g,h\}$

It is possible to discuss such things as

$$(P \cup Q) \cap (Q \cup R) = \{c,d,e,f\}$$

or

$$(P \cup R) \cap S = \{a,d,f\}$$

or

$$(P \cap Q) \cup (S \cap Q) = \{c,d,f\}$$

or

$$P \cup (Q \cap R) = \{a,b,c,d,e,f\}$$

or any other combination of these two operations. The reader should verify these four examples.

Our third operation on sets is known as "complementation" and is a unary operation. Unlike the operations of union and intersection, the complement is thought of as being formed from a single set. Suppose that we are considering sets that have a defined universal set U. From the definition of U, we understand that any set A being discussed is a subset of U. But we may also wish to discuss the set of elements in the universal set U that are not elements of set A. For example, if we are considering only sets of single-digit numerals, U is the set $\{0,1,2,3,4,5,6,7,8,9\}$, and if $A = \{2,3,4,5,6\}$, then the set of elements of the universal set that are not in set A is $\{0,1,7,8,9\}$. The notation for this set is $\tilde{A}$, read A complement. If $B = \{1,3,5,7,9\}$, then $\tilde{B} = \{0,2,4,6,8\}$.

In general, then, if A is a subset of a universal set U, then $\tilde{A}$ is the set of all elements of U that are not elements of A, or in symbols:

$\tilde{A}$ = the set of all objects x for which it is true that $x \in U$ and $x \notin A$.

Combinations of union, intersection, and complement can be used in a single expression. For example, if we say that a universal set is $U = \{a,b,c,d,e,f,g,h,i,j\}$ and use P, Q, R, and S as above, we have

$$\widetilde{(P \cap Q)} = \{a,b,e,f,g,h,i,j\}$$
$$\widetilde{(P \cup R)} = \{i,j\}$$
$$\widetilde{(P \cup R)} \cap S = \emptyset$$
$$\widetilde{(P \cup R)} \cup S = \{i,j,a,d,f\}$$
$$\widetilde{P \cup (Q \cap R)} = \{g,h,i,j\}$$

Again, the reader should verify these results.

In 1880, the English mathematician John Venn developed a method of visualizing set operation with drawings now known as "Venn diagrams." Suppose that we let an area in Fig. 7-1 represent the objects in a universal set. Then we indicate the set A as a portion of the universal set as a shaded circular area. Then $\tilde{A}$ is represented by the area outside the circle. This follows from the definition of complement, which says that all members of the universe that are not in set A are in set $\tilde{A}$. Because the area outside

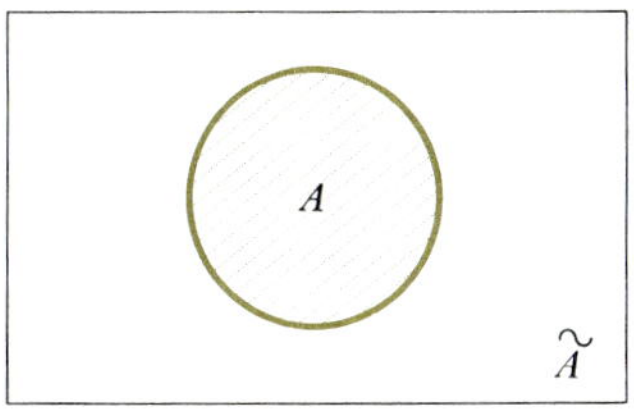

FIG. 7-1 SIMPLE VENN DIAGRAM.

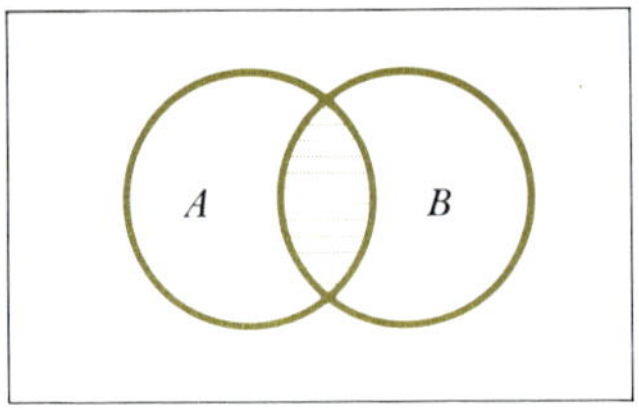

FIG. 7-2 VENN DIAGRAM OF $A \cap B$.

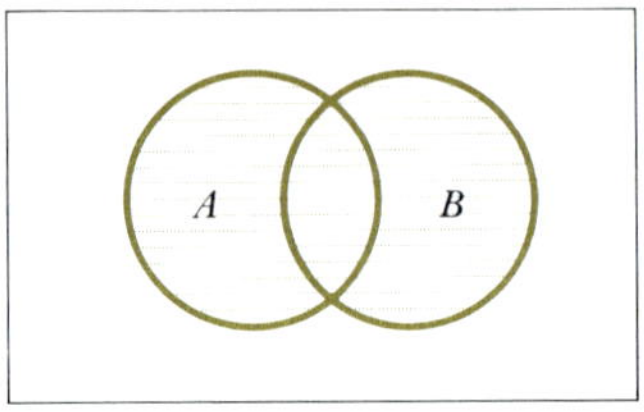

FIG. 7-3 VENN DIAGRAM OF $A \cup B$.

the circle represents the elements of U and these elements are not in A, they are elements of $\tilde{A}$.

In Fig. 7-2, the portion of the diagram within circle A represents set A, and that within circle B set B. The shaded portion represents elements in both set A and set B and is the representation of $A \cap B$.

Figure 7-3 shows $A \cup B$ shaded. This follows directly from the definition of union. We can illustrate many other sets and set notation with these diagrams. For example, in Fig. 7-4, the shaded portion is $Q \cap \tilde{P}$, because it represents those elements which are in set Q but not in set P and thus are in set $\tilde{P}$. This set is sometimes called the "relative" complement of P with respect to Q and is denoted by $Q - P$. In more general terms, the set $A - B$ is the set of all those elements of set A which are not elements of set B.

Another variation occurs when two sets are related through the concept of subset. Suppose that $R \subset S$. What would the Venn diagram of R and S look like? Figure 7-5 shows this relationship. Because all the elements of set R are elements of set S, it follows that the area representing set R is contained within that representing set S. If we shade the overlapping

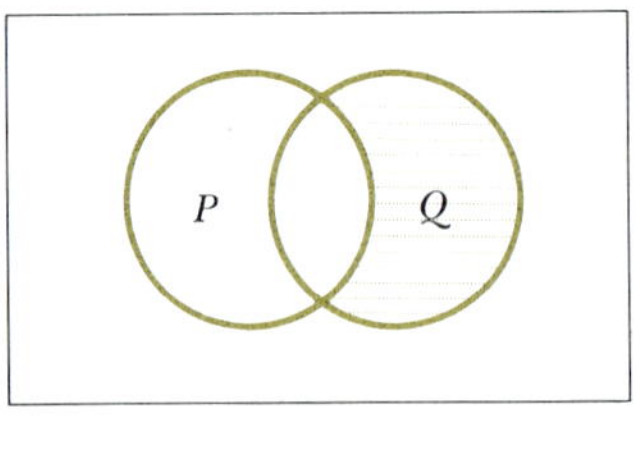

FIG. 7-4 VENN DIAGRAM OF $Q \cap \tilde{P}$ OR $Q - P$.

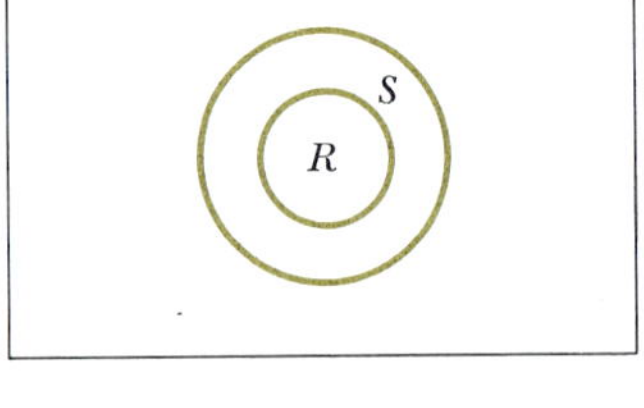

FIG. 7-5 VENN DIAGRAM OF $R \subset S$.

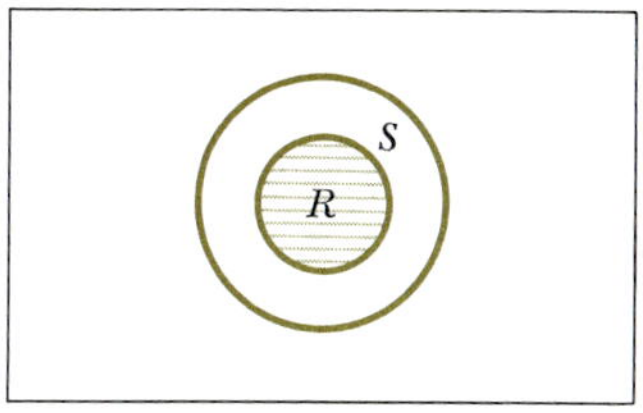

FIG. 7-6 VENN DIAGRAM OF $R \cap S$ WHEN $R \subset S$.

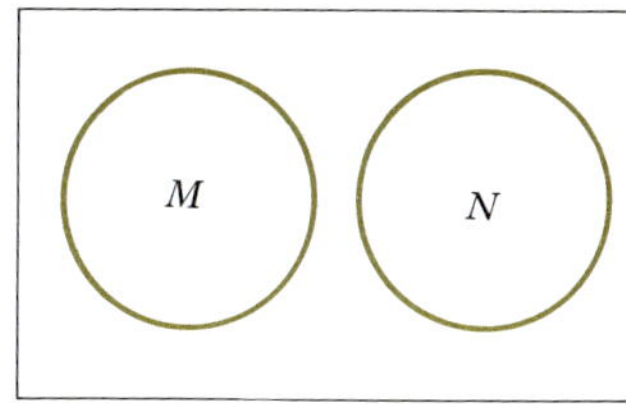

FIG. 7-7 VENN DIAGRAM OF $M \cap N = \emptyset$.

area for sets R and S in Fig. 7-5, we get Fig. 7-6. But we have shaded all the area of set R and nothing else; therefore $R \cap S = R$ if $R \subset S$. A similar line of reasoning leads to the fact that $R \cup S = S$ if $R \subset S$.

Venn diagrams can also show that two sets do not share any elements. Figure 7-7 shows that $M \cap N = \emptyset$. What would the diagram of $M \cup N$ look like? All that is required is to shade those areas which represent elements of M or N. This results in Fig. 7-8.

There is nothing about the definition of union to imply that the sets involved must share any elements. When two sets do not have any elements in common, that is, when their intersection is empty, we say that the sets are "disjoint." Saying that M and N are disjoint sets is another way of saying that $M \cap N = \emptyset$, or that M and N do not have any elements in common.

We can make Venn diagrams of complex combinations of sets by first finding the diagrams for the components of the final result. For example, how do we make the Venn diagram for $(\widehat{A \cup B}) \cap C$? First, we consider the fact that there are three sets. This is not a problem, because we can use a drawing like Fig. 7-9. The diagram for $A \cup B$ would be Fig. 7-10,

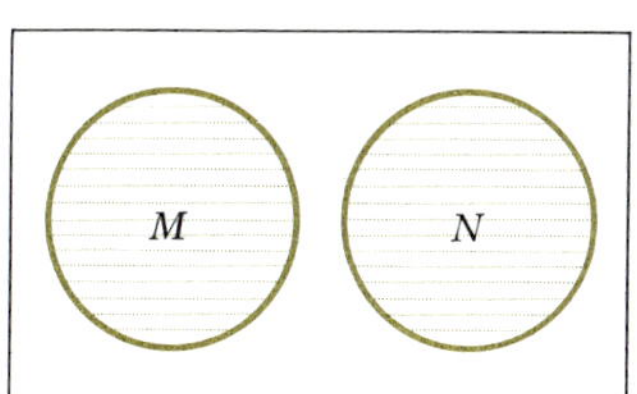

FIG. 7-8 VENN DIAGRAM OF $M \cup N$ WHEN $M \cap N = \emptyset$.

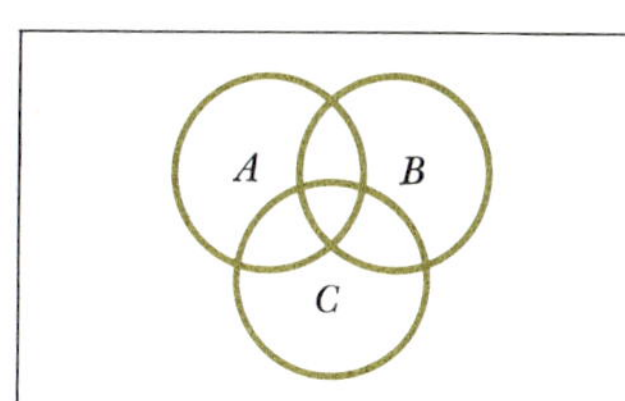

FIG. 7-9 VENN DIAGRAM OF THREE SETS.

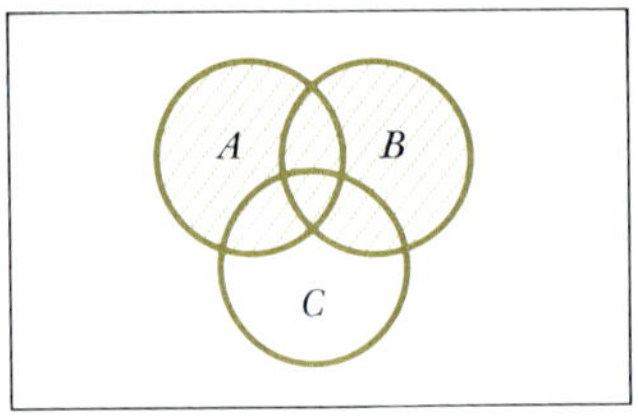

FIG. 7-10 VENN DIAGRAM OF $A \cup B$.

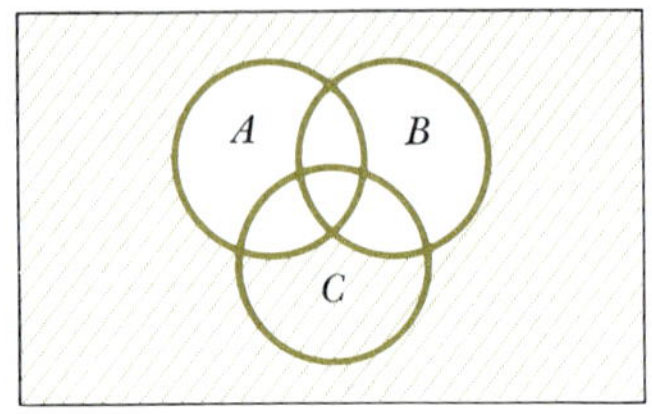

FIG. 7-11 VENN DIAGRAM OF $\widetilde{A \cup B}$.

where we have shaded the portion representing $A \cup B$. Then $\widetilde{A \cup B}$ would be illustrated by Fig. 7-11. Because we are looking for the intersection of the shaded set with set C, the final result would be Fig. 7-12, where the shaded portion corresponds to $(\widetilde{A \cup B}) \cap C$.

We have used several steps to get to the final result. With a little practice, many of these steps can be omitted. For example, in Figs. 7-13 to 7-17, the shaded portion of the diagram corresponds to the set identified in the legend. The reader should verify each result. Note that some of the set combinations have the same diagram. If we examine Figs. 7-14 and 7-17, we conclude that $A \cap (B \cup C) = (A \cap B) \cup (A \cap C)$; and by examining Figs. 7-15 and 7-16, we see that $(\widetilde{A \cap \tilde{B}}) = \tilde{A} \cup B$. These equalities are true for any sets A, B, and C, because no specific identification of their nature is given, and the only way two sets can have the same diagram is if they are the same set written with different symbols. And so the sets are equal. In one sense, these could be theorems in the algebra of sets. It should be understood that Venn diagrams do not represent formal proofs in the sense in which the term was used in Chaps. 5 and 6.

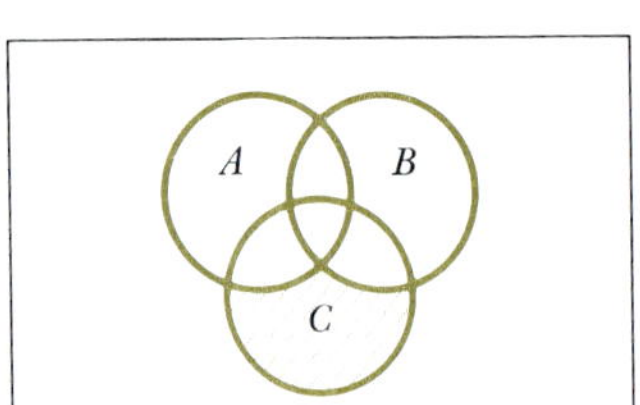

FIG. 7-12 VENN DIAGRAM OF $(\widetilde{A \cup B}) \cap C$.

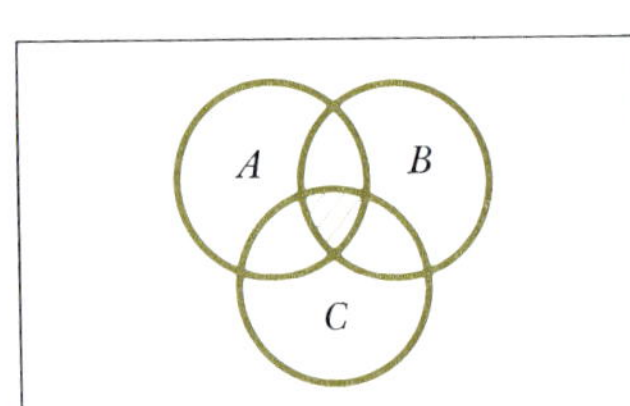

FIG. 7-13 VENN DIAGRAM OF $(A \cap B) \cap C$.

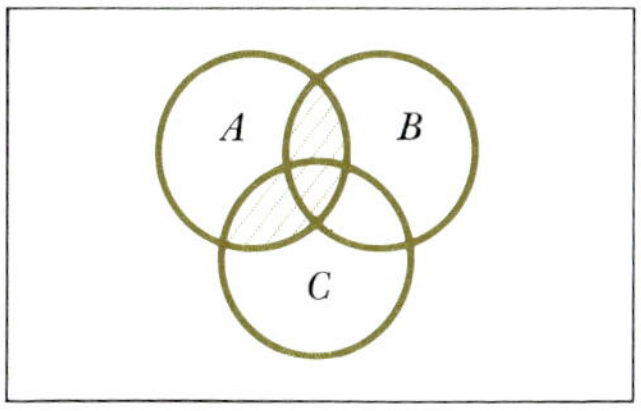

FIG. 7-14 VENN DIAGRAM OF $A \cap (B \cup C)$.

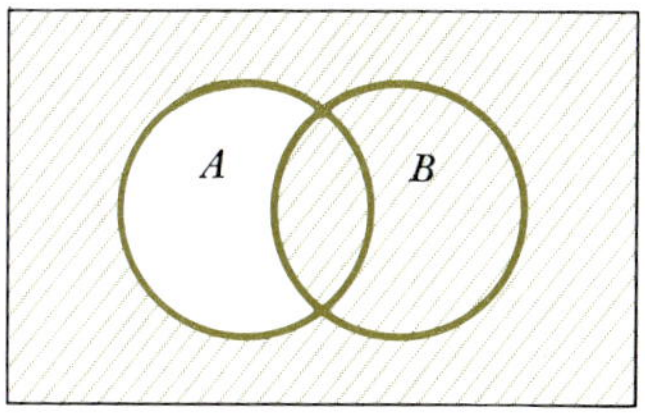

FIG. 7-15 VENN DIAGRAM OF $\widetilde{(A \cap \tilde{B})}$.

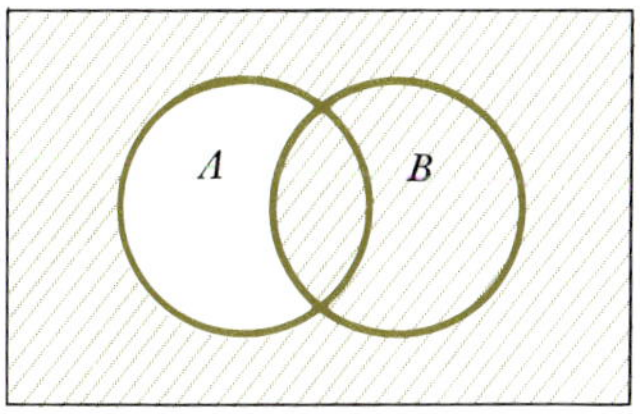

FIG. 7-16 VENN DIAGRAM OF $\tilde{A} \cup B$.

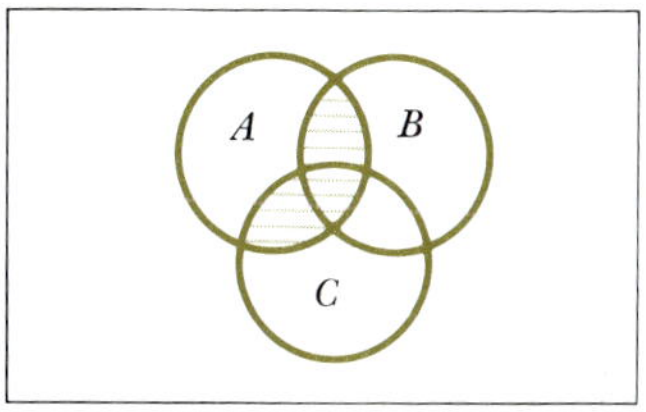

FIG. 7-17 VENN DIAGRAM OF $(A \cap B) \cup (A \cap C)$.

7-2 EXERCISES

For Exercises 1 to 4 use the following information:

$U = \{0,1,2,3,4,5,6,7,8,9\}$
$A = \{1,3,5,7\}$
$B = \{2,4,6,8\}$
$C = \{0,1,2,3,4\}$
$D = \{5,6,7,8,9\}$
$E = \{3,6,9\}$
$F = \{2,5,8\}$
$G = \{2,4,6\}$
$H = \{4,6,8\}$

1. Find the following sets.

   (*a*) $A \cup B$ (*b*) $A \cap C$ (*c*) $\tilde{D}$ (*d*) $E \cup F$ (*e*) $C - B$

2. Find the following sets.

   (*a*) $\widetilde{(D \cup B)}$ (*b*) $(A \cup B) \cap C$ (*c*) $\tilde{A} \cup B$
   (*d*) $(E \cap D) \cup (F \cap C)$ (*e*) $\widetilde{(E \cap D)} \cup \widetilde{(F \cap C)}$

3. Show, by direct calculation, that the following statements are true for the sets given.

   (*a*) $E \cup (B \cap E) = (E \cup B) \cap (E \cup E)$ (*b*) $\widetilde{A \cap C} = \tilde{A} \cup \tilde{C}$
   (*c*) $\widetilde{(\tilde{B})} = B$

4. Which of the following statements are true of the sets given?

   (*a*) $A \subseteq U$ (*b*) $A \cap B = \emptyset$ (*c*) B and D are disjoint
   (*d*) $G \subset C$

5. Draw Venn diagrams of the following.

   (*a*) $P \cap Q$ (*b*) $P \cup Q$ (*c*) $Q \cup P$ (*d*) $Q \cap P$

6. If $P \subset Q$, draw Venn diagrams of the following.

   (*a*) $\tilde{P} \cap Q$ (*b*) $\widetilde{(\tilde{P} \cup \tilde{Q})}$ (*c*) $\widetilde{P \cap \tilde{Q}}$ (*d*) $Q - P$

7. Draw Venn diagrams of the following.

   (*a*) $\widetilde{A \cap B} = \tilde{A} \cup \tilde{B}$
   (*b*) $A \cap (B \cup C) = (A \cap B) \cup (A \cap C)$
   (*c*) $A \cup (B \cap C) = (A \cup B) \cap (A \cup C)$
   (*d*) $(A \cap B) \subseteq A$

8. A manufacturer tested customer reaction to a new product offered in red, green, and blue boxes. As part of his research, he asked how many people liked the various colors, with the following results:

Number who liked red	138
Number who liked blue	123
Number who liked green	150
Number who liked both red and green	90
Number who liked both red and blue	37
Number who liked both blue and green	64
Number who liked all three	22

   (*a*) How many people liked red and green but not blue?
   (*b*) How many people liked green only?
   (*c*) If 250 people were surveyed, how many did not like any of the three colors?

   (*Hint:* Draw a Venn diagram with three overlapping areas—see the figure—and fill in the number of people who liked each color. In each area write the number of people who have the characteristics of that area. From the completed figure you should be able to answer the questions.)

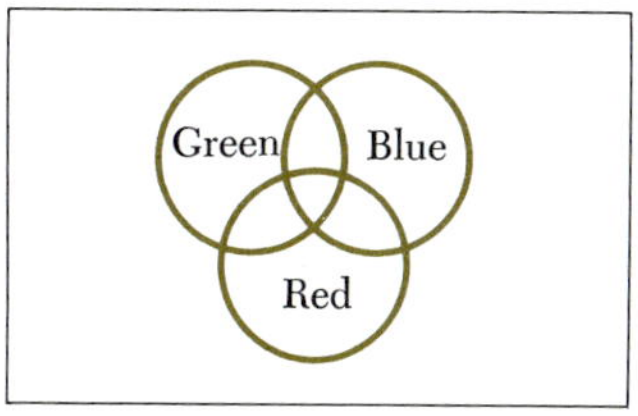

9. A poll was taken of 75 college students to determine the number of science courses in which each student was enrolled, with the following results:

Geology only	15
No science	17
Geology but not biology	21
Zoology and geology	15
Geology	40
Biology and zoology	15
Biology	30

(*a*) How many students took all three sciences?
(*b*) How many students took zoology?

10. A total of 101 housewives tested three brands of detergent for effectiveness and irritation to the skin, with the following results:

Brand	*Effective and non-irritating*	*Not effective and non-irritating*	*Effective and irritating*	*Not effective and irritating*
x	2	13	11	3
y	9	8	3	6
z	20	10	9	7

Let brand x = X Effective cleaner = E
brand y = Y Irritation to skin = I
brand z = Z

How many housewives are in each of the following classes?

(*a*) $X \cap \tilde{I}$ (*b*) $Y \cap \tilde{E} \cap I$ (*c*) $(Z \cup Y) \cap X$

(*d*) $(X \cup Z) \cap (E \cup \tilde{I})$ (*e*) $\widetilde{(\tilde{I} \cup E)}$ (*f*) $Z \cup \widetilde{(I \cap E)}$

(*g*) $[\widetilde{Y \cap (Z \cup I)}]$ (*h*) $X - \widetilde{(I \cup \tilde{E})}$

7-3 SET BUILDING

There are many ways to build new sets from old ones. In the present section we examine more "set-building" procedures. They are especially useful in the development of models in other mathematical systems. Like union, intersection, and complementation, these operations produce new sets; but although the elements of the union, intersection, or complement of two sets are the same kind of objects as in the original sets, the results of these new operations contain a completely new kind of element.

The first set-building operation is the construction of the "power set" of a given set. We begin with an illustration. Suppose $A = \{a,b,c\}$, a set containing three English letters. What are the subsets of set A? By inspection of set A, we discover that the following subsets of set A exist:

$\emptyset$
$\{a\}$
$\{b\}$
$\{c\}$
$\{a,b\}$
$\{a,c\}$
$\{b,c\}$
$\{a,b,c\}$

Because sets are objects, according to our meaning of the term, we can form a set of these subsets: $\{\emptyset,\{a\},\{b\},\{c\},\{a,b\},\{a,c\},\{b,c\},\{a,b,c\}\}$. This set of sets, all of which are subsets of set A, is known as the power set of A. The definition of the power set of any possible set S, is:

The power set of set S is the set of all possible subsets of the set S.

Consider $B = \{0,1\}$; then the power set of set B is $\{\{\ \},\{0\},\{1\},\{0,1\}\}$.

One aspect of power sets is the relationship between the number of elements in the original set and the number of elements in the power set of that set. Suppose that the original set has a finite number n of elements, compared with a set like the set of even numbers, which has an unlimited number of elements. The power set of this finite set has $2 \times 2 \times 2 \times \cdots \times 2 = 2^n$ elements, where n is the number of elements in the original set. Suppose that we add an element to set $B = \{0,1\}$. Call this result $B_2 = \{0,1,2\}$. The power set of B_2 is found by examining the power set of B, which is $\{\emptyset,\{0\},\{1\},\{0,1\}\}$. The power set of B_2 contains all the members of the power set of B and, in addition, the same four sets modified by the addition of the element 2; that is, the power set of B_2 is $\{\{\ \},\{0\},\{1\},\{0,1\},\{2\},\{0,2\},\{1,2\},\{0,1,2\}\}$. The last four elements of this set were formed by adding the element 2 to each of the sets in the power set of B. If we consider $\{0,1,2,3\}$, its power set has twice as many sets in it—the eight listed above and these eight union set $\{3\}$, a total of 16. We then reason that the addition of one more element to the generating set doubles the number of elements in the power set. If there were, say, 64 possible subsets for a set, and we added one new element to the set, there would be 64×2 subsets of the new set—the original 64 subsets and 64 new subsets containing one new element. This result is in agreement with our examples. Set A has three elements, and the power set of A has $2 \times 2 \times 2 = 8$ elements. Set B has two elements, and the power set of B has four elements.

The second set-building operation is the construction of a "cartesian-product set." Suppose that we have two sets of people, one of men and one of women. A natural set to form from this pair of sets is a set of pairs, or couples. The objects in such a set are not individuals but pairs with a special property. The objects in any pair are linked together in a special order. When we consider a pair of objects and insist that the order of the objects must be considered, we are dealing with ordered pairs. To handle the concept of an ordered pair, a special notation has been developed. Let $A = \{1,2,3\}$ and $B = \{a,b,c\}$. Then the cartesian product $A \times B$ is written:

$$A \times B = \{(1,a),(1,b),(1,c),(2,a),(2,b),(2,c),(3,a),(3,b),(3,c)\}$$

Note that the first element of each ordered pair in parentheses is taken from the first set in the cross product and the second element is taken from the second set. The pair $(a,1)$ is not an element of $A \times B$ but is an element of $B \times A$. Note also that every two-element combination of one element from A and one from B appears in the cartesian product $A \times B$.

The set $B \times A$ equals

$$\{(a,1),(a,2),(a,3),(b,1),(b,2),(b,3),(c,1),(c,2),(c,3)\}$$

This illustrates the notation, as $B \times A$ indicates, that B is to be considered the "first" set and A the "second" set. From the definition of ordered pair, $(A \times B) \cap (B \times A) = \emptyset$. This result follows, although the pairs of objects in $A \times B$ and $B \times A$ look alike; their orders are different, so that they are different objects.

Let us consider another example. Suppose $R = \{1,2\}$ and $S = \{2,3\}$. What is the form of $R \times S$?

$$R \times S = \{(1,2),(1,3),(2,2),(2,3)\}$$

It is a simple matter to determine the number of pairs in a cartesian-product set. If we want to form $P \times Q$, and set P has n elements and set Q has m elements, the number of possible pairs, and thus the number of pairs in the product set, is clearly n times m. We use the cross-product notation when using this product set to study the multiplication of numbers.

Finally, we can form product sets like $R \times R = \{(1,1),(1,2),(2,1),(2,2)\}$, where $R = \{1,2\}$. This kind of cartesian product is the one we mentioned in Sec. 2-6. We said that the operations of arithmetic were mappings of $S \times S$ into S, and we meant that S was to stand for the set of numbers and that the operations of addition, subtraction, multiplication, and division were associations of pairs of numbers—members of $S \times S$ with single numbers. For example, we can think of the division $20 \div 4$ as associating the pair (20,4) with the number 5 if we agree that for division the first object in the pair is divided by the second. Then (4,20) is associated with $\frac{1}{5}$. Multiplication associates both pairs (20,4) and (4,20) with the number 80. Mathematicians often study these operations from this point of view. In Sec. 7-4 we examine another use of the cartesian-product set, in studying the operations of arithmetic.

7-3 EXERCISES

Consider the following sets in doing the exercises in this section:

$A = \{a,b\}$ $\quad$ $C = \{4,5,6\}$ $\quad$ $E = \{6,7\}$

$B = \{b,c,d\}$ $\quad$ $D = \{1\}$

1. (*a*) Find the power set of set C.
   (*b*) Is 5 an element of the power set of C?

2. Are the power sets of A and E disjoint?
3. How many elements does the power set of the set $A \cup E$ have?
4. Does set D have a power set? If so, find it.
5. If a set has five elements, how many elements are in its power set?
6. Find $A \times E$ and $A \times D$.
7. (*a*) How many elements does $B \times C$ have?
   (*b*) How many elements does $B \times E$ have?
   (*c*) How many elements does $E \times B$ have?
8. What do the elements of $(A \times B) \times C$ look like? (Remember that parentheses usually mean do this first.)
9. In many problems, it is necessary to talk about ordered triples of objects. How could the notation for ordered pairs be extended or modified to handle ordered triples?

7-4 RELATIONSHIPS BETWEEN SET AND NUMBERS

There are many ways in which the concept of sets and the concept of numbers are related. Since the development of the theory of sets, mathematicians have been able to gain information about numbers by constructing "models" of the various parts of the number system from sets. In the present section we examine one of these set operations and the properties that can help us in the study of the arithmetic operations of addition and multiplication.

To begin, we classify sets into nonoverlapping groups, or, in the language of sets, into disjoint sets of sets. There are, of course, many bases we may use for such a classification. A simple one, and the one we use here, involves the idea of matching or equivalent sets.

Consider the sets $R = \{1,2,3,4,5\}$ and $S = \{a,b,c,d,e\}$. These two sets are not equal, because the word *equal* is reserved for those cases in which we have two different symbols for the same set. They seem equal, because they both have five members. Because we cannot use *equal,* we use the word *matching* or *equivalent* to identify their relationship.

We resist the natural tendency to identify the relationship between R and S as that of having the same number of elements and, instead, describe it in more primitive terms:

$$\begin{array}{ccccc} \{1, & 2, & 3, & 4, & 5\} \\ \updownarrow & \updownarrow & \updownarrow & \updownarrow & \updownarrow \\ \{a, & b, & c, & d, & e\} \end{array}$$

This same relationship has been shown before. In Sec. 1-2 we talked about matching, or the one-to-one pairing of objects, in reference to primitive counting. Two sets of objects are matching, or equivalent, if their elements can be placed in one-to-one correspondence.

We can use the idea of a one-to-one correspondence to divide or classify sets. Imagine a set of sets, described as the set of all the sets that match set R above. If set $A = \{a,e,o\}$, we can discuss the set of sets that match set A. Because A and R do not match and because we can clearly see that, if two sets both match a third, they match each other, the two sets of sets are disjoint. In other words, the set of sets that match set R and the set of sets that match set A do not overlap; they do not share any elements. It takes only a little imagination to observe that any set of objects that does not have an unlimited number of elements "fits" into only one set of matching sets. In this manner, we can use matching to classify sets. What is the common characteristic of all the members of such a set of sets? Formally, the only answer is that they match or are equivalent. Informally, we say that they have the same number of elements.

Putting all these ideas together, we arrive at the following formal definition of the abstract concept of number. Denote $N(B)$ as the number property of set B, so that $N(B)$ is the common property shared by all the sets that match set B. For example, 5 is the symbol for $N(R)$ where R is the same set as above, and 3 is the number property of set A above, that is, $N(A) = 3$ where $A = \{a,e,o\}$. In effect, we are using the concept of a one-to-one pairing of objects to define the more complex and more abstract concept of number.

The idea of the number property of set and set operations can be used to examine the operations of addition and multiplication of whole numbers. Suppose that we have two disjoint sets, say, $P = \{a,b,c\}$ and $Q = \{e,f,g,h\}$. Because $N(P) = 3$ and $N(Q) = 4$, we define the sum of 3 and 4, written, of course, as $3 + 4$, as $N(P \cup Q)$; that is, because P and Q are disjoint, $N(P) + N(Q) = N(P \cup Q)$. In general, suppose that we have two unspecified but disjoint sets A and B. Further, suppose that $n = N(A)$ and $m = N(B)$; we can then define $n + m$ as $N(A \cup B)$.

How does defining addition in this way help in studying it? Consider some of the things known about addition, for example, the property of the sum of two numbers that was called the commutative law for addition in Sec. 2-6. We can demonstrate this property of addition by using the set definitions of numbers and addition (and perhaps, through sets, seek new properties of numbers).

The set operation of union is commutative; that is, for two sets A and B, the definition of union assures us that $A \cup B = B \cup A$. Because $A \cup B$ and $B \cup A$ are alternative "names" for the same set, $N(A \cup B) = N(B \cup A)$. From the definition of addition, if $n = N(A)$ and $m = N(B)$ and A and B are disjoint, then $n + m = N(A \cup B)$ and $m + n = N(B \cup A)$. Therefore, $n + m = m + n$. Because A and B are not specified, it follows that $n + m = m + n$ for any whole number. In other words, the order in which two numbers are added does not matter; addition is a commutative operation on whole numbers.

Another property of whole numbers that can be demonstrated by the use of sets is the associative property of addition. Suppose that we have three unspecified but disjoint sets A, B, and C. We can form $A \cup B$ and then find its union with set C. Denote this result by $(A \cup B) \cup C$. On the other hand, we can find $B \cup C$ and then form its union with set A. We symbolize this result by $A \cup (B \cup C)$. A brief examination of Venn diagrams convinces us that these two results are the same set. We also observe that, if A and B are disjoint, and A and C are disjoint, and B and C are disjoint, then $(B \cup C)$ and A are disjoint, and so are $(A \cup B)$ and C.

Suppose that $N(C) = p$. Because

$$n + m = N(A \cup B)$$

we have

$$\begin{aligned} N((A \cup B) \cup C) &= N(A \cup B) + N(C) \\ &= (n + m) + p \end{aligned}$$

However

$$\begin{aligned} N(A \cup (B \cup C)) &= N(A) + N(B \cup C) \\ &= n + (m + p) \end{aligned}$$

because B and C are disjoint. Thus

$$(n + m) + p = N((A \cup B) \cup C) = N(A \cup (B \cup C)) = n + (m + p)$$

or

$$(n + m) + p = n + (m + p)$$

This result is known as the associative property of addition. In effect, it says that, when looking for the sum of three numbers, we may add the first to the second and then add the third to this result or add the second to the third and then add this result to the first. In either case we always get the same result.

These properties are obvious only in the sense that the rules of English grammar are obviously the only reasonable set of rules to use in a language. Many people say that it is clear that the rules of English grammar are the only natural way to link words in meaningful patterns. Most of us have grown up with English grammar, and most of us have grown up working with the number operations. As anyone who has studied another language knows, the rules of English are definitely not the only way to construct sentences. Alternatives do exist, both in language and in mathematics, but at a level that most students never reach. It is not our goal to examine any of these; rather we point out that, if we define numbers as we have, in terms of sets, we are able to demonstrate that the rules of operation are indeed the only correct ones, logically. We also point out that our proofs or arguments are not complete, in the sense of a sequence of statements, as discussed in Chap. 6. We have only indicated how we might begin to construct such formal proofs.

The concept of a cartesian-product set can be used to define the multiplication of whole numbers. Suppose that A, B, and C are disjoint sets with the number properties n, m, and p as above. We can then define n times m, or the product of n and m, as $N(A \times B)$. In other words, $n \cdot m = N(A \times B)$. Although $A \times B$ is not equal to $B \times A$, it can be shown that these two sets do match. A simple one-to-one correspondence between them matches each $(a,b) \in A \times B$ with the element (b,a) that is an element of $B \times A$. Thus, although $A \times B \neq B \times A$, it is true that $N(A \times B) = N(B \times A)$. This leads us to a chain of equalities that states that

$$n \cdot m = N(A \times B) = N(B \times A) = m \cdot n$$

that is, $n \cdot m = m \cdot n$ for any two whole numbers n and m. This is an argument for a commutative law for the multiplication of whole numbers.

A similar but more complex proof can be constructed for an associative law for the multiplication of two numbers, namely, that $m \cdot (n \cdot p)$ is the same number as $(m \cdot n) \cdot p$ for any whole numbers m, n, and p. We can also prove, with the aid of the set definitions for addition and multiplication, that for any three whole numbers, $m \cdot (n + p) = (m \cdot n) + (m \cdot p)$.

Mathematicians think of two kinds of numbers, cardinal and ordinal. Cardinal numbers say how many. Ordinal numbers indicate position, like second or fiftieth, or say which one. If we say that a set contains six objects, we are using a cardinal number. If we talk about the third object in a set, we are referring to the position of a particular object and are using an ordinal number. The elementary concept that numbers can be arranged in an order is the basis of many of the most advanced principles and many of the most useful concepts in mathematics. As with other concepts we have encountered or shall encounter, we do not explore further but hope that, if the reader is interested, he will explore these powerful concepts himself.

7-4 EXERCISES

These sets are to be used in the Exercises 1 to 7:

$A = \{1,3,5,7,9\}$
$B = \{1,2,3,4,5,6,7,8,9\}$
$C = \{2,4,6,8,10\}$
$D = \{$☆,○,□,△,✳$\}$
$E \times \{1,+,$✳,□,◪,⊠,⊞,△,⚠$\}$

1. Which of these sets are equivalent?
2. Construct two sets equivalent to set A.
3. Find $N(A)$ and $N(D)$; also find $N(A \cup D)$.
4. Does $N(A) + N(B) = N(A \cup B)$?
5. Why must the sets used in defining a sum be disjoint?
6. What is $N(A \times B)$?
7. Sets A and B are not disjoint. Because $N(A) = 5$ and $N(B) = 9$, is $N(A \times B) = 9 \cdot 5 = 45$? What do you think it is?
8. Must the sets used to define the multiplication of a product be disjoint?
9. Addition can be defined in terms of the union of disjoint sets and multiplication in terms of the cross product. What set operation can be used to define subtraction?
10. Which are cardinal and which are ordinal numbers?

    (*a*) First base
    (*b*) Seven days in May
    (*c*) May 7
    (*d*) 25 lb
    (*e*) Butterfield 8-3341

11. Give a formula for $N(R \times S)$.
12. There is a train T and a bus B and a plane P providing transportation between towns D and C. How many ways can one make a round trip between these two towns?
13. How is the answer to Exercise 12 related to the cartesian product?

Let set $A = (a,b,c)$ and $B = (1,2,3,4)$ and $C = (\square,\triangle)$. Give the number property and list the elements of the following.

14. $(A \cup B) \times C$
15. $(A \cap B) \times C$
16. $(A \times C) \times B$

7-5 SETS AND LOGIC

Consider any set of simple statements p, q, r, and s. From Chaps. 5 and 6, we can identify 16 distinct possible arrangements of the truth values of these four statements:

Outcome	p	q	r	s
1	T	T	T	T
2	T	T	T	F
3	T	T	F	T
4	T	T	F	F
5	T	F	T	T
6	T	F	T	F
7	T	F	F	T
8	T	F	F	F
9	F	T	T	T
10	F	T	T	F
11	F	T	F	T
12	F	T	F	F
13	F	F	T	T
14	F	F	T	F
15	F	F	F	T
16	F	F	F	F

To relate sets and logic, let these logically possible truth values be a universal set U. Then one element of U is outcome ⑫, (FTFF), another is outcome ⑤, (TFTT). Thus

$U = \{$①, ②, ③, ④, . . . , ⑯$\}$

where the numerals represent the 16 distinct truth-value arrangements for four statements of logic. A "truth set," then, is a subset of the universal set and contains all the elements of the universal set for which the truth value of a particular statement is T. Truth sets are denoted by capital letters. Thus the truth set for statement p is:

$P = \{$①, ②, ③, ④, ⑤, ⑥, ⑦, ⑧$\}$

Set P, then, is made up of all the elements of U for which statement p is true. In the same manner, we list the truth sets Q, R, and S:

$Q = \{$①, ②, ③, ④, ⑨, ⑩, ⑪, ⑫$\}$
$R = \{$①, ②, ⑤, ⑥, ⑨, ⑩, ⑬, ⑭$\}$
$S = \{$①, ③, ⑤, ⑦, ⑨, ⑪, ⑬, ⑮$\}$

As a more complicated example, consider any of the infinite number of compound statements we can form from the four statements p, q, r, and s, for example, $(p \wedge q) \rightarrow s$. For convenience identify this compound with a single symbol, say, z; that is, $z \equiv (p \wedge q) \rightarrow s$. Let Z equal the set of outcomes for which statement z is true. By examining the list of outcomes, we arrive at $Z = \{$①, ③, ⑤, ⑥, ⑦, ⑧, ⑨, ⑩, ⑪, ⑫, ⑬, ⑭, ⑮, ⑯$\}$. What is the set of outcomes for which statement z is false or, in terms of negations, the set of outcomes for which $\sim z$ is true? Again, by inspection of the list of outcomes, we find that this set of outcomes is $\{$②, ④$\}$. What relationship exists between this set and the set Z? By inspection, we see that $\{$②, ④$\} = \tilde{Z}$. By indicating the set of outcomes for which a statement is true as the truth set of the statement, then, although Z is the truth set of z, the truth set of $\sim z$ is $\tilde{Z}$. This shows that there is a close connection between the negation of logic and the complementation of sets. To complete the parallel between sets and logic, we need the truth sets for $p \wedge q$ and $p \vee q$. To find the truth set of $p \vee q$, note that $p \vee q$ is true if p is true or if q is true or if both are true. Then the truth set for $p \vee q$ is $P \cup Q = \{\textcircled{1}, \textcircled{2}, \textcircled{3}, \textcircled{4}, \textcircled{5}, \textcircled{6}, \textcircled{7}, \textcircled{8}, \textcircled{9}, \textcircled{10}, \textcircled{11}, \textcircled{12}\}$ for P and Q as defined above. The statement $p \wedge q$ is true only when both p and q are true. Thus the truth set for $p \wedge q$ is $P \cap Q = \{\textcircled{1}, \textcircled{2}, \textcircled{3}, \textcircled{4}\}$. The connection between sets and logic is reinforced if you recall that the definition of union depended on the word *or* and that of intersection on the word *and.*

From these examples we can generalize the relationship between set operations and logical connectives. Suppose that a and b are any two statements. If U stands for all the combinations of truth values, define A as the set of outcomes where a is true and B as the set of outcomes where b is true; that is, A is the truth set for a, and B the truth set for b. Although we cannot make a list of the members of A and B, we can use a Venn diagram, as in Fig. 7-18, to visualize this idea.

The two circles divide the universe into four distinct portions, labeled 1, 2, 3, and 4. Area 1, which is unshaded, represents those outcomes where neither a nor b is a true statement. Area 2, shaded only vertically, represents those outcomes or conditions where a is a true statement but b is a false statement. Area 3, shaded both vertically and horizontally, represents those outcomes where both a and b are true, and area 4 those where b is true and a is false.

Area 3 also represents the truth set for the statements $a \wedge b$. We know this, because a conjunction is true only when both statements are true. This area also represents $A \cap B$. We can conclude that, in general, the truth set for $a \wedge b$ is $A \cap B$. What happens to $a \vee b$? The similarity between symbols hints at the answer. The truth set for $a \vee b$ is $A \cup B$. We can also observe that the truth set for $\sim a$ is $\tilde{A}$, and for $\sim b$ is $\tilde{B}$.

How can we express the truth set for $a \rightarrow b$ and $a \leftrightarrow b$ in terms of the

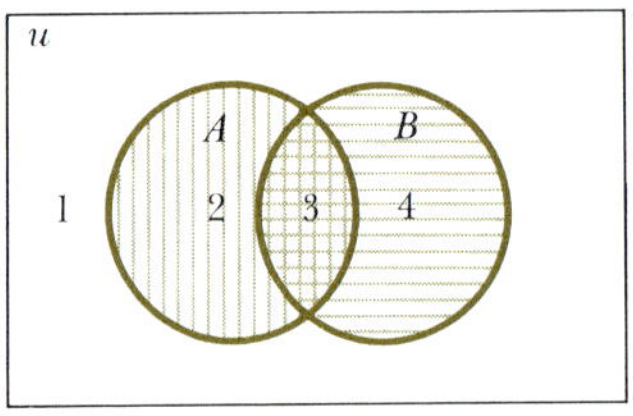

FIG. 7-18 VENN DIAGRAM OF TRUTH SETS.

truth set of a and b? Because, as we saw in Chap. 5, $a \rightarrow b$ and $\sim a \vee b$ are equivalent statements, they have identical truth sets. We can find the truth set for $a \rightarrow b$ by finding the truth set for $\sim a \vee b$. This is $\tilde{A} \cup B$. The reader can verify that the truth set for $a \leftrightarrow b$ is $(\tilde{A} \cup B) \cap (\tilde{B} \cup A)$.

Because the basic connectives of logic correspond to the basic set operations, we can study the concepts of logic in terms of the concepts of sets, and vice versa. Sets seem like a simpler concept than logic, but from our analysis we conclude that in a basic sense they are the same concept. This "sameness" is the basis of Boolean algebra, which we shall examine in Chap. 8.

The question of logically equivalent statements has already been answered. Because logically equivalent statements are true under the same conditions and false under the same conditions, they have the same truth sets; that is, if statement a is logically equivalent to statement b, then truth set A equals set B.

Statements that are true under all conditions are called "logically true" statements, and those which are always false are known as "logically false" statements. "Tautology" is another name for a logically true statement. Most statements are neither logically true nor logically false but are sometimes true and sometimes false, depending on various conditions. Logically true and logically false statements are easily seen to be parallel to the notions of the universal and empty sets. A logically true statement, by definition, is a statement that is true in every case. In terms of truth sets, the truth set for a logically true statement is U, the universal set. A similar line of reasoning leads us to the conclusion that the truth set for a logically false statement is $\emptyset$, the empty set, because there are no cases where such a statement is true.

Much of logic involves valid arguments. To relate arguments to sets, we need a special kind of conditional. This is one that is always true or saying the same thing, a conditional that is logically true. Thus we say that a implies b if and only if $a \rightarrow b$ is logically true. The symbol for *a implies b* is $a \Rightarrow b$. The "implication" is a conditional that is always true. To see how the parts of the implication are related, we have only to examine a conditional to determine the conditions under which it is true. If the antecedent is false, the conditional is true, no matter what the truth value of the consequent. However, if the antecedent is true, the consequent is also true to make the conditional true. Thus we say that $a \Rightarrow b$ means that, whenever a is true, b is true. This is very close to the definition of a valid argument. An argument is valid if the conclusion is true

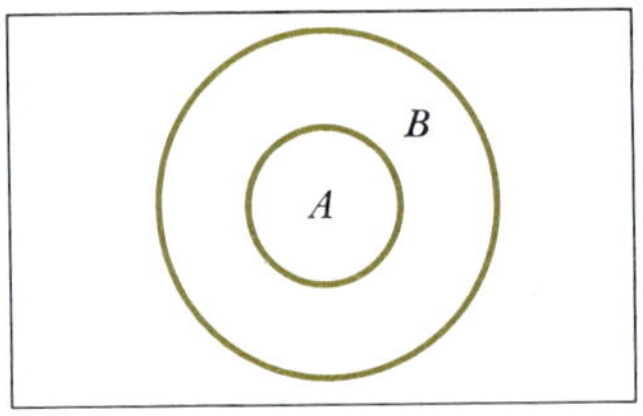

FIG. 7-19 VENN DIAGRAM OF $A \subseteq B$.

whenever all the premises are true. There is a small difficulty. An argument may contain many premises, but an implication contains only one antecedent. To avoid this difficulty, consider the single statement formed by joining all the premises by *and.* This statement, perhaps very long, is true only when all the premises are true. Thus an argument is valid if, whenever the "conjunction" of the premises is true, the conclusion is also true. But this is the same as saying that an argument is valid if the conjunction of the premises implies the conclusion.

Thus any argument can be reduced to two statements. If these two statements form an implication, the argument is valid; if they do not, the argument is invalid. Thus, to parallel valid arguments in set theory, all that is needed is set representation for implication. Because the truth set for $a \rightarrow b$ is $\tilde{A} \cup B$, we conclude that a implies b if $\tilde{A} \cup B = U$. We indicated that this is the necessary requirement for the truth set of a logically true statement. This statement is easier to analyze if we understand that, if set C is equal to U, then $\tilde{C} = \emptyset$. In other words, saying that a set is equal to the universal set is the same thing as saying that its complement is the empty set. For our problem, this means that $\widetilde{\tilde{A} \cup B} = \emptyset$, but in Sec. 7-2 we showed that $\widetilde{\tilde{A} \cup B} = A \cap \tilde{B}$. Therefore, a implies b if $A \cap \tilde{B} = \emptyset$. Under what conditions would this be true? If we draw a Venn diagram in which this is true, we must make it in such a way that the area outside B, that is, the area representing $\tilde{B}$, and the area inside A do not overlap. A little trial and error results in Fig. 7-19. This is the only way we can draw the Venn diagram to meet the given conditions. How do we describe the relationship this diagram indicates? It is $A \subseteq B$. We have to use $\subseteq$ rather than $\subset$, because we have no way of showing that $A \neq B$. In summary, a implies b if $A \subseteq B$.

To illustrate this result, we construct a valid and an invalid argument of the p, q, r, and s statements from the example at the beginning of the present section. For example, we check the validity of *modus ponens:*

$$\begin{array}{l} p \\ \underline{p \to q} \\ \therefore q \end{array}$$

To show that this argument is valid, we show that $(p \land (p \to q)) \Rightarrow q$. Working with sets, this is equivalent to showing that the truth set of $(p \land (p \to q))$ is contained in the truth set of q. $(p \land (p \to q))$ has truth set $H = \{\textcircled{1}, \textcircled{2}, \textcircled{3}, \textcircled{4}\}$ and q has truth set $Q = \{\textcircled{1}, \textcircled{2}, \textcircled{3}, \textcircled{4}, \textcircled{9}, \textcircled{10}, \textcircled{11}, \textcircled{12}\}$. It is clear that $H \subseteq Q$, so that the argument is valid. Now the argument:

$$\begin{array}{l} q \\ \underline{p \to q} \\ \therefore p \end{array}$$

$(q \land (p \to q))$ has truth set $W = \{\textcircled{1}, \textcircled{2}, \textcircled{3}, \textcircled{4}, \textcircled{9}, \textcircled{10}, \textcircled{11}, \textcircled{12}\}$, and p has truth set $P = \{\textcircled{1}, \textcircled{2}, \textcircled{3}, \textcircled{4}, \textcircled{5}, \textcircled{6}, \textcircled{7}, \textcircled{8}\}$. Because $W \not\subseteq P$, the argument is invalid.

Our model of logic built from sets is useful in learning about either logic or sets. This model and the similar one for numbers are only two examples of the use of sets in studying other areas of mathematics.

7-5 EXERCISES

1. Construct Venn diagrams for the truth sets for the following compound statements.

   (*a*) $p \land \sim q$ (*b*) $\sim p \land q$ (*c*) $\sim p \to q$ (*d*) $(\sim p \land \sim q)$

2. Construct Venn diagrams for the truth set for the following compound statements.

   (*a*) $p \to q$ (*b*) $q \to p$ (*c*) $\sim q \to \sim p$ (*d*) $\sim p \to \sim q$

3. From the diagrams in Exercise 2, is the relationship between conditional statements, converse, inverse, and so on, confirmed? Explain.
4. Consider the following sets. How would you use truth tables to show that they are empty sets? Carry out the truth-table analysis, and determine which are empty sets.

   (*a*) $P \cap \tilde{P}$ (*b*) $(P \cap Q) \cap (\widetilde{P \cup Q})$ (*c*) $(P \cup \tilde{Q}) \cup (\tilde{P} \cup Q)$

5. List some advatages of being able to construct a set model of logic.
6. We say that a and b are equivalent statements if a implies b and b implies a. What is the corresponding relationship among sets?
7. Use Venn diagrams of truth sets to check the validity of the following arguments.

(a)
$$\begin{array}{l} p \to q \\ \underline{\sim q \qquad} \\ \therefore \sim p \end{array}$$

(b)
$$\begin{array}{l} p \wedge q \\ \underline{\sim p \to q} \\ \therefore \sim q \end{array}$$

(c)
$$\begin{array}{l} p \to q \\ \underline{\sim r \to \sim q} \\ \therefore \sim r \to \sim p \end{array}$$

(d)
$$\begin{array}{l} p \leftrightarrow q \\ q \vee r \\ \underline{\sim r \qquad} \\ \therefore \sim p \end{array}$$

8. Construct a compound statement whose truth set is:

   (a) Empty
   (b) The universal set

7-6 MODELS

We saw how we can use sets to construct models of other mathematical systems. These examples illustrate some of the power of the theory of sets as a model-building medium. They also demonstrate the more general concept of a symbolic model. In our discussion of computers we mentioned that symbolic models are used to determine the results of experiments or to predict the outcome of an event, such as an election. In the present section we examine the concept of model building, especially mathematical model building.

We can only conjecture where and when physical models were first used. Perhaps, at the dawn of history, a primitive man observed a twig floating down a stream and knew that a larger version could be used to provide water transportation. Perhaps a rolling stone served as a model of the wheel. We do know that early records indicate that craftsmen in stone and metal used wood or clay to model a building before attempting construction. But what is a model? What characterizes models and model building?

A dictionary uses such descriptions as a small copy built to scale or a preliminary representation of something larger to characterize a model. But must a model be smaller? Must it be to scale? Must it even be material? A plastic model of a DNA molecule is not smaller than the molecule it represents. Galileo's clockwork model, built to show the rotation of the planets around the sun, was hardly to scale. Relative size, or whether a model is to scale or not, does not determine whether we have a model or not. As to the material nature of models, the history of mathematics is almost a history of the construction of nonmaterial models, as the examples in Secs. 7-4 and 7-5 are illustrations.

Many, if not most, developments in mathematics resulted from attempts to model, in symbols, some principle or pattern of nature. Even most of the advanced topics in mathematics result from model-building experiments. In these cases, mathematicians attempt to construct symbolic models by using one system of symbols for other mathematical systems. The earliest kind of formal mathematics, numeral systems, was probably developed to symbolize or model the concept of a year and the passage of time. We saw that the Mayan numeral system was based on the number 360, which is a primitive estimate of the number of days in a year. These systems were also used in the study of astronomy, which may be man's oldest area of study. In these same early periods we find records of the construction of symbolic tools for use in measurement. These served as the beginning of geometry. Although Greek scholars considered geometry a pure subject, free from problems related to considerations of the physical world, the word *geometry* itself means earth measure.

If we jump from these early developments twenty-odd centuries ago to the seventeenth and eighteenth centuries, we find that most of Isaac Newton's many discoveries were made while trying to describe the underlying principles of physics. In Chaps. 5 and 6, we mentioned that symbolic logic was developed to symbolize, or to be a model of, everyday logic. We could go on to list thousands of individual examples of mathematical methods or concepts being developed to describe specific physical phenomena. Some of the great works in the history of mathematics have titles that suggest that they belong to the science of physics or astronomy rather than the art of mathematics. Examples of these include Joseph Fourier's "Mathematical Theory of Heat" and Pierre Simon de Laplace's "Mécanique céleste."

Many useful and interesting results occur when a model of one mathematical system is constructed with another mathematical system as its basis. An example of this kind of model building is the construction of a numerical model of geometry. Or is it a geometric model of numbers? This is called "analytic geometry" by mathematicians. In analytic geometry one is able to study the properties of geometric figures by studying equations related to these figures, and one is also able to study the properties of equations and numbers by studying their geometric "pictures." In this we could not say which is the model and which is the system being modeled.

In effect, in Secs. 7-4 and 7-5 we examined models of numbers and logic constructed of sets: we could also take the view that we examined models of set theory constructed of numbers and propositions. Sets are the most valuable model-building material of mathematics. In Chap. 9 we construct still another model built of sets, this time of the laws of chance. Sets are, of course, more than just a model-building device. They are also a fundamental concept underlying more complex systems. Geometry could be viewed as the study of sets of objects called points. Algebra, the study of the rules and properties of numbers, could be considered the study of sets of objects known as numbers.

As scholars have developed more and more results from the theory of sets, they have come to the conclusion that, in this relatively simple idea of a collection of objects, is the one central concept common to all areas of mathematics. The theory of sets is viewed as the concept that links all the various branches of mathematics into one unified subject.

We have used sets to construct symbolic models, but an economic theory is a symbolic model, because it is constructed of words, which, as we have pointed out, are simply arbitrary symbols. In one sense, almost all our thinking is done with such symbolic models. The idea of models and model building is one of man's most powerful, if not his most powerful, tool. In mathematics, model building has reached an advanced state. Mathematicians have perfected it to such a degree that, in some cases, a distinction between the system and the model is almost impossible to make. These are, naturally, in symbolic models of symbolic systems. In some cases, this is carried to such a degree that we can conclude only that rather than working with two distinct sets of objects, we are dealing with two different sets of symbols for the same set of objects. This kind of result has led us to the discovery of many relationships among physical phenomena.

Sometimes we find that two physical phenomena have the same mathematical model. A good example is the similarity among the manner in which heat passes through an object, the nature of electricity in space, and the manner in which gravity affects objects. In the process of working out mathematical equations to model these phenomena, it has been found that, under certain conditions, the same equations describe all three. This helped physicists and mathematicians in their search for basic patterns to describe the physical nature of the universe. They reasoned that, because the same equation described the three distinct phenomena, these three were examples of a more basic physical principle.

7-6 EXERCISES

1. Give some examples of the practical use of physical models.
2. Give some examples of nonmathematical symbolic models used by man.
3. Give some examples of physical models that are smaller than the objects they represent.
4. What do we mean when we say that a model is to scale? Give some examples of physical models that are to scale.
5. Suggest some reasons why the relationships between mathematical models are called mappings?
6. What differences exist between reasoning by analogy and using symbolic models?

REVIEW EXERCISES

1. What is the difference between a well-defined collection of objects and a collection that is not well defined?
2. Two ways of specifying sets are given in the present chapter. What are they, and what are the advantages of each?
3. Why is it impossible to restrict completely the use of capital letters as symbols for sets with lowercase letters used only as elements of sets?
4. One of the following statements is true and one false. Which is which, and why?

   (*a*) For every set A, $\emptyset \subset A$. F

   (*b*) For every set A, $\emptyset \subseteq A$. T

5. The operation of complementation involves two sets, that whose complement we are looking for and the universal set. Why is it not considered a binary operation?

For Exercises 6 to 8 consider $U = \{m,a,t,h,e,i,c,s\}$, $A = \{m,a,t,h\}$, $B = \{t,h,e,i\}$, $C = \{e,i,c,s\}$.

6. Find:

   (*a*) $A \cup B$ (*b*) $A \cap B$ (*c*) $A \cap C$ (*d*) $A \cup C$

7. Find:

   (*a*) $A \cap (B \cup C)$ and $(A \cap B) \cup (A \cap C)$
   (*b*) $A \cup (B \cap C)$ and $(A \cup B) \cap (A \cup C)$

8. Find:

   (*a*) $\widetilde{A \cup B}$ (*b*) $\tilde{A} \cup \tilde{B}$ (*c*) $A \cup \tilde{B}$ (*d*) $\tilde{A} \cup B$
   (*e*) $\tilde{A} \cap \tilde{B} \cap \tilde{C}$

9. Draw Venn diagrams to illustrate the following:

   (*a*) $P \subset Q$ (*b*) $\tilde{P} \cap Q = \emptyset$ (*c*) $A \subset B$ and $B \subset C$

10. Find the power set of the power set of $\{1\}$.
11. Find $A \times B$ if $A = \{0,1\}$ and $B = \{1,2\}$; also find $B \times A$.
12. Describe a one-to-one correspondence between $A \times B$ and $B \times A$ in Exercise 11.
13. Why are sets a useful way to model numbers?
14. Why is the concept of a set of ordered pairs needed when we try to model such arithmetic operations as division?
15. Make a list of the set-logic pairs of similar concepts.
16. One mathematician described a poem as a verbal model. What did he mean?
17. Explain why you believe or disbelieve that models are physical.
18. What problems exist when we deduce results from a model rather than the objects themselves?
19. If a model produces results that conflict with observed results, what steps should we take if we still want to use the model for our studies?

SELECTED REFERENCES

Byrne, J. Richard: "Modern Elementary Mathematics," McGraw-Hill Book Company, New York, 1966.

Easily paced development of elementary set theory with extensive exercises. Heavy emphasis on notation.

Dinkines, Flora: "Elementary Concepts of Mathematics," Appleton-Century-Crofts, Inc., New York, 1964.

Complete discussion of sets on an elementary level, including an introduction to advanced set theory.

Kemeny, J., J. L. Snell, and G. L. Thompson: "Introduction to Finite Mathematics," Prentice-Hall, Inc., Englewood Cliffs, N.J., 1965.

Has some good examples and some harder exercises for set operations.

Meserve, B. E., and M. A. Sobel: "Introduction to Mathematics," Prentice-Hall, Inc., Englewood Cliffs, N.J., 1964.

This presentation of sets is more elementary than the others in this list.

Stoll, Robert: "Sets, Logic and Axiomatic Theories," W. H. Freeman and Company, San Francisco, 1961.

This is a rigorous treatment of set theory.

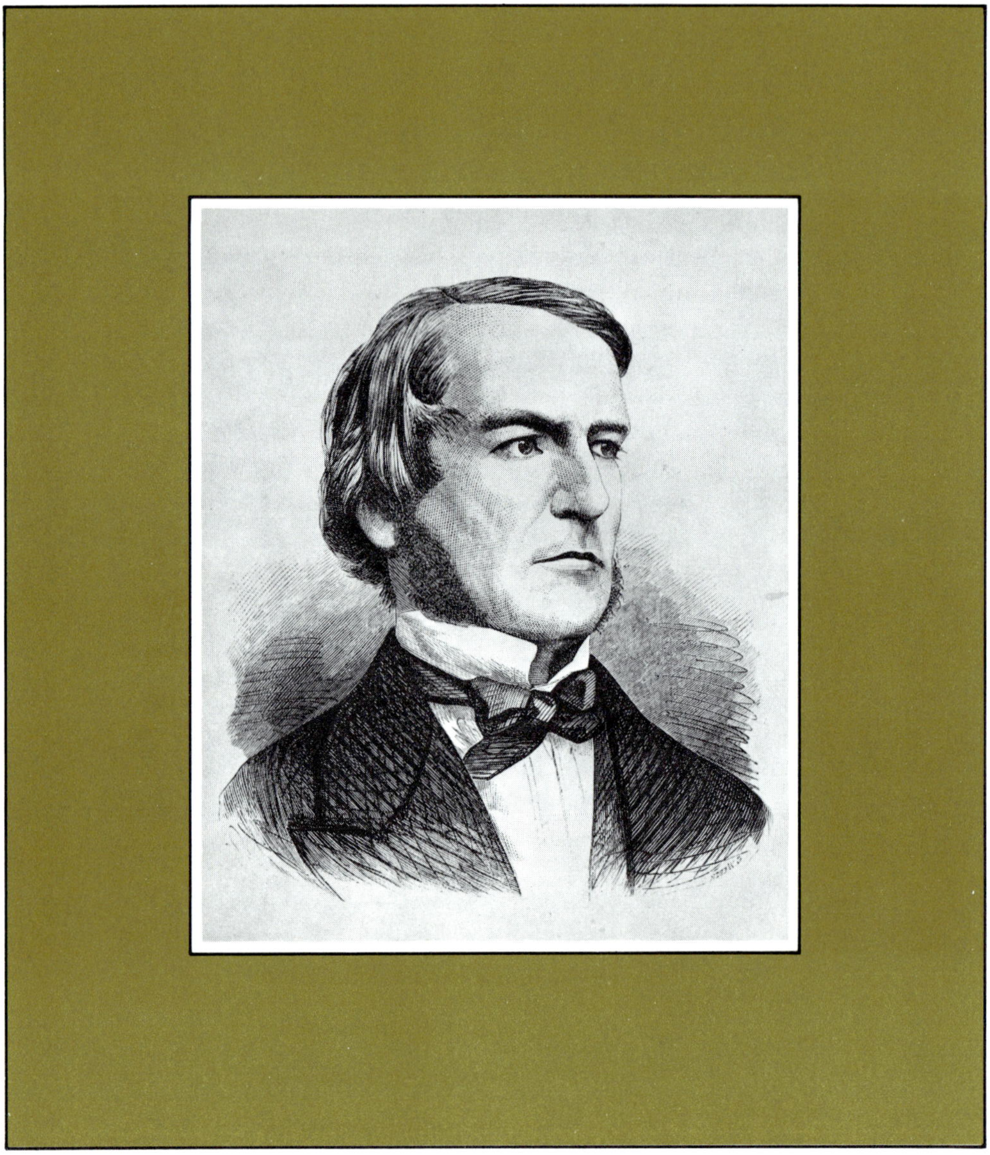

George Boole (1815–1864).
COURTESY BETTMAN ARCHIVES.

chapter eight

BOOLEAN ALGEBRA

The similarities between logic and sets, as indicated in Sec. 7-5, are surprising at first. Seemingly unrelated, the two subjects prove to have many common aspects when examined closely. To the mathematician they are not merely similar but are two examples of the same mathematical system. In the present chapter this system is examined formally. You will find that sets and logic are not the only structures that are a part of it. In particular we examine an electrical structure with the same characteristics.

8-1 A FORMAL MATHEMATICAL SYSTEM

The word *algebra* is used in many ways in mathematics. For almost all students the word brings to mind unknowns and equations and the other topics of the high school study of real numbers. For mathematicians the word has a much broader meaning and is used in connection with almost any organized system. There is the algebra of propositions, which was considered in Chaps. 5 and 6. There are also the algebra of sets, modern algebra, and linear algebra. One of the most interesting and important of the algebras is that named in honor of the English mathematician and logician George Boole.

The study of Boolean algebra has many valuable ends. It gives us an excellent example of a formal axiomatic system. It also gives us examples of formal mathematical proof and the power of symbolic model building by showing how other systems are identical to abstract Boolean algebras.

To begin, a description and a set of postulates of a Boolean algebra are needed. A Boolean algebra consists of a set B with elements identified as a, b, c, and so on. The elements of set B combine using two binary operations, one denoted by $\bigcirc$ and called "circle," and one denoted by $\star$ and called "star." In other words if we take any two elements of set B, say, a and b, then $a \bigcirc b$ (read a circle b) is some new element of B and $a \star b$ (read a star b) is also some new element of B. $\bigcirc$ and $\star$ obey the following postulates.

POSTULATE 1 For any elements a and b of set B, $a \bigcirc b = b \bigcirc a$ and $a \star b = b \star a$. This property is known as "commutativity."

POSTULATE 2 There exist two distinct elements of B, $i_\bigcirc$ and $i_\star$, which have the following properties. For any element a of set B, $a \bigcirc i_\bigcirc = a$ and $a \star i_\star = a$. The elements $i_\bigcirc$ and $i_\star$ are known as "identity" elements. $i_\bigcirc$ is the identity element with respect to circle, and $i_\star$ with respect to star.

POSTULATE 3 The operations $\bigcirc$ and $\star$ are related in the following manner. For any elements a, b, and c of B, the following holds:

$$a \bigcirc (b \star c) = (a \bigcirc b) \star (a \bigcirc c)$$

and

$$a \star (b \bigcirc c) = (a \star b) \bigcirc (a \star c)$$

The operations are said to distribute with respect to each other.

POSTULATE 4 For every element a of B there is an element a' also of B such that $a \bigcirc a' = i_\star$ and $a \star a' = i_\bigcirc$. a' is called the "inverse" of a.

There are any number of examples of sets of objects that follow these postulates. For instance, consider the set of all subsets of the set $\{1,2,3\}$. This is $B = \{\emptyset,\{1\},\{2\},\{3\},\{1,2\},\{1,3\},\{2,3\},\{1,2,3\}\}$. Instead of $\bigcirc$ and $\star$ let us use the more familiar union $\cup$ and intersection $\cap$. If a, b, and

c denote any of the elements of set B, it is not hard to verify that the postulates of a Boolean algebra hold:

1. $a \cup b$ and $a \cap b$ are elements of B.
2. $a \cup b = b \cup a$ and $a \cap b = b \cap a$ no matter what sets a and b are.
3. $a \cap \{1,2,3\} = a$ and $a \cup \emptyset = a$, so that $\{1,2,3\}$ is the identity element with respect to $\cap$, and $\emptyset$ is the identity element with respect to $\cup$.
4. $a \cap (b \cup c) = (a \cap b) \cup (a \cap c)$, $a \cup (b \cap c) = (a \cup b) \cap (a \cup c)$.
5. If we use $\{1,2,3\}$ as the universal set, $a \cup \tilde{a} = \{1,2,3\}$ and $a \cap \tilde{a} = \emptyset$.

Although the symbols $\cup$, $\cap$, and $\sim$ have been used instead, ○, ☆, and ′, it is easy to see that B with $\cap$, $\cup$ and $\sim$ is a Boolean algebra.

Using only the basic properties that any Boolean algebra has, it is possible to deduce a number of interesting theorems, one of the most important of which is not about the set and operations but about Boolean algebras. One of the most obvious characteristics of the postulates of Boolean algebra is the perfect symmetry between ○ and ☆. If any statement is made about ○ and $i_{○}$, exactly the same statement can be made about ☆ and $i_{☆}$, and vice versa. This is summarized in theorem 1, which is also known as the "principle of duality."

THEOREM 1 Any theorem of Boolean algebra remains valid if the operations of ○ and ☆ and the identity elements $i_{○}$ and $i_{☆}$ are interchanged throughout the theorem.

The powerful advantage this theorem provides is easy to see. One gets two theorems for the price of one in a Boolean algebra. Once one of the pair is established, the other "dual" theorem is also proved. For example, the dual of a ☆ $i_{○}$ is a ○ $i_{☆}$.

8-1 EXERCISES

Write the dual of each of the following.

1. Commutative law: a ☆ $b = b$ ☆ a
2. Associative law: a ☆ (b ☆ c) = (a ☆ b) ☆ c
3. Idempotent law: a ☆ $a = a$
4. Distributive law: a ☆ (b ○ c) = (a ☆ b) ○ (a ☆ c)
5. De Morgan's law: $(a$ ☆ $b)' = a'$ ○ b'

6. Absorption law: $a \bigcirc (a \star b) = a$
7. Identity: $a \star i_{\star} = a$
8. Inverse: $a \star a' = i_{\bigcirc}$
9. Let $A = \{a,b\}$, and $\oplus$ and $\odot$ represent binary operations defined by the following tables:

$\oplus$	a	b
a	a	b
b	b	b

$\odot$	a	b
a	a	a
b	a	b

Show that A with $\oplus$ and $\odot$ forms a Boolean algebra.

10. Show that the set $A = \{0,1,2\}$ and the following binary operations do not form a Boolean algebra:

$\oplus$	0	1	2
0	0	1	2
1	1	1	1
2	2	1	2

$\odot$	0	1	2
0	0	0	0
1	0	1	2
2	0	2	1

11. When doing arithmetic with the set of whole numbers and the operations of addition and multiplication, what are the identity elements?
12. Do the operations of additions and multiplication, as you learned them in elementary school, and the set of whole numbers comprise a Boolean algebra? Explain.

8-2 FORMAL THEOREMS

With the postulates of a Boolean algebra it is easy to prove a number of interesting theorems about the way objects behave. As indicated in Chaps. 5 and 6, a formal proof should consist of a list of statements that link the beginning information and the final conclusion. They should be of two kinds. The first are those which are restatements of earlier statements in the proof; that is, they are either stated postulates or previously established theorems adapted to the form needed in the proof. The second kind of statement is a conclusion logically derived from the postulates and previous theorems or earlier statements in the proof. Several sample proofs will clarify these ideas.

THEOREM 2 For any element a of B, $a \bigcirc a = a$.
(This is a statement of the theorem to be established.)

Step 1 $a = a \bigcirc i_{\bigcirc}$ postulate 2
(As a starting point in the proof we quote postulate 2 of the Boolean algebra postulates.)

Step 2 $a = a \bigcirc (a \star a')$ postulate 4
(Because postulate 4 says that $i_{\bigcirc} = a \star a'$, the $i_{\bigcirc}$ in step 1 is replaced by $a \star a'$ in step 2.)

Step 3 $a = (a \bigcirc a) \star (a \bigcirc a')$ postulate 3
(Postulate 3 states that $a \bigcirc (b \star c) = (a \bigcirc b) \star (a \bigcirc c)$, where a, b, and c are any elements of B. Because a and a' are indeed elements of B, this postulate applies to them. If we identify a in step 2 with a and b in the statement of postulate 3 and a' in step 2 with c in the postulate, then

$$a \bigcirc (b \star c) = (a \bigcirc b) \star (a \bigcirc c)$$

becomes

$$a \bigcirc (a \star a') = (a \bigcirc a) \star (a \bigcirc a'))$$

Step 4 $a = (a \bigcirc a) \star i_{\star}$
(By postulate 4, $a \bigcirc a' = i_{\star}$.)

Step 5 $a = a \bigcirc a$ postulate 2
(Postulate 2 states that $a \star i_{\star} = a$ for any element of B. $a \bigcirc a$ must be some element of B; therefore $(a \bigcirc a) \star i_{\star} = a \bigcirc a$.)

Step 5 concludes the proof, because it was the indicated theorem. Without the discussion the proof would look like this.

THEOREM 2 For any element a of B, $a \bigcirc a = a$.

Proof

Step 1	$a = a \bigcirc i_{\bigcirc}$	postulate 2
Step 2	$a = a \bigcirc (a \star a')$	postulate 4
Step 3	$a = (a \bigcirc a) \star (a \bigcirc a')$	postulate 3
Step 4	$a = (a \bigcirc a) \star i_{\star}$	postulate 4
Step 5	$a = a \bigcirc a$	postulate 2

This theorem has its dual, which, thanks to the principle of duality, requires no proof.

THEOREM 2′ For any element a of B, $a \star a = a$.

Theorems 2 and 2′ are important enough to Boolean algebra so that mathematicians have given them a special name. They are known as the "idempotent laws."

THEOREM 3 $a \star i_{\bigcirc} = i_{\bigcirc}$

Proof

Step 1	$i_{\bigcirc} = a \star a'$	postulate 4
Step 2	$i_{\bigcirc} = a \star (a' \bigcirc i_{\bigcirc})$	postulate 2
Step 3	$i_{\bigcirc} = (a \star a') \bigcirc (a \star i_{\bigcirc})$	postulate 3
Step 4	$i_{\bigcirc} = i_{\bigcirc} \bigcirc (a \star i_{\bigcirc})$	postulate 4
Step 5	$i_{\bigcirc} = (a \star i_{\bigcirc}) \bigcirc i_{\bigcirc}$	postulate 1
Step 6	$i_{\bigcirc} = a \star i_{\bigcirc}$	postulate 2

The dual of this system is

THEOREM 3′ $a \bigcirc i_{\star} = i_{\star}$.

In the proof of theorem 3 only the steps and the related postulates are given. Each step is a modified form of the step preceding it. Of course, this is not always the case, as often two or more steps are combined to produce a result. The first two sample proofs use postulates only. Any further proofs can use theorems 1, 2, and 3. For example, consider the following proof of a result known as the "absorption law."

THEOREM 4 For any elements a and b of B, $a \star (a \bigcirc b) = a$.

Proof

Step 1	$a = a \bigcirc i_{\bigcirc}$	postulate 2
Step 2	$a = a \bigcirc (b \star i_{\bigcirc})$	theorem 3
Step 3	$a = (a \bigcirc b) \star (a \bigcirc i_{\bigcirc})$	postulate 3
Step 4	$a = (a \bigcirc b) \star a$	postulate 2
Step 5	$a = a \star (a \bigcirc b)$	postulate 1

Step 2 uses a theorem rather than one of the postulates. Because theorem 3 says that $a \star i_{\bigcirc} = i_{\bigcirc}$, no matter what element a represents, then it is equally true that $b \star i_{\bigcirc} = i_{\bigcirc}$. Then $i_{\bigcirc}$ of step 1 replaced by $b \star i_{\bigcirc}$ in step 2.

The dual of this theorem is

THEOREM 4′ For any element a of B, $a \bigcirc (a \star b) = a$.

Naturally as the theorems get more and more complex, so do the corresponding proofs. For this reason the following theorems are stated without proof.

THEOREM 5 For each element a of B, a' is unique.

THEOREM 6 For any elements a, b, and c of set B, $a \bigcirc (b \bigcirc c) = (a \bigcirc b) \bigcirc c$.
THEOREM 6′ For any elements a, b, and c of set B, $a \star (b \star c) = (a \star b) \star c$.

Theorems 6 and 6′ are known as the "associative law."

For a final example of proofs of this kind, let us consider a proof for De Morgan's laws.

THEOREM 7 $(a \bigcirc b)' = a' \star b'$

To establish this result, it is necessary to show that $a' \star b'$ has the properties of $(a \bigcirc b)'$. Theorem 5 says that, whatever element $(a \bigcirc b)'$ is, it is unique; that is, there is only one element with the required properties. What are these properties? From postulate 4 it is clear that two results must be shown.

PART 1 $(a \bigcirc b) \bigcirc (a' \star b') = i_{\star}$

PART 2 $(a \bigcirc b) \star (a' \star b') = i_{\bigcirc}$

The formal proof of theorem 7 then takes the following form.

Proof (part 1) $(a \bigcirc b) \bigcirc (a' \star b') = i_{\star}$

Step 1 $(a \bigcirc b) \bigcirc (a' \star b')$
$= [(a \bigcirc b) \bigcirc a'] \star [(a \bigcirc b) \bigcirc b']$ postulate 3

Step 2 $= [a' \bigcirc (a \bigcirc b)] \star [(a \bigcirc b) \bigcirc b']$ postulate 1

Step 3 $= [(a' \bigcirc a) \bigcirc b] \star [a \bigcirc (b \bigcirc b')]$ theorem 6

Step 4 $= [b \bigcirc (a' \bigcirc a)] \star [a \bigcirc (b \bigcirc b')]$ postulate 1

Step 5 $= (b \bigcirc i_{\star}) \star (a \bigcirc i_{\star})$ postulate 4

Step 6 $= i_{\star} \star i_{\star}$ theorem 3′

Step 7 $= i_{\star}$ postulate 2

Proof (*part 2*) $(a \bigcirc b) \star (a' \star b') = i_{\bigcirc}$

Step 1	$(a \bigcirc b) \star (a' \star b')$ $= (a' \star b') \star (a \bigcirc b)$	postulate 1
Step 2	$= [(a' \star b') \star a] \bigcirc [(a' \star b') \star b]$	postulate 3
Step 3	$= [a \star (a' \star b')] \bigcirc [(a' \star b') \star b]$	postulate 1
Step 4	$= [(a \star a') \star b'] \bigcirc [a' \star (b' \star b)]$	theorem 6
Step 5	$= [b' \star (a \star a')] \bigcirc [a' \star (b \star b')]$	postulate 1
Step 6	$= (b' \star i_{\bigcirc}) \bigcirc (a' \star i_{\bigcirc})$	postulate 4
Step 7	$= i_{\bigcirc} \bigcirc i_{\bigcirc}$	theorem 3
Step 8	$= i_{\bigcirc}$	postulate 2

Now part 1 and part 2 can be combined. Because $a' \star b'$ behaves in the same way as $(a \bigcirc b)'$ and because by theorem 5 the "complement" elements are unique, the only possible conclusion is that

$$(a \bigcirc b)' = a' \star b'$$

The principle of duality holds, so we may write

THEOREM 7′ $(a \star b)' = a' \bigcirc b'$

There are a number of aspects of proofs that these examples illustrate, of which an important one is the sterility of this kind of proof. There is nothing about these proofs that show, in any way, the thinking that led De Morgan and others to arrive at these theorems.

Once the desired conclusion has been arrived at, there is also no indication as to the thinking that led to the particular steps involved. We can follow the steps, once someone else has thought them up. Man's knowledge of creativity, as well as all the other higher mental functions, is limited.

8-2 EXERCISES

The theorems needed to prove the following statements are the first eight exercises of Sec. 8-1 and their duals.

1. Show that $A \bigcirc (B \star C)' = (A \bigcirc B') \bigcirc C'$.

2. What is the set equivalent of the Boolean algebra inverse?

3. Show that $A \star (A' \bigcirc B) = A \star B$.

4. Prove that A star the inverse of the quantity A star B, is A star inverse of B.
5. Prove $A = (A \star B) \bigcirc (A \star B')$.
6. Prove that the inverse of the star identity element is the circle identity element.
7. Prove that the inverse of the circle identity element is the star identity element.
8. Show that $(A \star B) \star [A \star (B \bigcirc C)] = A \star B$.
9. Prove that $(a')' = a$. We sometimes drop the parentheses and write $(a')'$ as a''.

8-3 EXAMPLES OF BOOLEAN ALGEBRAS

The study of Boolean algebra may seem abstract and useless, interesting to "pure" mathematicians only. It is certainly true that it is abstract. However, it is not useless. The present section examines some of the ways it can be applied to save work in proving related results.

In Sec. 8-1 it was pointed out that the power set of the set $\{1,2,3\}$ and the operations of $\cup$, $\cap$, and $\sim$ formed a Boolean algebra; that is, if the set $\{\{\ \},\{1\},\{2\},\{3\},\{1,2\},\{1,3\},\{2,3\},\{1,2,3\}\} = C$ is used as the set mentioned in the Boolean postulates and the $\bigcirc$ operation is replaced with $\cup$, the $\star$ operation with $\cap$, and the $'$ with $\sim$, then the postulates are still satisfied; that is,

POSTULATE 1 For any elements a and b, $a \bigcirc b = b \bigcirc a$ and $a \star b = b \star a$.

becomes

POSTULATE 1 For any elements a and b, $a \cup b = b \cup a$ and $a \cap b = b \cap a$.

POSTULATE 2 There exist two distinct elements of B, $i_{\bigcirc}$ and $i_{\star}$, with the following properties: for every element a of B, $a \bigcirc i_{\bigcirc} = a$ and $a \star i_{\star} = a$.

becomes

POSTULATE 2 For every element a of C, $a \cup \emptyset = a$ and $a \cap \{1,2,3\} = a$.

POSTULATE 3 For any elements a, b, and c of B

$$a \bigcirc (b \star c) = (a \bigcirc b) \star (a \bigcirc c)$$

and

$a \star (b \bigcirc c) = (a \star b) \bigcirc (a \star c)$

becomes

POSTULATE 3 For any elements a, b, and c of C

$a \cup (b \cap c) = (a \cup b) \cap (a \cup c)$

and

$a \cap (b \cup c) = (a \cap b) \cup (a \cap c)$

POSTULATE 4 For every element a of B there is an element a' of B such that

$a \bigcirc a' = i_{\star}$ and $a \star a' = i_{\bigcirc}$

becomes

POSTULATE 4 For every element a of C there is an element $\tilde{a}$ of C such that

$a \cup \tilde{a} = \{1,2,3\}$ and $a \cap \tilde{a} = \{\ \}$

The validity of the first three of the postulates in this new form was established in Chap. 7. To show that all the necessary $\tilde{a}$ elements exist, a list is made in Table 8-1.

Table 8-1 shows that the required $\tilde{a}$ elements exist by finding them for each possible element a of C. For C the elements $\emptyset$ and $\{1,2,3\}$ are the $i_{\bigcirc}$ and $i_{\star}$ elements. The set $\{1,2,3\}$ is the universal set of the example. However, this can be generalized to state that the identity with respect to $\cup$ is $\emptyset$ and with respect to $\cap$ is U. The table verifies the more abstract method of establishing postulate 4 by observing that every element of C has a complement with respect to the set $\{1,2,3\}$ that is also an element of C. Using a to denote any element of C and $\tilde{a}$ to denote the complement of a with respect to $\{1,2,3\}$, then $a \cup \tilde{a} = \emptyset$ and $a \cap \tilde{a} = \{1,2,3\} = U$. In this way the required property of Boolean postulate 4 is again verified.

Why go to all the trouble to establish that C with $\cup$, $\cap$, and $\sim$ form a Boolean algebra? Because once it is shown that we are dealing with a Boolean algebra, all the theorems established in Sec. 8-2 are equally valid when stated in terms of the new operations and set. This means that each

TABLE 8-1 DIRECT LISTING OF $\tilde{a}$ ELEMENTS REQUIRED OF THE SET TO SATISFY BOOLEAN ALGEBRA POSTULATE 4

Element a	*Element* $\tilde{a}$	$a \cup \tilde{a} = U$	$a \cap \tilde{a} = \emptyset$
{ }	{1,2,3}	$\emptyset \cup \{1,2,3\} = \{1,2,3\}$	$\emptyset \cap \{1,2,3\} = \emptyset$
{1}	{2,3}	$\{1\} \cup \{2,3\} = \{1,2,3\}$	$\{1\} \cap \{2,3\} = \emptyset$
{2}	{1,3}	$\{2\} \cup \{1,3\} = \{1,2,3\}$	$\{2\} \cap \{1,3\} = \emptyset$
{3}	{1,2}	$\{3\} \cup \{1,2\} = \{1,2,3\}$	$\{3\} \cap \{1,2\} = \emptyset$
{1,2}	{3}	$\{1,2\} \cup \{3\} = \{1,2,3\}$	$\{1,2\} \cap \{3\} = \emptyset$
{1,3}	{2}	$\{1,3\} \cup \{2\} = \{1,2,3\}$	$\{1,3\} \cap \{2\} = \emptyset$
{2,3}	{1}	$\{2,3\} \cup \{1\} = \{1,2,3\}$	$\{2,3\} \cap \{1\} = \emptyset$
{1,2,3}	$\emptyset$	$\{1,2,3\} \cup \emptyset = \{1,2,3\}$	$\{1,2,3\} \cap \emptyset = \emptyset$

theorem of Sec. 8-2 has a parallel theorem involving the set C and $\cup$ and $\cap$. For example,

PARALLEL THEOREM For any element a of C, $a \cup a = a$ and $a \cap a = a$.

PARALLEL THEOREM For any element a of C

$$a \cap \emptyset = \emptyset$$
$$a \cup U = U$$

PARALLEL THEOREM For any elements a and b of C

$$a \cap (a \cup b) = a$$
$$a \cup (a \cap b) = a$$

PARALLEL THEOREM $\tilde{a}$ is unique.

PARALLEL THEOREM

$$a \cup (b \cup c) = (a \cup b) \cup c$$
$$a \cap (b \cap c) = (a \cap b) \cap c$$

PARALLEL THEOREM For any elements a and b of C

$\widetilde{(a \cup b)} = \tilde{a} \cap \tilde{b}$
$\widetilde{(a \cap b)} = \tilde{a} \cup \tilde{b}$

These theorems do not require any kind of formal proof. They were proved in Sec. 8-2. Also, if these theorems were unknown prior to an investigation of the properties of C, they could be found by the comparisons made here. Here, then, is a powerful tool for the discovery of new results. If it is possible to construct an exact model of one system in another, the results discovered in the first lead to parallel ones in the second.

A more interesting Boolean algebra was discussed in Chaps. 5 and 6. It was the one formed by the set of all possible statements of logic and the connectives $\wedge$, $\vee$, and $\sim$. Instead of equality, the relationship of logical equivalence is used. The symbol $\equiv$ was used in Chaps. 5 and 6 and is used here.

Let D be the set of all possible statements. Then D, $\wedge$, $\vee$, and $\sim$ relative to $\equiv$ form a Boolean algebra. To establish this, it is necessary to show that the postulates of Boolean algebra are satisfied.

Relative to Boolean postulate 1, it is necessary to show that for any two statements p and q

$p \vee q \equiv q \vee p \qquad \text{and} \qquad p \wedge q \equiv q \wedge p$

This result was shown in Chap. 5. For Boolean postulate 2 an $i_{\bigcirc}$ and $i_{\star}$ must be found. Let T stand for any logically true statement and F for any logically false one; that is, T stands for a statement that is true in every logically possible case and F stands for one that is false in every logically possible case. The statements T and F are the $i_{\bigcirc}$ and $i_{\star}$ for this Boolean algebra. The reader can verify this by using truth-table analysis to establish that for any statement p

$p \wedge \mathrm{T} \equiv p \qquad \text{and} \qquad p \vee \mathrm{F} \equiv p$

Postulates 3 and 4 can easily be verified. To verify postulate 3, it is necessary to show that for any statements p, q, and r

$p \vee (q \wedge r) \equiv (p \vee q) \wedge (p \vee r)$

and

$p \wedge (q \vee r) \equiv (p \wedge q) \vee (p \wedge r)$

A truth table can be used to establish these, as in Table 8-2.

TABLE 8-2 TRUTH TABLE ANALYSIS FOR $p \wedge (q \vee r)$ AND $(p \wedge q) \vee (p \wedge r)$

$p\ q\ r$	$p \wedge (q \vee r)$	$(p \wedge q) \vee (p \wedge r)$
T T T	T T T T T	T T T T T T T
T T F	T T T T F	T T T T T F F
T F T	T T F T T	T F F T T T T
T F F	T F F F F	T F F F T F F
F T T	F F T T T	F F T F F F T
F T F	F F T T F	F F T F F F F
F F T	F F F T T	F F F F F F T
F F F	F F F F F	F F F F F F F

Because Table 8-2 shows that $p \wedge (q \vee r)$ and $(p \wedge q) \vee (p \wedge r)$ have the same truth table, they are logically equivalent. The truth-table analysis showing $p \vee (q \wedge r)$ equivalent to $(p \vee q) \wedge (p \vee r)$ is left as an exercise.

The $'$ elements required in Boolean postulate 4 are easily identified as being the negations in Chaps. 5 and 6. $p \vee \sim p$ is a logically true statement, and $p \wedge \sim p$ a logically false one. Then $p \vee \sim p \equiv \mathrm{T}$ and $p \wedge \sim p \equiv \mathrm{F}$. Thus $\sim p$ is the inverse element needed to satisfy Boolean postulate 4.

Because this shows that D, $\wedge$, $\vee$, $\sim$, and equivalence form a Boolean algebra, all the theorems in Sec. 8-2 hold; that is,

THEOREM For any statement p

$$p \vee p \equiv p$$

and

$$p \wedge p \equiv p$$

THEOREM For any statement p

$$p \vee \mathrm{T} \equiv \mathrm{T}$$

and

$$p \wedge \mathrm{F} \equiv \mathrm{F}$$

THEOREM For any statements p and q

$$p \wedge (p \vee q) \equiv p$$

and

$$p \vee (p \wedge q) \equiv p$$

THEOREM $\sim p$ is unique.

THEOREM For any statements p, q, and r

$$p \wedge (q \wedge r) \equiv (p \wedge q) \wedge r$$

and

$$p \vee (q \vee r) \equiv (p \vee q) \vee r$$

THEOREM For any statements p and q

$$\sim (p \vee q) \equiv \sim p \wedge \sim q$$

and

$$\sim (p \wedge q) \equiv \sim p \vee \sim q$$

These six theorems do not require proof. Their validity was established in Sec. 8-2. Suppose that the present chapter had preceded those on logic and sets. Then much of them could have been omitted, because the results would have already been established. If economy of effort is to be a major criterion in judging a method of learning, then going from a general concept like Boolean algebra to a specific one like the algebra of propositions and sets would be more common. However, almost everyone is more comfortable when learning takes place from specific examples to general concepts. Is this because we are familiar with this method or because of some intrinsic mechanism in the operation of learning? If there is nothing in this method to make it easier, then economy of effort would dictate a change to learning from general concepts to specific examples whenever possible. This is especially true in light of the knowledge explosion of the twentieth century.

8-3 EXERCISES

1. State the theorem of duality for sets.
2. State the theorem of duality for logic.

State the following theorems in set notation.

3. Idempotent laws

4. De Morgan's laws
5. Identities
6. Absorption laws
7. Inverses
8. Double negation

9 to *14*. State the above theorems (Exercises 3 to 8) in logic notation.
15. Use truth tables to prove $p \vee (q \wedge r) \equiv (p \vee q) \wedge (p \vee r)$.
16. Use Boolean algebra to prove $P \cap (\tilde{P} \cup Q) = P \cap Q$.
17. Prove $P \cap (\widetilde{P \cap Q}) = P \cap \tilde{Q}$ using:

(*a*) Boolean algebra
(*b*) Venn diagrams

18. Repeat Exercise 17 for $(P \cap Q) \cap [P \cap (Q \cup R)] = P \cap Q$.
19. Use Boolean algebra to prove $p \wedge (\sim p \vee q) \equiv p \wedge q$.
20. Prove $p \wedge \sim (p \wedge q) \equiv p \wedge \sim q$ using:

(*a*) Boolean algebra
(*b*) Truth tables

21. Repeat Exercise 20 for $(p \wedge q) \wedge [p \wedge (q \vee r)] \equiv p \wedge q$.
22. What is the difference among sets, logic, and Boolean algebra?

8-4 PARTIAL ORDERING

For the development of more complex theorems it is necessary to define a new relationship among the elements of a Boolean algebra.

DEFINITION If a and b are elements of B, then $a -\!\!\!\exists\, b$ (read a contained in b) if and only if $a \bigcirc b = b$.

Some theorems about $-\!\!\!\exists$ are easily developed.

THEOREM 8 For every element a of B, $a -\!\!\!\exists\, a$.

Proof

Step 1 $a \bigcirc a = a$ theorem 2

Step 2 $a -\!\!\!\exists\, a$ definition

This proof is somewhat different from those in Sec. 8-2 in that it establishes a result by showing that the requirements of a definition have been met. There is another concept that was used in earlier proofs but not formally stated. In the discussion of Boolean algebra the equal sign has been used as a primitive notation in that it has been used without having first been given a formal meaning. The primitive understanding of the meaning of $=$ has included the notion that the two items linked with it could always be interchanged; that is, in any expression where one of the two items linked with $=$ is used, the other can be substituted without changing the truth value of the expression. This is called the principle of "substitution." This property of $=$ has been used but not stated in earlier sections. It is an advantage to state this principle now so that it can be used as a justification in a proof, such as that of theorem 9.

THEOREM 9 If a and b are elements of B and if $a \mathrel{-\!\!\ni} b$ and $b \mathrel{-\!\!\ni} a$, then $a = b$.

Proof

Step 1	$a \bigcirc b = b$	definition of $a \mathrel{-\!\!\ni} b$
Step 2	$b \bigcirc a = a$	definition of $b \mathrel{-\!\!\ni} a$
Step 3	$b \bigcirc a = a \bigcirc b$	postulate 1
Step 4	$a = b$	substitution

Substitution is used several times in the proof of theorem 10.

THEOREM 10 If a, b, and c are elements of B and if $a \mathrel{-\!\!\ni} b$ and $b \mathrel{-\!\!\ni} c$, then $a \mathrel{-\!\!\ni} c$.

Proof

Step 1	$a \bigcirc b = b$ and $b \bigcirc c = c$	definition of $a \mathrel{-\!\!\ni} b$ and $b \mathrel{-\!\!\ni} c$
Step 2	$a \bigcirc c = a \bigcirc (b \bigcirc c)$	substitution
Step 3	$a \bigcirc c = (a \bigcirc b) \bigcirc c$	theorem 6
Step 4	$a \bigcirc c = b \bigcirc c$	substitution
Step 5	$a \bigcirc c = c$	substitution
Step 6	$a \mathrel{-\!\!\ni} c$	definition of $\mathrel{-\!\!\ni}$

The results of theorems 8 and 10 are similar to those we could establish using $=$ as a relationship instead of $\mathrel{-\!\!\ni}$. It is certainly true that for any a an element of B, $a = a$ (corresponding to $a \mathrel{-\!\!\ni} a$) and that if $a = b$ and $b = c$, then $a = c$ (corresponding to $a \mathrel{-\!\!\ni} b$ and $b \mathrel{-\!\!\ni} c$, then $a \mathrel{-\!\!\ni} c$). However, $a = b$, then $b = a$, a result not true for the $\mathrel{-\!\!\ni}$ relationship. (In general if $a \mathrel{-\!\!\!3} b$, then $b \not\mathrel{-\!\!\!3} a$.) This property of equality is known as the "symmetric property." Because $\mathrel{-\!\!\!3}$ does not have this property, but rather follows the rule established in theorem 9, $\mathrel{-\!\!\!3}$ is said to be "antisymmetric." The properties of a relationship indicated in theorems 8 and 10 are also given special names. When a relationship has the property proved for $\mathrel{-\!\!\!3}$ in theorem 8, it is said to be "reflexive." If a relationship has the property proved for $\mathrel{-\!\!\!3}$ in theorem 10, it is said to be "transitive." A relationship that has the properties of antisymmetry, reflexivity, and transitivity is known as "partial ordering." In other words, the relationship of $\mathrel{-\!\!\!3}$ on the elements of Boolean algebra is a partial ordering on them.

The identity elements $i_{\bigcirc}$ and $i_{\star}$ also have a special connection with $\mathrel{-\!\!\!3}$. This is established in the following theorems.

THEOREM 11 For any element a of B, $i_{\bigcirc} \mathrel{-\!\!\!3} a$ and $a \mathrel{-\!\!\!3} i_{\star}$.

Proof (part 1) $i_{\bigcirc} \mathrel{-\!\!\!3} a$

Step 1 $i_{\bigcirc} \bigcirc a = a \bigcirc i_{\bigcirc}$ postulate 1

Step 2 $i_{\bigcirc} \bigcirc a = a$ postulate 2

Step 3 $i_{\bigcirc} \mathrel{-\!\!\!3} a$ definition of $\mathrel{-\!\!\!3}$

Proof (part 2) $a \mathrel{-\!\!\!3} i_{\star}$

Step 1 $a \bigcirc i_{\star} = i_{\star}$ theorem 3

Step 2 $a \mathrel{-\!\!\!3} i_{\star}$ definition of $\mathrel{-\!\!\!3}$

THEOREM 12 If $a \mathrel{-\!\!\!3} x$ and $b \mathrel{-\!\!\!3} x$, then $(a \bigcirc b) \mathrel{-\!\!\!3} x$.

Proof

Step 1 $a \bigcirc x = x$ and $b \bigcirc x = x$ definition of $a \mathrel{-\!\!\!3} x$ and $b \mathrel{-\!\!\!3} x$

Step 2 $(a \bigcirc b) \bigcirc x = a \bigcirc (b \bigcirc x)$ theorem 6

Step 3 $(a \bigcirc b) \bigcirc x = a \bigcirc x$ substitution

Step 4 $(a \bigcirc b) \bigcirc x = x$ substitution

Step 5 $(a \bigcirc b) \mathrel{-\!\!\!3} x$ definition of

There are many other theorems concerning a Boolean algebra that could be established. There are also other sets of postulates that could have been used in place of those in Sec. 8-1 to structure a Boolean algebra. Why bother with all this formality? Sometimes formality seems to be a distinct handicap to learning—unnecessary and dull. But formality is often vital, especially when intuition fails.

Uses for these theorems are examined in Sec. 8-5. They are applied to the models of Boolean algebra in Sec. 8-3. There are, however, several aspects of the proofs in the present section that should be noted. Three distinct kinds of thinking seem to be needed in building a proof. First one must find out what to prove. Then one must decide how to establish the validity of the result; that is, one must discover a chain of reasoning that links the known information to the forecasted result. Finally, one must be able to translate a line of reasoning into a sequence of verifiable steps. The proofs in the present section serve mostly to illustrate this last kind of thinking rather than the other two. Once a result is chosen and a line of reasoning is found, almost everyone can, with a little practice, identify the formal steps required.

It is the techniques of discovery, the first kind of thinking, that have escaped analysis. It is not hard to find examples of discovery, but we do not know how the process of discovery works. Perhaps we never shall.

8-4 EXERCISES

1. Use the numbers 3, 4, and 5 and show that they are partially ordered by the relation *less than or equal to.*

2. Define *less than* in a manner similar to the definition of $\mathrel{-\!\!\!\ni}$.

3. In the present section, it is shown that $a \mathrel{-\!\!\!\ni} i_{\star}$ and $i_{\bigcirc} \mathrel{-\!\!\!\ni} a$. Do the additive identity element and the multiplicative identity element of the real numbers behave in a similar manner under the operation *less than*? Explain.

4. Name an operation other than *less than or equal to* that also partially orders the set of whole numbers.

5. Definition: $a \mathrel{\in\!\!\!-} b$ if and only if $a \star b = b$. Show that $\mathrel{\in\!\!\!-}$ partially orders set B, where a and b are members of set B.

6. Prove $i_{\star} \mathrel{\in\!\!\!-} a$ for all a in B.

7. Prove $a \mathrel{\in\!\!\!-} i_{\bigcirc}$ for all a in B.

8. Show that, if $a \star b' = i_{\bigcirc}$, then $a \mathrel{-\!\!\!\ni} b$.

9. Show that, if $x \mathrel{-\!\!\!\ni} a$ and $x \mathrel{-\!\!\!\ni} b$, then $x \mathrel{-\!\!\!\ni} (a \star b)$.

10. Prove that, if $a \mathrel{-\!\!\!\ni} b$, then $b' \mathrel{-\!\!\!\ni} a'$.

8-5 PARTIAL ORDERING, SETS, AND LOGIC

Because the systems in Sec. 8-3 are Boolean algebras, they also have partial-ordering relations. What is the form of this relation? Have we already considered the partial-ordering relation but under another name? Consider the properties a partial-ordering relation has. First let us examine this topic with respect to the power set of $\{1,2,3\}$.

Restating the defining condition of a partial-ordering relation in terms of sets $\cup$, $\cap$, and $\sim$, this becomes:

DEFINITION If a and b are elements of C, then a ?? b if and only if $a \cup b = b$. The ?? are used in place of the —3 of the original form of the definition in the hope that an appropriate symbol already exists. In Chap. 7 several examples showed that, if $a \cup b = b$, then all the elements of a are elements of b. Care must be used here, because a and b are at once elements of the Boolean algebra formed from C and sets, which themselves contain elements. Which of these two roles a and b play is usually clear from the context. The usual way $a \cup b = b$ is described is that a is a subset of b; that is, when every element of a is also an element of b, then a is a subset of b. Then the correct form of the definition of partial ordering in this case reads:

DEFINITION If a and b are elements of C, then $a \subseteq b$ if and only if $a \cup b = b$. Here the —3 has been replaced with the usual symbol for subset, $\subseteq$. The properties of —3 proved in Sec. 8-4 hold when —3 is replaced with $\subseteq$. In particular,

THEOREM $a \subseteq a$

This theorem, of course, does not hold for the proper subset relation.

THEOREM If a and b are elements of C and $a \subseteq b$ and $b \subseteq a$, then $a = b$. A more general version of this theorem is true for any two sets a and b, although they are not elements of C.

THEOREM If a, b, and c are elements of C and $a \subseteq b$ and $b \subseteq c$, then $a \subseteq c$.

THEOREM For any element a of C

$\emptyset \subseteq a$ and $a \subseteq \{1,2,3\} = U$

As in Sec. 8-3 these theorems do not require any proof. They were established for all Boolean algebras in Sec. 8-4.

To determine if a partial-ordering relation has already been defined on the set of propositions, consider two statements p and q that are related in the manner required by the definition of partial orderings, namely,

$$p \vee q \equiv q$$

In terms of truth values this condition means that the statement $p \vee q$ is true whenever q is true and false whenever q is false. This places a severe restriction on p. If one assumes that p is true, then $p \vee q$ is also true; and if $p \vee q$ is true, then q is true. On the other hand, if one assumes that p is false, then $p \vee q$ can be either true or false and no restriction is placed on q. In other words, whenever p is true, q is also true. This relation between two statements was given a name in Chap. 7; p was said to imply q. In Chap. 7 the symbol $\Rightarrow$ was used to indicate implication. It is reasonable to continue using this symbol here; that is, if in every case that p is true, so is q, than p implies q, written $p \Rightarrow q$. Thus, to define a partial-ordering relation in logic, $p \Rightarrow q$ if and only if $p \vee q \equiv q$. Each theorem of Sec. 8-4 also has a parallel theorem for implication. As in the example of sets above, these theorems do not require proof but are necessary properties of any Boolean algebra.

THEOREM For any statement p, $p \Rightarrow p$.

THEOREM If p and q are any statements and $p \Rightarrow q$ and $q \Rightarrow p$, then $p \equiv q$.

THEOREM If p, q, and r are statements and $p \Rightarrow q$ and $q \Rightarrow r$, then $p \Rightarrow r$.

THEOREM For any statement p, $\mathrm{F} \Rightarrow p$ and $p \Rightarrow \mathrm{T}$.

The last theorem is interesting. Translated into everyday words, it states that any logically false statement can be used to imply any desired conclusion. On the other hand, any statement can be used to imply a logically true statement. Suppose that one begins with a logically false statement. It is still possible to reach valid conclusions without detecting any problems within the logical structure of the reasoning. On the other hand, if one wants to prove any logically true statement, he can use any statement or group of statements as premises of the argument. This theorem assures us that both these approaches will be successful.

How does the usual form of a valid argument, where several statements are used to reach a conclusion, fit into logic as a Boolean algebra? By connecting all the premises in an argument with $\wedge$, one arrives at a single statement that implies the desired conclusion if the original argument is valid. From this point of view, logic is a neat Boolean package.

8-5 EXERCISES

1. Show that, if $A \cap \tilde{B} = \emptyset$, then $A \subseteq B$.
2. Prove that if $C \subseteq A$ and $C \subseteq B$, then $C \subseteq (A \cap B)$.
3. Prove that, if $A \subseteq B$, then $\tilde{B} \subseteq \tilde{A}$.
4. Prove that, if $p \wedge \sim q \equiv \mathrm{F}$, then $p \Rightarrow q$.
5. Prove that, if $p \Rightarrow q$ and $p \Rightarrow r$, then $p \Rightarrow (q \wedge r)$.
6. Prove that, if $p \Rightarrow q$, then $\sim q \Rightarrow \sim p$.
7. What is the relationship between the following pairs of the above exercises: 1 and 4, 2 and 5, 3 and 6?
8. An argument is valid if the conjunction of the premises implies the conclusion. Use Boolean algebra to show that the following is valid:

$$\begin{array}{l} p \vee q \\ \underline{\sim p} \\ \therefore q \end{array}$$

9. Use Boolean algebra to show that the following is valid:

$$\begin{array}{l} p \rightarrow q \\ \underline{\sim q} \\ \therefore \sim p \end{array}$$

8-6 AN ELECTRICAL BOOLEAN ALGEBRA

Of all the Boolean algebras we can study perhaps none is more interesting than that based on the idea of an electrical switch. For this purpose an electrical switch is considered any electrical device with two specific states relative to the flow of current. When the switch is in the "on" state, then electrical current can flow through it. When the switch is in the "off"

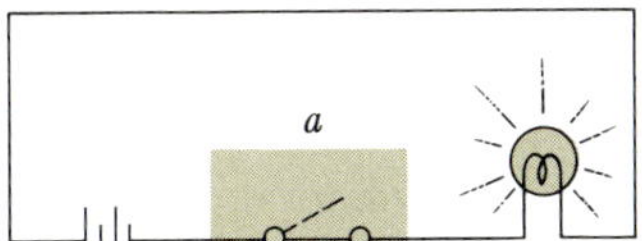

FIG. 8-1 SWITCH a IN THE ON STATE.

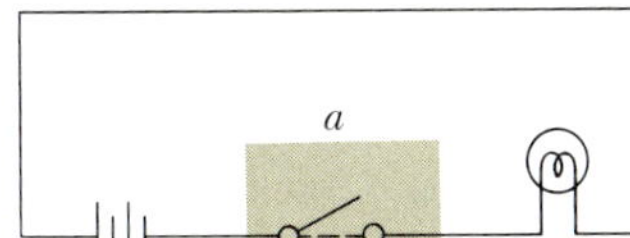

FIG. 8-2 SWITCH a IN THE OFF STATE.

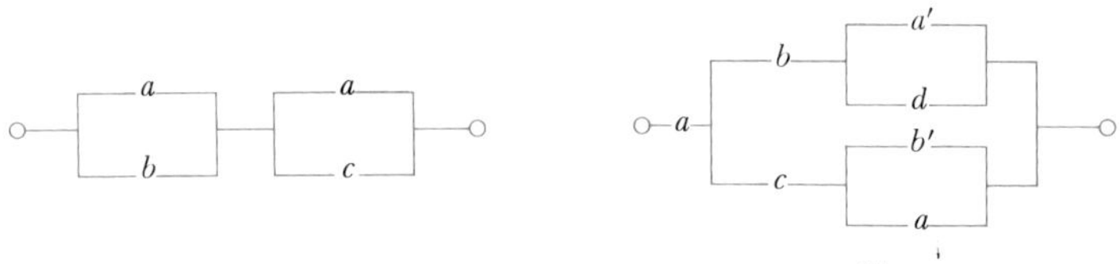

(a) (b)

FIG. 8-3 SAMPLE SWITCHING NETWORKS.

state, then the flow of current is stopped. Figures 8-1 and 8-2 are simple diagrams of such switches.

From such switches, it is possible to build more complex switching networks, as in Fig. 8-3, where letters like a, b, c, a', and so on, stand for switches that may be in either state. If a letter appears more than once in a network, each time it appears it is assumed to be mechanically linked to all the other switches identified by the same letter. This is done in such a way that all the switches with the same identification are always in the same state.

If a switch is identified with a', such as a' and b' in Fig. 8-3, then it is mechanically linked to switches marked with the same identifying letter without the $'$ in such a way that it always has the opposite state. For example, if switch a is on, then switch a' is off; if switch a is off, then switch a' is on.

A simple relation between switching networks can be defined. If two networks allow current to flow under exactly the same conditions (same switch settings for the individual switches), then they are called "equivalent," and the symbol $\sim$ is used to denote the relationship. See, for example, F_1 and F_2 below.

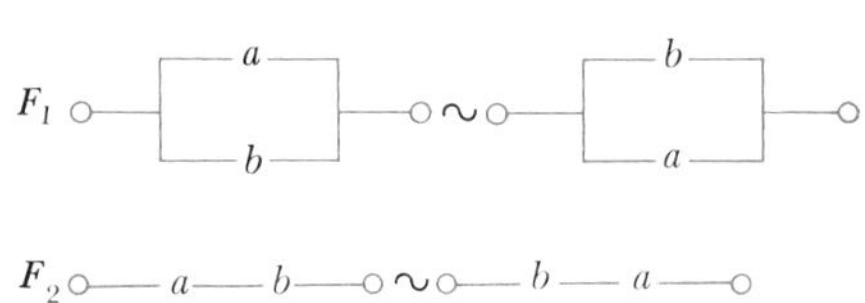

The two switches in F_1 are said to be connected "in parallel," and those in F_2 "in series." The parallel and series connection of switches and switching networks and the equivalence of switching networks form an electrical Boolean algebra. Note that there is no real difference between a switch and a switching network; they both are simply devices with two states, on and off. It is reasonable to talk about something like (o—*a*—*b*—o)′ as a network that is off if o—*a*—*b*—o is on, and on if o—*a*—*b*—o is off. And it is reasonable to think of letters like *a* standing for switching networks as it is to restrict their meaning to simple switches. To demonstrate that this systems of switches forms a Boolean algebra, it has to be shown that they satisfy the four postulates in Sec. 8-1. Let us replace the ○ operation with the phrase *wired in parallel* and the ☆ operation with *wired in series;* that is, *a* ○ *b* is to be replaced by 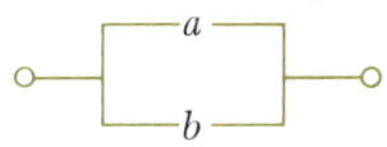 and *a* ☆ *b* by o—*a*—*b*—o

Postulate 1 then becomes:

POSTULATE 1 (COMMUTATIVE LAW FOR SWITCHES) For any two switching networks *a* and *b*

and o—*a*—*b*—o ∼ o—*b*—*a*—o

These are F_1 and F_2 above and are easily shown to be true by trying all possible cases, as in Table 8-3, which shows that the two networks are equivalent; that is, they allow current to flow under the same conditions. Thus o—*a*—*b*—o ∼ o—*b*—*a*—o, because current flows when both switches are on and at no other time. To establish postulate 2, we find a switching network that can play the roles of $i_{○}$ and $i_{☆}$. Let o—*O*—o stand for an unbroken piece of wire through which current can always flow, and let o—*F*—o stand for a broken piece of wire through which current can never flow. Then o—*O*—o and o—*F*—o are easily seen to be the required identity elements of postulate 2; that is,

TABLE 8-3 POSSIBLE CASES FOR NETWORKS 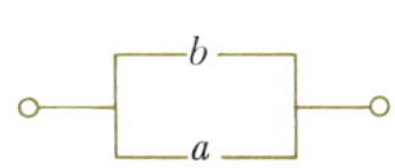 AND 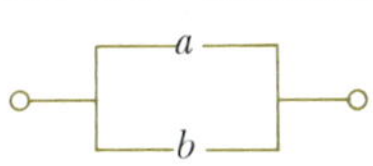

Switch a	*Switch b*	*Network* 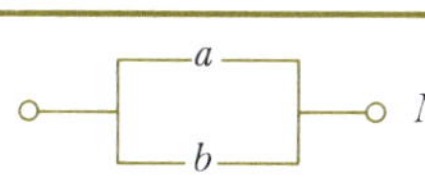	*Network* 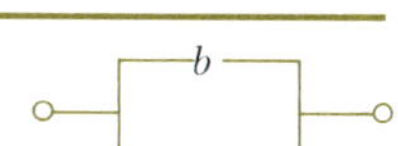
On	On	On	On
On	Off	On	On
Off	On	On	On
Off	Off	Off	Off

POSTULATE 2 For any switching network a

identity for series

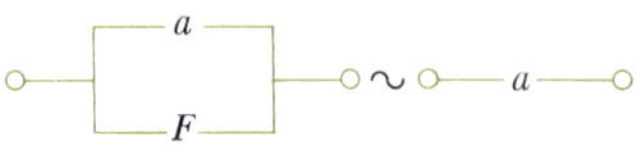

identity for parallel

Using ○—O—○ and ○—F—○, it is easy to show that postulates 3 and 4 are also satisfied.

POSTULATE 3 For any switching networks a, b, and c

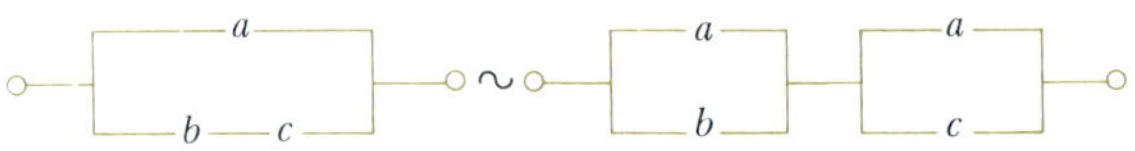

distributive law for parallel

and

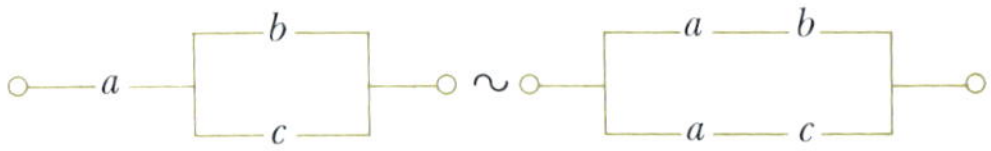

distributive law for series

POSTULATE 4 For any switching network a

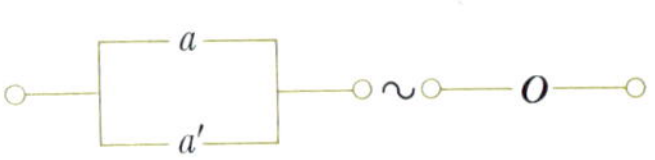

inverse for parallel

and

inverse for series

These postulates can be more formally verified with switching tables, such as Table 8-3.

As in the other examples of Boolean algebras, once the postulates have been shown to be satisfied, the theorems in Sec. 8-2 can be quoted without further comment.

THEOREM 1

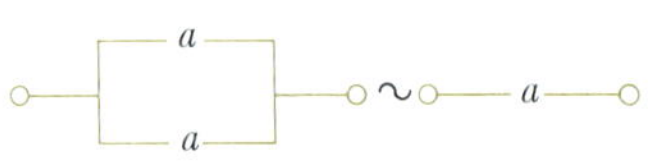

and

THEOREM 2

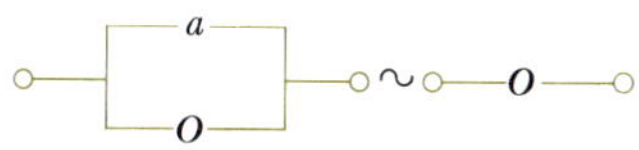

and

THEOREM 3

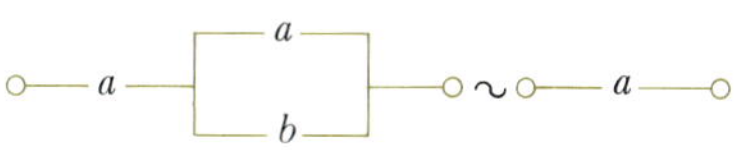

and

THEOREM 4

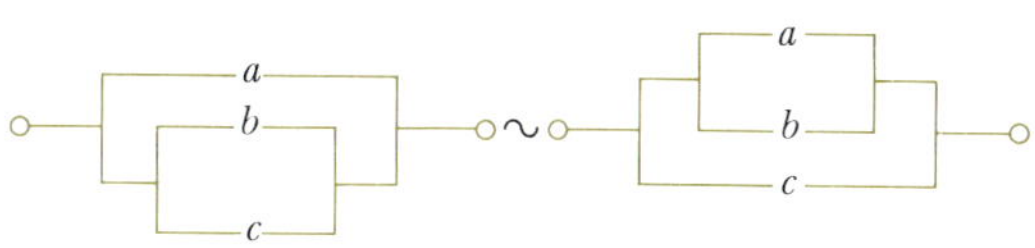

and

THEOREM 5

and

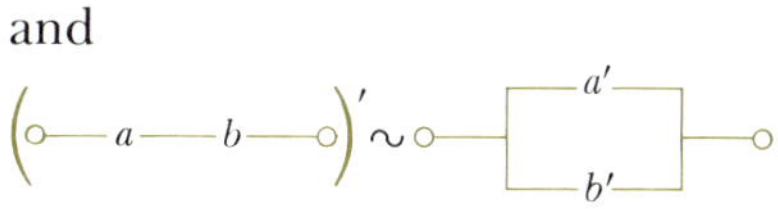

These theorems can be put to a practical use. For example, consider the following network:

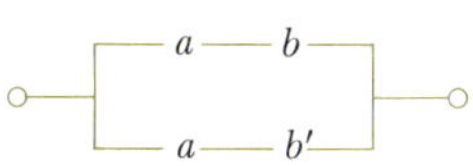

If this network is compared with those in postulate 3 above, it is easy to see that

However

from postulate 4. Any network can be replaced by an equivalent network; thus

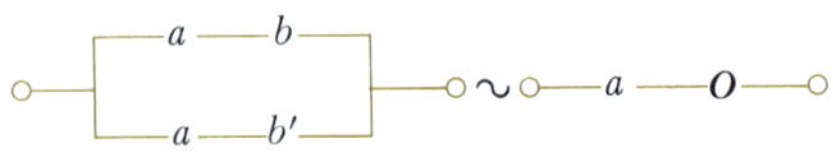

However

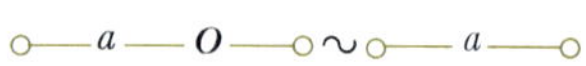

therefore

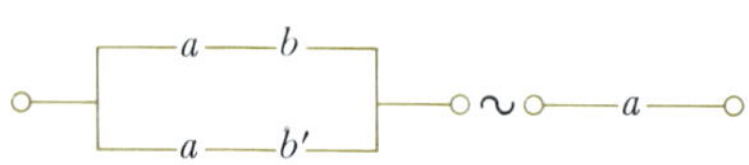

Problems of this kind are usually attacked with the use of a "Boolean polynomial" rather than the switching networks themselves. A Boolean polynomial is any expression made up of the elements and operations of Boolean algebra. For example, the network

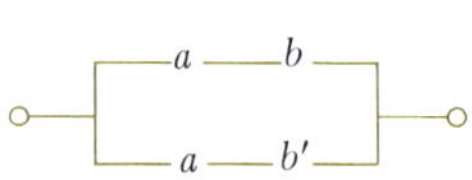

corresponds to the Boolean polynomial

$(a \star b) \bigcirc (a \star b')$

In a similar manner the polynomial

$a \star (b \bigcirc a')$

corresponds to

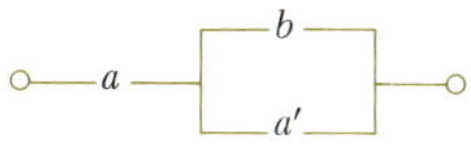

and

$(a \star b') \bigcirc ((a' \bigcirc c) \star b)$

corresponds to

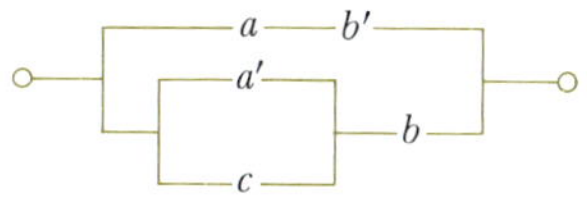

This translation of switching networks into Boolean polynomials helps in designing and simplifying networks for selected jobs. Consider the switching network

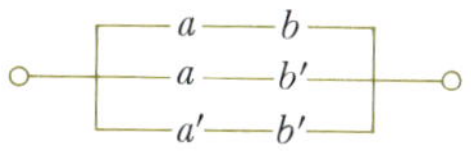

The corresponding Boolean polynomial is

$(a \star b) \bigcirc (a \star b') \bigcirc (a' \star b')$

Because by postulate 3

$(a \star b) \bigcirc (a \star b') = a \star (b \bigcirc b')$

then

$(a \star b) \bigcirc (a \star b') \bigcirc (a' \star b') = (a \star (b \bigcirc b')) \bigcirc (a' \star b')$

But by postulate 4

$b \bigcirc b' = i_{\star}$

therefore

$(a \star b) \bigcirc (a \star b') \bigcirc (a' \star b') = (a \star i_{\star}) \bigcirc (a' \star b')$

By postulate 2

$a \star i_{\star} = a$

so that

$$(a \star b) \bigcirc (a \star b') \bigcirc (a' \star b') = a \bigcirc (a' \star b')$$

Applying postulate 3 again,

$$a \bigcirc (a' \star b') = (a \bigcirc a') \star (a \bigcirc b')$$

then

$$(a \star b) \bigcirc (a \star b') \bigcirc (a' \star b) = (a \bigcirc a') \star (a \bigcirc b')$$

By postulate 4

$$a \bigcirc a' = i_{\star}$$

and

$$i_{\star} \star (a \bigcirc b') = a \bigcirc b'$$

Therefore

$$(a \star b) \bigcirc (a \star b') \bigcirc (a' \star b') = a \bigcirc b'$$

To what switching network does $a \bigcirc b'$ correspond? It should be

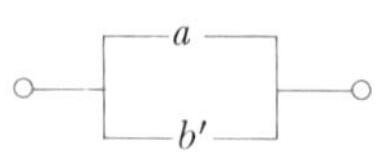

Then

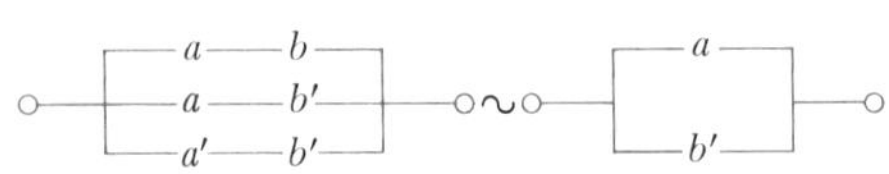

As an interesting variation, consider the following problem. Suppose that in a certain three-member committee, a majority vote carries an issue, but the chairman has a veto; that is, his "no" vote prevents the passage of an issue. Assume that each member of the committee controls a switch. The problem is to design a switching network that is on when an issue passes and off when it does not, according to the votes of the members. Then once the network is designed, it must be simplified.

A basic network meeting these requirements is relatively easy to design. There are only three ways an issue can pass. The chairman and both other members of the committee can vote in favor of an issue, or the chairman and either of the other members can be in favor of the issue and the third member opposed. Let the switch controlled by the chairman be denoted by c and those controlled by the other committee members by a and b. If one could find a network that represented each of the favorable kinds of votes, then these could be linked in parallel so that current flowed if any favorable vote occurred. In the case of all three committee members being in favor, the following network passes current:

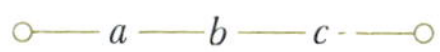

Networks corresponding to the other two favorable conditions are

and

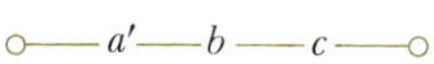

Linking these produces

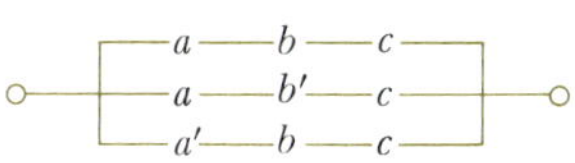

Can this network be simplified? The corresponding Boolean polynomial for this network is

$$(a \star b \star c) \bigcirc (a \star b' \star c) \bigcirc (a' \star b \star c)$$

Applying postulate 3, it becomes

$$(a \star ((b \star c) \bigcirc (b' \star c))) \bigcirc (a' \star b \star c)$$

Then by postulate 1, it is

$$(a \star ((c \star b) \bigcirc (c \star b'))) \bigcirc (a' \star b \star c)$$

Then by postulate 3, it is

$$(a \star (c \star (b \bigcirc b'))) \bigcirc (a' \star b \star c)$$

Then applying postulate 4, it is

$$(a \star (c \star i_{\star})) \bigcirc (a' \star b \star c)$$

Postulate 2 reduces this to

$$(a \star c) \bigcirc (a' \star b \star c)$$

Then by postulate 1, it is

$$(c \star a) \bigcirc (c \star (a' \star b))$$

which postulate 3 changes to

$$c \star (a \bigcirc (a' \star b))$$

which by postulate 3 is

$$c \star ((a \bigcirc a') \star (a \bigcirc b))$$

Postulate 4 reduces this to

$$c \star (i_{\star} \star (a \bigcirc b))$$

Then applying postulate 1, it becomes

$$c \star ((a \bigcirc b) \star i_{\star})$$

Finally, postulate 2 changes this to

$$c \star (a \bigcirc b)$$

Translating this final form into a switching network, we have

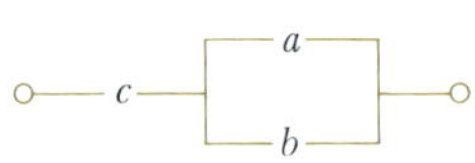

Summarizing in terms of switching networks, this result is

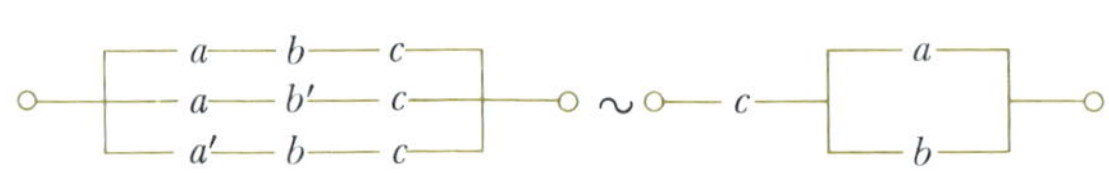

The largest application of switching circuitry and Boolean algebra is computer design. Computers contain logic circuits. It is not difficult to create circuits for the basic logic operations of *and* and *or*. We stated that $\star$ corresponds to series and $\bigcirc$ to parallel. However, $\wedge$ also corresponds to $\star$ and $\vee$ to $\bigcirc$. Thus if we allow two switches to represent statements of logic, with the on position meaning true and the off position false, $p \wedge q$ are represented by switches in series and $p \vee q$ by switches in parallel:

$p \wedge q$ $\qquad$ $p \vee q$

Note that $p \wedge q$ is true only when both statements are true and that current flows through a series of two switches only when both are on. Similarly, $p \vee q$ is true when either or both statements are true and is false only when both are false, and current flows through a circuit of two parallel switches when one or both are on and is blocked only when both are off. Thus more complicated statements of logic can be translated into switching circuits which can be simplified with Boolean algebra. The simplification of logic circuits is important for two reasons. It can save a great deal of time and money for computer manufacturers, and it can save space and weight in computers for aircraft and space vehicles.

8-6 EXERCISES

Design switching circuits to represent the following.

1. $(p \wedge q) \vee (r \wedge s)$
2. $p \vee (p \wedge q) \vee (\sim p \wedge \sim q)$
3. $[p \vee (p \wedge q) \vee (\sim p \wedge \sim q)] \wedge p$
4. $[p \wedge (q \vee r)] \vee [q \wedge (p \vee \sim r)] \vee [q \wedge r]$

Translate the following into statements of logic.

5. 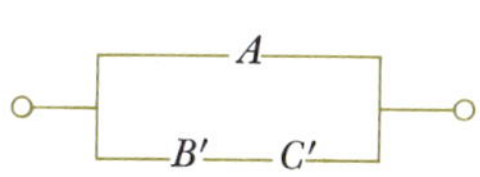

6. 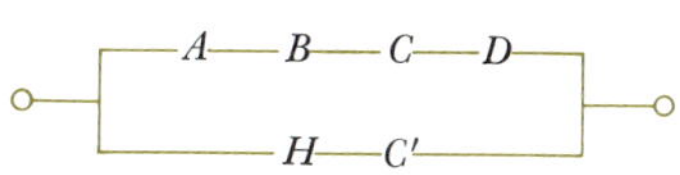

7. 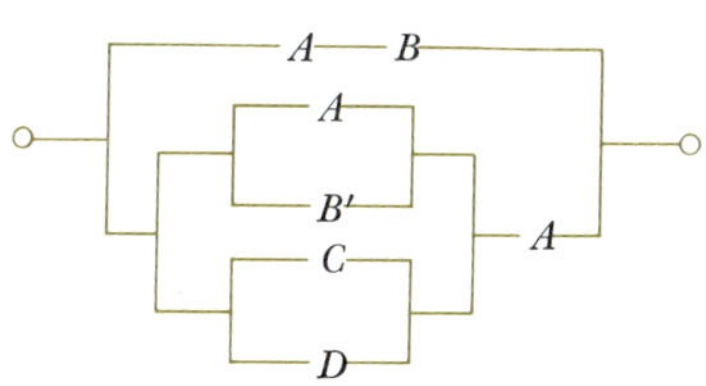

Use Boolean algebra to simplify the following circuits.

8. 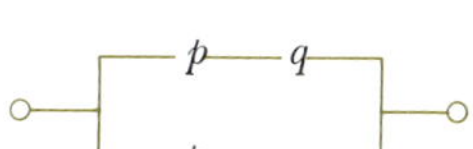

9.

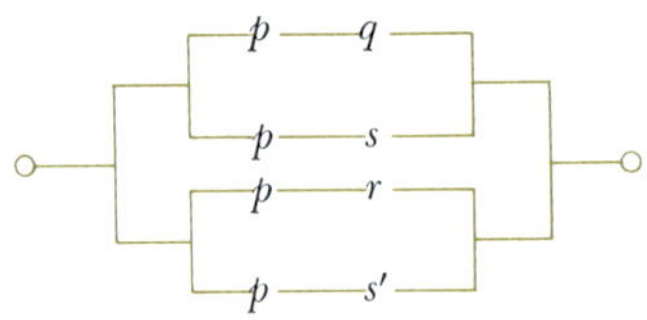

10.

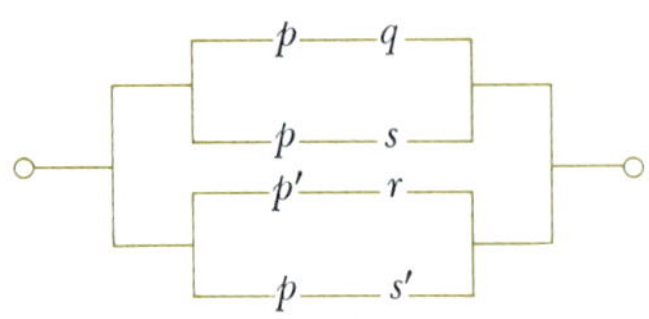

11.

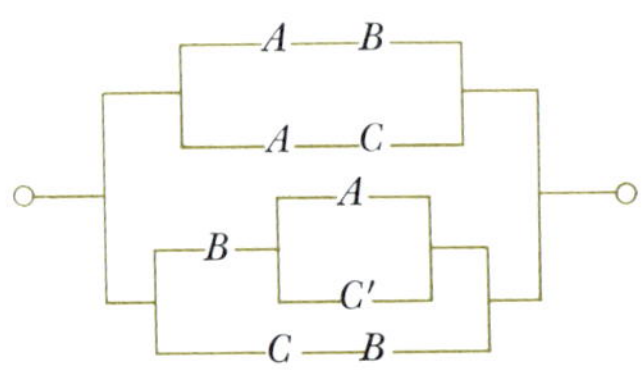

12.

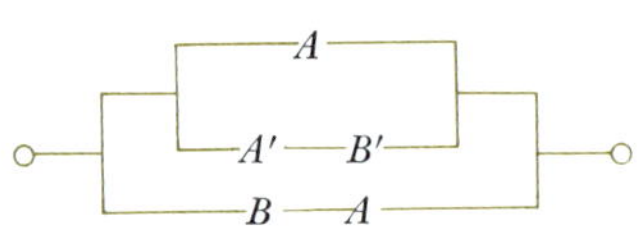

13. Translate $p \rightarrow q$ into an electric circuit.

14. Design a circuit that a three-man committee can use to indicate a simple majority for an issue. The members push a button for a "yes" vote and do not push it for a "no" vote.

15. In a committee of four the chairman has veto power; otherwise a majority of three carries an issue. The issue does not carry in the event of tie votes. Design the simplest possible switching circuit that turns on a light to indicate that an issue has passed (as above, to vote "yes," a member pushes a button; otherwise the vote is "no").

REVIEW EXERCISES

1. Prove $A \star (A' \star B)' = A$.

2. Use Boolean algebra to prove $(A \cap \tilde{B}) \cap [A \cap (\widetilde{B \cap \tilde{C}})] = A \cap \tilde{B}$.

3. Show that, if $A \star B = i_{\bigcirc}$, then $B \multimap A'$.

4. Show that, if $p \wedge q \equiv \mathrm{F}$, then $p \Rightarrow \sim q$.
5. Show that, if $A \cap B = \emptyset$, then $B \subseteq \tilde{A}$.
6. With respect to the concept of a Boolean algebra, what is the difference among Exercises 3, 4, and 5 above?
7. Simplify:

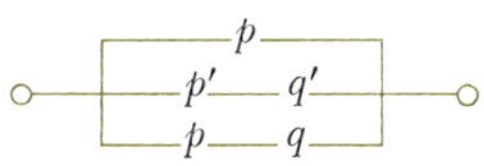

SELECTED REFERENCES

Eves, Howard: "An Introduction to the Foundations and Fundamental Concepts of Mathematics," New York: Holt, Rinehart and Winston, Inc., New York, 1965.
Good discussion and problems related to sets and Boolean algebra.

Lipschutz, Seymour: "Schaum's Outline Series: Set Theory and Related Topics," Schaum Publishing Co., New York, 1964.
Many problems worked and theorems proved in Boolean algebra. Also a discussion of partial-ordering relations.

Young, Frederick, H.: "Digital Computers and Related Mathematics," Ginn and Company, Boston, 1961.
Many circuitry problems solved using Boolean algebra.

"Les Joueurs de Carte," from a sixteenth-century wood engraving.
COURTESY NEW YORK PUBLIC LIBRARY.
According to mathematical legend, the study of probability began in an analysis of gambling odds.

chapter nine

THE LAWS OF CHANCE

The chapters on logic and sets contained two examples of mathematical models. The structures examined were considered limited models of real occurrences; that is, we viewed the algebra of propositions as a symbolic model of logical thinking. In sets we found models for logic and other mathematical systems. In the present chapter we examine another mathematical model. This model is built of numbers; it is more complex than the others we have looked at, but it is also more interesting. We are going to examine games of skill and chance, such as card and dice games, and a numerical model mathematicians and statisticians have constructed for lady luck, or probability.

9-1 PROBABILITY

Almost every important decision we make is based on our feeling that the action we take will produce a favorable result; that is, we base our actions on a probable result. We know that we cannot be sure of all the details of the future, but we usually act on the chance a desirable outcome will take place. You are studying this text because you feel that going to college increases your chances for success. When you cross an intersection with a green light, you feel that the chances are good that traffic coming the other way will stop. In the back of your mind you know that an oncoming car can have a brake failure or the driver may not see the light, but you think it unlikely. When you buy something, such as a new television set, you choose the one you do because you think that the chances are that you are getting the best buy.

In the world of mathematics and science, feeling that the chances are good for a favorable outcome is not enough. When a manned spacecraft is sent into space, we cannot afford to know only that there is a good chance for a safe flight; we must make the decisions involved on a more accurate basis. Or suppose that we are testing a new drug. Suppose that under normal circumstances 8 out of 10 victims of a disease die. During a certain period the new drug is administered to 20 disease victims. Out of the 20, 15 recover, but 4 of them have a bad reaction to the drug. Does the drug do any good? Suppose that there have been cases where a group of 10 persons who contracted the disease all recovered. How does this fact change our interpretation of the experimental results? In these cases the feeling that the drug is effective has to be replaced by more positive evidence. The mathematical models needed for these problems are considered in a course in statistics. Let us examine some of the underlying concepts on which these models are built.

The study of the mathematical nature of chance began in the seventeenth century with two French mathematicians, Blaise Pascal and Pierre de Fermat. We met Pascal before as the inventor of an early computating machine. He and Fermat set about to study a gambling problem posed to Pascal by a French nobleman, the Chevalier de Méré. Ideas outlined in correspondence between Pascal and Fermat concerning this problem evolved into the modern theory of probability. As with many of the ideas we have examined, it was not until the late nineteenth and early twentieth century that these ideas of probability found a fertile ground to grow in. In the present section we examine how we might quantify the feelings we have about chance. In the following sections we shall see how this numerical or quantified notion of probability can be used to study more complex problems. We shall also examine how it may be used to help us make decisions. To begin, we examine three new concepts: mutually exclusive outcomes for an experiment, independent experiments, and equally likely outcomes of an experiment.

By "experiment" we mean the event whose outcomes are examined. First, let us consider what we mean by "mutually exclusive" outcomes. Formally, two or more outcomes of an experiment are mutually exclusive if the occurrence of one precludes that of any of the others; that is, if one of a set of mutually exclusive outcomes of an experiment takes place, we can be sure that none of the others does. Let us consider a few examples. If we toss a coin and it comes up heads, we know that the coin does not come up tails. We assume that the coin does not split in two or come up on edge; only normal outcomes are possible. Getting a head and getting a tail on a single toss of a coin are, then, an example of two mutually exclusive outcomes. As a second example, consider the birthday of the next person you meet. His birthday may be on a Sunday, or on a Monday, or on neither. Having his birthday fall on a Sunday, or on a Monday, or on neither are three mutually exclusive outcomes. For an example of a larger set of mutually exclusive outcomes, consider the set of possible sums when one tosses a pair of dice, that is, when the experiment is tossing a pair of dice. On a single toss of the dice we can have sums of 2, 3, 4, 5, 6, 7, 8, 9, 10, 11, or 12; whichever sum occurs, the others are eliminated. An example of two outcomes that are not mutually exclusive is drawing a card from a standard deck of 52 playing cards and getting an ace and drawing a card and getting a spade. We can draw the ace of spades, a card that is at once an ace and a spade.

The second term is "independent experiments." Again formally, two or more experiments are independent if the outcome of any one of them does not affect that of any of the others. For example, consider the tossing of a pair of dice. Suppose one is red, and the other green. The outcome on the green die does not affect that on the red one. For another example, let us consider what may happen if one draws a single card from a standard deck of 52 playing cards. Suppose that a king is drawn. If it is replaced, the cards mixed, and then a second card drawn, the second draw is an independent experiment with respect to the first draw. If one did not replace the first card, the two draws or experiments would not be independent, for what could take place on the second draw would depend on what was removed on the first.

"Equally likely" outcomes are outcomes of an experiment that we believe to have the same chance of occurring. We believe that the physical structure of the experiment of tossing a fair coin is of such a nature that getting a head and getting a tail are equally likely. The same thing may be said of any one of the six possible outcomes of 1, 2, 3, 4, 5, and 6 of tossing a single fair die. If we toss a pair of dice and consider a sum of 2 or a sum of 7 as outcomes, we do not have equally likely outcomes, for to get a 2 we must have a 1 on each die, but we can get a 7 with a 1 and a 6 or a 2 and a 5 or a 3 and a 4. We cannot say that equally likely outcomes have the same mathematical probability, because we have not yet defined *probability.* Rather, we are using *equally likely* as a primitive concept with which we can define the numerical probability of an outcome of an experiment.

We now have the terms with which to construct a formal definition for the numerical probability of a specific outcome. Let us suppose that we have an experiment with a specific set of possible outcomes. Let us assume that the outcomes in this set are mutually exclusive. Let us also suppose that all the outcomes are equally likely and, further, that we are sure that one of them takes place. The probability of a favorable outcome is defined as the quotient obtained when the number of favorable outcomes is divided by the total number of possible outcomes. Perhaps the definition can be clarified as follows. Let M be the total number of equally likely, mutually exclusive possible outcomes for an experiment. Let N be the number of these which are considered favorable. The probability of a favorable outcome is then defined as N/M.

For example, consider the toss of a single fair die as the experiment. As indicated, the outcomes of 1, 2, 3, 4, 5, and 6 are a set of equally likely, mutually exclusive outcomes. We are also sure that, if the die is tossed, one of these $M = 6$ possible outcomes occurs. If one wishes to calculate the probability of getting a 3 on the toss of the die, he observes that only $N = 1$ of the six outcomes is favorable, because only one side of the die has 3 spots. The probability of a 3 as an outcome is then $N/M = 1/6$.

For a more complex example, let us calculate the probability of drawing a picture card (king, queen, or jack) from a deck of playing cards on one draw. The 52 distinct cards are the $M = 52$ possible outcomes of the experiment. Because there are $N = 12$ possible picture cards (four kings, four queens, and four jacks), the probability of drawing a picture card is $N/M = \frac{12}{52}$, which reduces to $\frac{3}{13}$.

We note that it is not necessary to know the nature of each outcome; rather we need to know only how many outcomes we are dealing with. A still more complex example is the following problem. Let us assume that we live in one of the many states that issue automobile license numbers consisting of three letters followed by three numerals. What is the probability that a driver gets a license plate having his initials as the three letters? This problem requires some assumptions. First we assume that licenses are assigned in a random manner. Then we assume that all possible three-letter arrangements are used. How many arrangements of letters are possible? The first letter can be chosen in 26 ways. For each possible choice of the first letter there are 26 choices for the second. Thus a total of 26 × 26 or 676 choices are possible for the first two letters. For each of these 676 choices there are 26 choices for the third letter. Thus the total number of choices is 676 × 26 or 17,576. Only one of these is the initials of the driver; therefore the probability that a driver has his own initials on a license is $\frac{1}{17,576}$.

In this example, we were able to determine how many cases we had to consider without listing any of them. Using similar procedures for calculating arrangements and numbers of possibilities, we are able to estimate and calculate many complex probabilities.

Although the previous definition of probability can be used to calculate many probabilities, an alternative definition may help in using and interpreting the numerical values themselves. We can define the probability of an outcome as the fraction of the times (or frequency) the experiment is performed that the outcome occurs. Thus the probability of $\frac{1}{6}$ for getting a 3 when we toss a die means that $\frac{1}{6}$ of the times we toss the die we get 3. In the second example we can say that $\frac{3}{13}$ of the times we draw a card from a deck of 52 we get a picture card. In the third example we can say that $\frac{1}{17,576}$ of the drivers get their own initials on their license plates. This should not be interpreted as meaning that, if we tossed the die five times and a 3 had not come up, the next toss would be a 3 or that, if we drew and replaced 10 cards from the deck and none of them was a picture card, the next 3 would be picture cards. Rather, we mean that on the average or for a large number of trials these fractional occurrences are observed. If we repeat the card drawing 13,000 times, we expect to get about 3000 picture cards. However, we should be surprised to get exactly 3000. If we throw the die 6000 times or throw one group of 6000 dice once, we expect to get about 1000 threes.

The "frequency" definition is also useful when we are faced with a situation where the formal definition cannot be applied. For example, we may ask the probability that a man who is 40 years old will live to be 65. Clearly there is no way to list, or even estimate, the number of mutually exclusive, equally likely outcomes. However, we can examine records and determine what percentage of the men who were 40 twenty-five years ago are alive today. This would tell us what fraction of the time men lived twenty-five years longer and give us an estimate of the probability of a similar occurrence today. If, out of a group of 1000 men alive twenty-five years ago, 650 were still alive, we should say that 65 percent of the time a man who is 40 will live to be 65. Of course, this is not really accurate. The practice of medicine keeps on improving, and we have better life-saving techniques, keep better records, and so on. For these and similar reasons we have, at best, only a good estimate of the probability. Still this numerical estimate is more useful than saying only that the chances of living to age 65 are good.

9-1 EXERCISES

1. Give some advantages of having numerical probabilities over having only a feeling for an outcome.

2. Consider the experiment of drawing a card from a deck of 52 playing cards. For each of the sets of outcomes listed below determine if all the members of the set are (*a*) mutually exclusive or (*b*) equally likely.

(*a*) {An ace, a king, a four of spades}
(*b*) {An ace, a three}
(*c*) {A red card, a picture card}
(*d*) {A red picture card, a heart that is a 9, 10, jack, queen, king, or ace}
(*e*) {A red card, a black card}
(*f*) {The king of spades, the ace of clubs, the seven of diamonds, the queen of hearts}

3. Give some examples of outcomes of drawing cards from a deck that are:

(*a*) Equally likely
(*b*) Mutually exclusive
(*c*) Equally likely and mutually exclusive
(*d*) Equally likely and not mutually exclusive
(*e*) Not equally likely but mutually exclusive
(*f*) Neither equally likely nor mutually exclusive

4. Which of the pairs of experiments listed below are independent?

(*a*) Drawing a card and then replacing it and drawing a second
(*b*) Tossing a coin and then tossing it again
(*c*) Two rolls of a pair of dice

(*d*) One student's getting an A in math and another's getting a B in history
(*e*) An election for President of the United States and an election for governor of a state
(*f*) The sex of a baby and the sex of a second child born to the same parents 4 years later
(*g*) Your birthday and the birthday of the next person you meet

5. Give some examples of:

(*a*) Independent experiments
(*b*) Experiments that are not independent

6. If we consider a pair of dice as being made up of one red and one green die, we can list 36 distinct outcomes when we toss them:

red	*green*	*red*	*green*	*red*	*green*	*red*	*green*	*red*	*green*	*red*	*green*
1	1	2	1	3	1	4	1	5	1	6	1
1	2	2	2	3	2	4	2	5	2	6	2
1	3	2	3	3	3	4	3	5	3	6	3
1	4	2	4	3	4	4	4	5	4	6	4
1	5	2	5	3	5	4	5	5	5	6	5
1	6	2	6	3	6	4	6	5	6	6	6

(*a*) Are these 36 outcomes mutually exclusive?
(*b*) Are they equally likely?
(*c*) Must one of them take place under normal conditions?

7. Using the results of Exercise 6, calculate the probabilities of all possible sums for a pair of fair dice on one roll. For example, out of the 36 outcomes only one, red—1 and green—1, results in a sum of 2; therefore, the probability of a 2 is $\frac{1}{36}$. Calculate the probabilities of the other 10 outcomes.

8. What is the probability that the next person you meet was born:

(*a*) On your birthday?
(*b*) In January?
(*c*) On a Tuesday?
(*d*) Not on your birthday?
(*e*) In the same year as you?

9. Give some examples of outcomes of experiments that have to be esti-

mated by using the fraction-of-the-time concept rather than the formal definition given.

10. When we calculated the probability of someone's getting license plates with his initials, we made two assumptions. Why were they required?

11. What assumptions are required if we say that the probability we found in the text for a man's living from 40 to 65 years of age was the exact numerical probability of such an outcome?

9-2 CALCULATION OF PROBABILITIES

In the present section we examine how probabilities combine. To do this, the concepts and procedures developed in the study of the algebra of sets are needed. We can use the same approach as that used in finding the truth values of compound statements. There it was found that the truth value of a compound depended on the truth value of the simple statements involved and on the manner in which the statements were linked. A similar result occurs in probabilities; that is, the probability of a complex outcome depends on the probabilities of the simple outcomes involved and the manner in which the simple experiments are connected to form the final complex experiment. Also a more compact notation, or set of symbols, to denote the probabilities is needed.

Let $P(A)$ stand for the probability of outcome A. For example, P(head on a toss of a coin) means the probability of getting a head as the outcome of the experiment of tossing a coin once. When it is clear what the experiment is, we may shorten this to P(heads) or $P(h)$. Because, in many cases, reference to specific outcomes brings only one experiment to mind, this notation presents few problems. P(sum 3) or $P(3)$ refers, then, to getting a sum of three spots if we are discussing the tossing of a pair of dice, and $P(6)$ to getting the sum of six spots. If we discuss living to age 65, we write P(age 65).

For the first relationship consider the following formula:

$$P(\text{not } A) = 1 - P(A)$$

The formula says that the probability of outcome A not taking place is 1 minus the probability that it does. This makes sense when we recall the fraction-of-the-time definition for probability. The fraction of the time or percentage of the time a certain outcome takes place and the fraction of the time it does not take place add up to the "whole" time or 1. If $\frac{1}{6}$ of

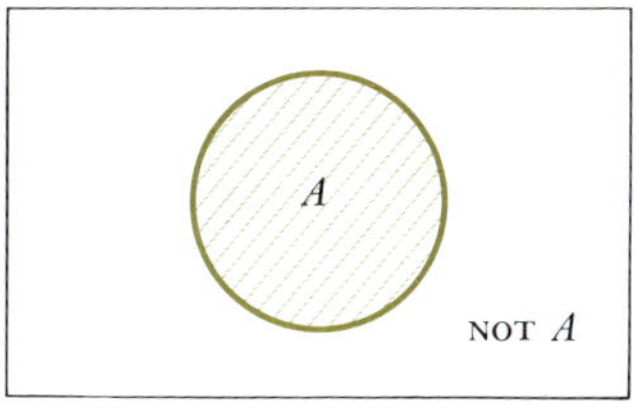

FIG. 9-1 PROBABILITY APPLICATION OF A VENN DIAGRAM.

the time we throw a die we get a 3, then $\frac{5}{6} = 1 - \frac{1}{6}$ of the time we do not. If 65 percent of the time a man who is 40 years old lives to be 65, then 35 percent of the time he does not. A variation of the Venn diagrams of previous chapters can be used to illustrate this point. Suppose that in Fig. 9-1 the whole diagram is used to represent the set of all possible outcomes. A portion of the diagram is shaded to represent the fraction of the time outcome A takes place and the unshaded portion then represents the fraction of the time outcome A does not take place.

We can use Venn diagrams in working out a formula to determine the probability of a sequence of outcomes of events when we know the probabilities of the individual outcomes making up the sequence. For example, the probability of drawing a king from a deck of playing cards is $\frac{4}{52}$ or $\frac{1}{13}$, and that of drawing a second king, assuming that we replace the first card we draw, is also $\frac{4}{52}$ or $\frac{1}{13}$. What is the probability of drawing two kings in a row? In effect, we have two distinct experiments—the drawing of the first card and the drawing of the second. We then have a compound experiment made up of two simple experiments, and we are trying to calculate the probability of a compound outcome made up of two simpler outcomes.

It may be helpful if we know the answer to this problem before attempting to develop a method of calculating it from this point of view. Although this experiment is more complex than those we are accustomed to, we can still use the basic definition to calculate an answer. First we need a set of outcomes. The first card is any of the 52, and for any selection of a first card there are 52 choices for the second. Therefore a set of outcomes includes 52×52 or 2704 distinct outcomes. How many of these consist of two kings? There are four ways to draw a king, so that with replacement there are $4 \times 4 = 16$ ways of drawing two kings in a row. Then the probability of drawing two kings is $N/M = \frac{16}{2704}$. This fraction reduces to $\frac{1}{169}$. Whatever formula we develop to combine probabilities should yield the same answer.

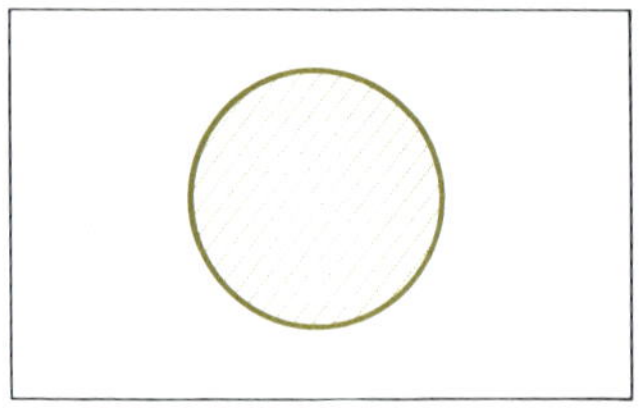

FIG. 9-2 VENN DIAGRAM WITH ONE-THIRTEENTH OF THE AREA SHADED.

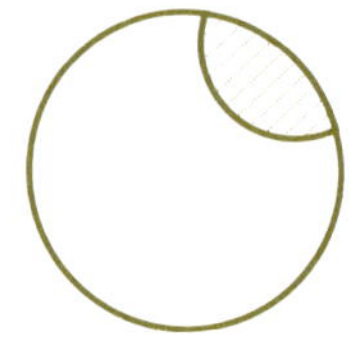

FIG. 9-3 ENLARGED VIEW OF PART OF FIG. 9-2.

A probability diagram for the first draw might look like Fig. 9-2, where the shaded portion represents $\frac{1}{13}$ of the total area, and the whole diagram the 2704 possible outcomes. This is because $\frac{1}{13}$ of the time we draw two cards, the first is a king. Now consider Fig. 9-3. Suppose that the circle is a magnified view of the shaded portion of Fig. 9-2. Then the shaded portion in Fig. 9-3 represents the fraction of the time we draw the second king. The draw of the two cards must be considered independent experiments, because there is no evidence the cards have "memories" they use to recall how many times they have been drawn. What fraction of the time is the second card a king? Clearly $\frac{1}{13}$ of the time. Then the shaded portion of Fig. 9-3 represents $\frac{1}{13}$ of the total area of Fig. 9-3. And because Fig. 9-3 is $\frac{1}{13}$ of the total area of Fig. 9-2, the shaded area of Fig. 9-3 is $\frac{1}{13}$ of $\frac{1}{13}$ of the total area of Fig. 9-2. $\frac{1}{13}$ of $\frac{1}{13}$ of a total is $\frac{1}{169}$ of the total. This result agrees with the probability we calculated from more basic concepts. Can this method be generalized?

Let us suppose that we have to independent experiments, 1 and 2. And suppose that we want to calculate the probability of outcome A of experiment 1 and outcome B of experiment 2, or $P(A$ of 1 and B of 2) or, in an even shorter form, $P(A$ and $B)$. We can follow the line of reasoning of the example. $P(A)$ represents the fraction of the time A occurs for experiment 1, and $P(B)$ the fraction of the time outcome B of experiment 2 occurs. What fraction of the time do both outcomes occur? The answer is $P(A) \cdot P(B)$. Examine Fig. 9-4 to confirm this. Suppose that the horizontally shaded area is the portion of the total outcome of the two experiments in which A occurs. The vertically shaded area is the portion of the total outcome in which B occurs. Because the experiments are independent, the area shaded both ways is $P(A)$ of the vertically shaded area and $P(B)$ of the horizontally shaded area. Viewed from either direction, it is either

$P(A)$ of $P(B)$ of the total or $P(B)$ of $P(A)$ of the total. Remember that $P(A)$ and $P(B)$ represent the actual numerical probabilities, and either represents $(A) \cdot P(B)$ of the total outcomes. In summary, we conclude that, if 1 and 2 are a pair of independent experiments, then

FORMULA 1 $P(A \text{ of } 1 \text{ and } B \text{ of } 2) = P(A \text{ and } B) = P(A) \cdot P(B)$

where $P(A)$ is the probability of outcome A of event 1 and $P(B)$ is the probability of outcome B of event 2.

What happens when the two experiments are not independent of each other? For example, how do we treat a problem of drawing two cards where we do not replace the first one? To solve this problem, it is necessary to introduce a new kind of probability, a "conditional" probability. The notation is $P(B/A)$ or $P(B \text{ of } 2/A \text{ of } 1)$. It is read "the probability of outcome B of experiment 2 given that outcome A of experiment 1 has taken place." Putting it another way, $P(B/A)$ is the fraction of the time outcome B of experiment 2 takes place, assuming that experiment 1 had outcome A. Again following the line of reasoning established in the example, we modify formula 1 to read

FORMULA 2 $P(A \text{ and } B) = P(A) \cdot P\left(\frac{B}{A}\right)$

As an example, suppose that we want to calculate the probability of drawing two kings if we do not replace the first card drawn. It should be $P(K_2/K_1) = \frac{3}{51}$; that is, the probability of drawing a king on the second draw, assuming that we drew a king on the first draw, is $\frac{3}{51}$. This follows by noting that out of the 51 cards remaining after a king has been removed, there are 3 kings. Then $\frac{3}{51}$ of the time we draw a second king after drawing a king on the first draw. The probability of drawing two kings without

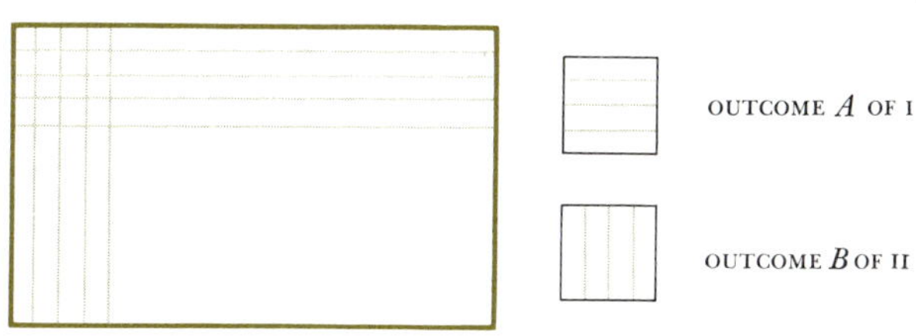

FIG. 9-4 VENN DIAGRAM OF A COMPOUND PROBABILITY OF OUTCOMES A AND B.

replacing the first draw is

$$P(K_1 \text{ and } K_2) = P(K) \cdot P\left(\frac{K_2}{K_1}\right) = \tfrac{4}{52} \times \tfrac{3}{51} = \tfrac{1}{13} \times \tfrac{1}{17} = \tfrac{1}{221}$$

The only formula we need to remember is formula 2, because if the experiments are independent, $P(B/A)$ and $P(B)$ are the same; that is, the probability of outcome B is not affected by the outcome of the first experiment.

A few more examples are necessary. Earlier we looked at a problem involving a 40-year-old man's chances of living to be 65. Although our answer has no basis in fact, that is, the figure $P(40 \text{ to } 65) = 0.65$ was simply a made-up value, we can use it. Let us assume that the probability of a 40-year-old woman's living to be 65 is 0.75; that is, $P(40 \text{ women to } 65) = 0.75$. Assuming further that the life span of a husband and wife are completely independent, what is the probability that a 40-year-old couple who have been married 20 years will live to celebrate 45 years of marriage? In reality we have two experiments. The man's living to be 65 is one, and the woman's living to be 65 another. We assumed them to be independent, and we make our calculations on this basis. In formula form we have

$$\begin{aligned} P(\text{both } 65) &= P(\text{man to } 65 \text{ and woman to } 65) \\ &= P(\text{man to } 65) \cdot P(\text{woman to } 65) = 0.65 \times 0.75 = 0.4875 \end{aligned}$$

or just a little less than $\frac{1}{2}$.

For a second example, what is the probability of the next two persons you meet being born on the same day of the week as you were? The experiments involved can again be assumed to be independent. The probability that the first person you meet was born on the same day is $\frac{1}{7}$. The probability that the second person was also born on this day is also $\frac{1}{7}$. Therefore

$$\begin{aligned} P(\text{first same day and second same day}) &= P(\text{first}) \cdot P(\text{second}) \\ &= \tfrac{1}{7} \times \tfrac{1}{7} = \tfrac{1}{49} \end{aligned}$$

For a third example, let us consider the following, using a conditional probability. Suppose that from state motor vehicle records the following probabilities are estimated. A driver who has not had an accident in the last year has a probability of 0.05 of having one within the next year. Further, assume that, if a driver has had an accident in the last year, his probability of having one within the next year is 0.08. If you are a driver who has not had an accident in more than a year, what is the probability of your having one this year and another one the following year? We wish to find:

P(accident this year and accident next year)

From the assumed data

P(accident this year) = 0.05

$$P\left(\frac{\text{accident following year}}{\text{accident this year}}\right) = 0.08$$

Therefore

$$P(\text{two accidents}) = P(\text{accident this year}) \cdot P\left(\frac{\text{accident following year}}{\text{accident this year}}\right)$$
$$= 0.05 \times 0.08 = 0.04$$

Next, let us try to develop a formula to determine the probability of either of two outcomes of a single experiment if we know the probability of the outcomes taken separately. To begin, assume that the outcomes are mutually exclusive. For example, suppose that we are going to toss a pair of dice. What is the probability that we get a 7 or an 11 as the sum. We can calculate $P(\text{sum } 7) = \frac{6}{36}$ and $P(\text{sum } 11) = \frac{2}{36}$ (see Exercise 6, Sec. 9-1 or 9-3). What is P(sum 7 or 11)? Because $\frac{6}{36}$ of the time we get a sum of 7 and $\frac{2}{36}$ of the time a sum of 11 and the outcomes are mutually exclusive, we reason that $P(7 \text{ or } 11) = \frac{6}{36} + \frac{2}{36} = \frac{8}{36}$ of the time.

In a more general analysis, suppose that we have two mutually exclusive outcomes, A and B of an experiment. Assume that we know $P(A)$ and $P(B)$. We might have a "probability diagram" such as Fig. 9-5. Here again, the whole rectangle represents the total set of possible outcomes. The shaded area labeled A represents the fraction of the whole for which A is the outcome; the shaded area labeled B, the fraction for B. What fraction of the time is A or B the outcome? The answer, a reasonable one is $P(A) + P(B)$. In effect, we are saying that the total area within the rectangle is assigned a value of one unit. $P(A)$ is some fraction of this unit corresponding to the fraction of the time A occurs. $P(B)$ is the corre-

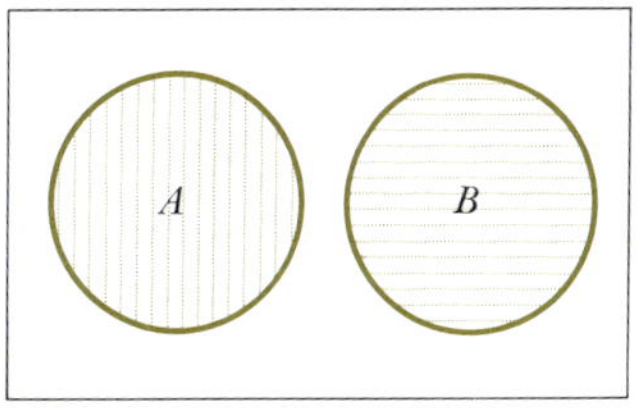

FIG. 9-5 VENN DIAGRAM OF THE PROBABILITY OF A OR B WHERE A AND B ARE MUTUALLY EXCLUSIVE.

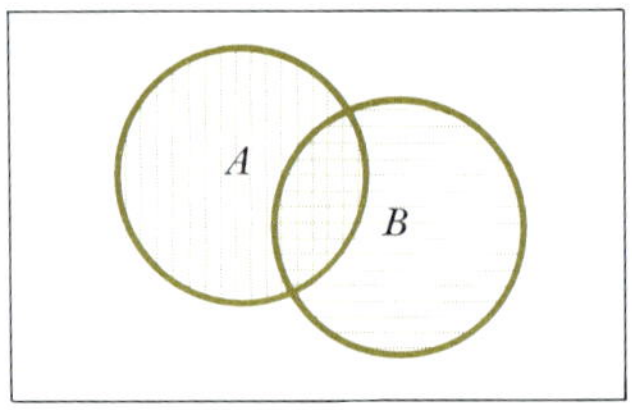

FIG. 9-6 VENN DIAGRAM OF THE GENERAL CASE OF A OR B.

sponding fraction for B. The sum of the two fractions correspond to the fraction of the time one or the other takes place. The basic formula is then

FORMULA 3 $P(A \text{ or } B) = P(A) + P(B)$

where A and B are mutually exclusive outcomes of the same event.

Next, it is natural to ask what happens to $P(A \text{ or } B)$ if A and B are not mutually exclusive outcomes. In this case, the probability diagram looks like Fig. 9-6. There is an overlap between the areas corresponding to the two outcomes. How do we interpret this? If we are talking about an outcome in the overlap, we are considering an outcome that is both an A and a B outcome. The overlap corresponds to $P(A \text{ and } B)$, because any outcome in the overlap has both A and B occurring. For example, if the experiment is drawing a card from a deck of playing cards and A is drawing a picture card and B is drawing a spade, the area in the overlap corresponds to drawing the king, queen, or jack of spades.

The $P(A \text{ or } B)$ is the fraction of the whole area the shaded area represents. If we said that this fraction is $P(A) + P(B)$, we would be making an error, because the overlapping area would be counted twice. We can correct this by subtracting the overlap from the total. Therefore, we get the following formula:

FORMULA 4 $P(A \text{ or } B) = P(A) + P(B) - P(A \text{ and } B)$

For the problem involving drawing a picture card or a spade from a deck of playing cards, we have the following:

$$P(\text{picture card}) = \tfrac{12}{52}$$
$$P(\text{spade}) = \tfrac{13}{52}$$
$$P(\text{picture and spade}) = \tfrac{3}{52}$$

Therefore

$$\begin{aligned} P(P \text{ or } S) &= P(P) + P(S) - P(P \text{ and } S) \\ &= \tfrac{12}{52} + \tfrac{12}{52} - \tfrac{3}{52} \\ &= \tfrac{22}{52} = \tfrac{11}{26} \end{aligned}$$

If A and B were mutually exclusive outcomes, formula 4 would still yield the correct answer, because $P(A$ and $B)$ would be 0. In effect, this is saying that from the basic definition of mutually exclusive outcomes the probability that two of them can both take place at the same time is zero. Because of this only formula 4 is really necessary, because it can be used in all cases, while formula 3 requires that the outcomes be mutually exclusive.

9-2 EXERCISES

1. Let W stand for Johnny winning a contest and T for Bill being taller than the next person he meets. Write the following in words.

   (*a*) $P(W \text{ and } T)$

   (*b*) $P(W \text{ or } T)$

   (*c*) $P\left(\frac{T}{W}\right)$

   (*d*) $P\left(\frac{W}{T}\right)$

   (*e*) $P(W \text{ and not } T)$

   (*f*) $P(\text{not } W)$

2. If, by use of a computer, careful study, and blind luck we discover that P (John making a million dollars) $= 0.005$ and P(George making a million dollars) $= 0.075$ and assuming these two are independent, calculate the following.

   (*a*) P(both make a million)
   (*b*) P(John does not make a million)
   (*c*) P(George does not make a million)
   (*d*) P(neither makes a million)

3. If two cards are drawn from a deck of 52 cards, and the first card is replaced before the second is drawn, calculate the following.

(*a*) *P*(both are queens)
(*b*) *P*(the first card is a heart or a king)
(*c*) *P*(neither card is a red card)
(*d*) *P*(the first card is red and the second is black)
(*e*) *P*(both cards are the same color)
(*f*) *P*(the first card is a picture card or is black)

4. Repeat Exercise 3, assuming that the first card drawn is not replaced after it is drawn.
5. If three fair coins are tossed on a table, calculate the probabilities of the following.

(*a*) *P*(no heads)
(*b*) *P*(only one head)
(*c*) *P*(exactly two heads)
(*d*) *P*(all three heads)
(*Hint:* See Exercise 6 below.)

6. What is the difference between tossing three coins at once and tossing one coin three times?

9-3 LOGICALLY POSSIBLE OUTCOMES

In order to use the basic definition of probability given in Sec. 9-1, it was necessary to determine a set of possible outcomes of an experiment. If we were unable to determine the exact nature of the set, we were able, at least, to calculate the number of members it had. To be useful in the calculation of probabilities, this set of possible outcomes had to be made up of mutually exclusive results, and every outcome had to fit the set. For almost all the problems we examined, such a set of outcomes was easy to find; in fact, this is why these examples were chosen. For tossing a pair of dice, we found 36 outcomes, as listed below. For clarity, we assumed that one of the pair was red, and the other green:

red	*green*	*red*	*green*	*red*	*green*	*red*	*green*	*red*	*green*	*red*	*green*
1	1	2	1	3	1	4	1	5	1	6	1
1	2	2	2	3	2	4	2	5	2	6	2
1	3	2	3	3	3	4	3	5	3	6	3
1	4	2	4	3	4	4	4	5	4	6	4
1	5	2	5	3	5	4	5	5	5	6	5
1	6	2	6	3	6	4	6	5	6	6	6

In this table each entry represents an outcome, with the face showing on each die given by the appropriate numeral in the pair. For example, the third entry in the fourth column is 4 and 3, indicating a four showing on the red die and a three on the green die. It seems reasonable to state that this set of outcomes is complete, because 1 out of the 36 occurs if we toss the dice. It also is reasonable that these outcomes are mutually exclusive and equally likely, so that this set could be, and was, used to calculate probabilities for various outcomes of tossing a pair of dice. We also examined another set of possible outcomes of tossing a pair of dice. This set might be denoted as 2, 3, 4, 5, 6, 7, 8, 9, 10, 11, and 12, where each outcome was represented by the sum of spots on the dice. Although this set was complete and mutually exclusive, the outcomes were not equally likely, so that the set could not be used to calculate probabilities directly. In Chap. 10 we shall find an interesting use for this set when we examine a total experiment. If we wished to examine what occurs if we toss a pair of dice, this set of outcomes, together with the probabilities assigned to each individual member of the set, gives us a clear picture of what is taking place.

We see that knowing something about a set of logically possible outcomes has at least two distinct and important uses. The first is the direct calculation of probabilities themselves, and the second the examination of the whole experiment. It is useful, then, to develop methods for finding such sets of outcomes.

We may divide the basic problem into two simpler ones. How can we determine a set of logically possible outcomes in those cases where a list of the outcomes is of a reasonable length? And what, if anything, can we discover about an experiment where a list of the outcomes is of so great a length as to be impossible to handle in any useful manner? In a previous section, to estimate the number of arrangements of letters on a license plate, we reasoned that the total selection was made up of three individual selections. Each of these had 26 possibilities, and we reasoned that this resulted in $26 \times 26 \times 26$ possible outcomes for the whole. A similar line of reasoning can be adopted for the following example. Suppose that we are asked to draw three chips from a bowl containing a red chip, a black chip, and a green chip. If we replace each chip after noting its color, we may ask what our chances are of drawing the same chip three times. To do this, we have to determine a set of logical possibilities. If we consider the first chip only, we note three outcomes. We could visualize this in the following way:

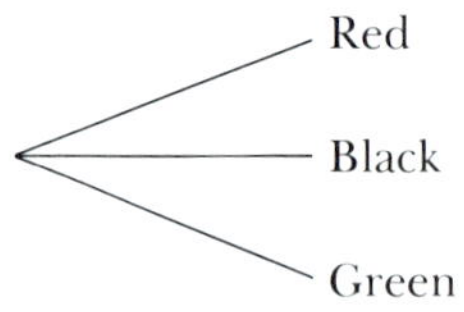

For each of these selections any of three outcomes is possible. This could be added to the diagram as follows:

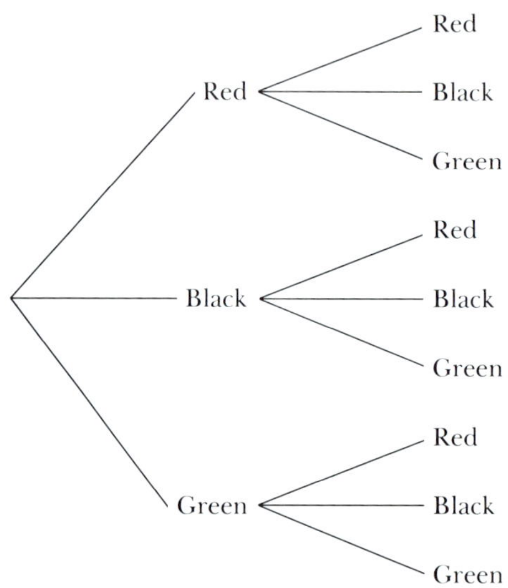

Then, for the third draw, we would obtain the diagram shown on the next page.

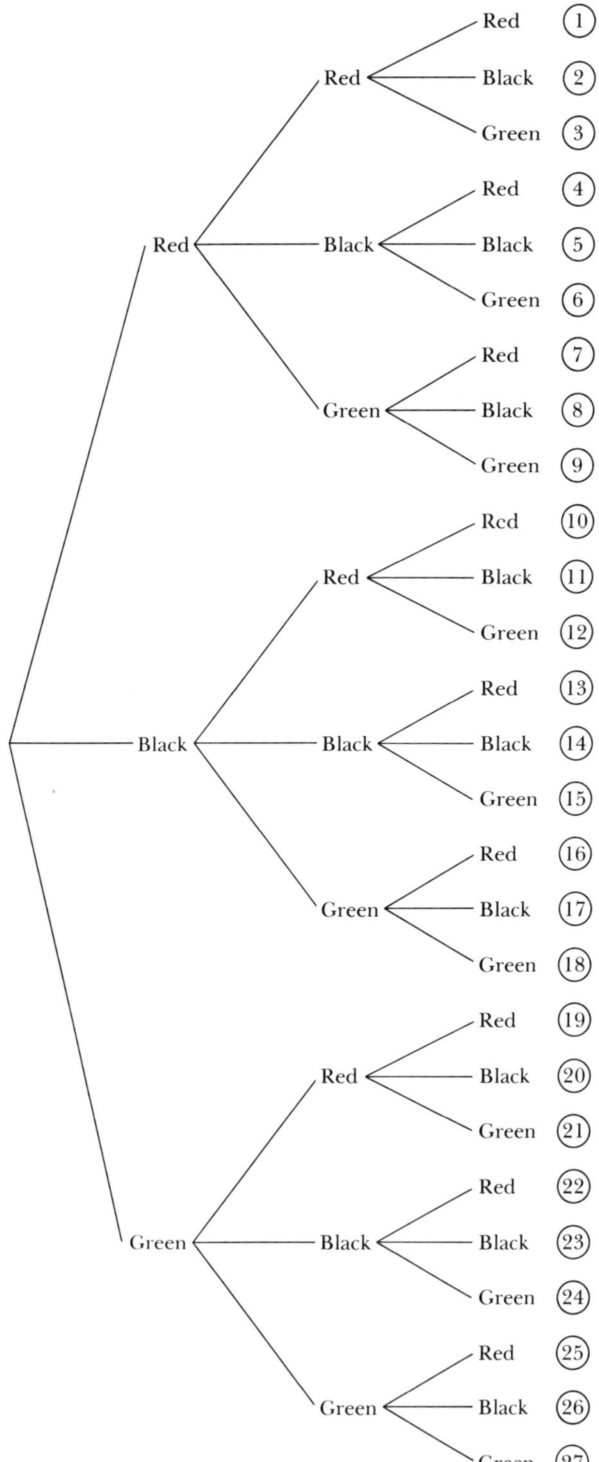

Red
Red
Red
Black
Green
Black
Red
Black
Green
Green
Red
Black
Green
Black
Red
Red
Black
Green
Black
Red
Black
Green
Green
Red
Black
Green
Green
Red
Red
Black
Green
Black
Red
Black
Green
Green
Red
Black
Green
1
2
3
4
5
6
7
8
9
10
11
12
13
14
15
16
17
18
19
20
21
22
23
24
25
26
27

This diagram, which is known as a "tree" diagram, because of its obvious resemblance to tree branches, indicates that there are 27 distinct outcomes. It also shows that outcomes 1, 14, and 27 involve three similar draws. If we feel that the 27 distinct outcomes meet the basic requirements, we calculate the probability of drawing the same chip all three times as $\frac{3}{27}$ or $\frac{1}{9}$. The tree-diagram method could be used to calculate the number of possible outcomes if we draw one chip at a time and do not replace it:

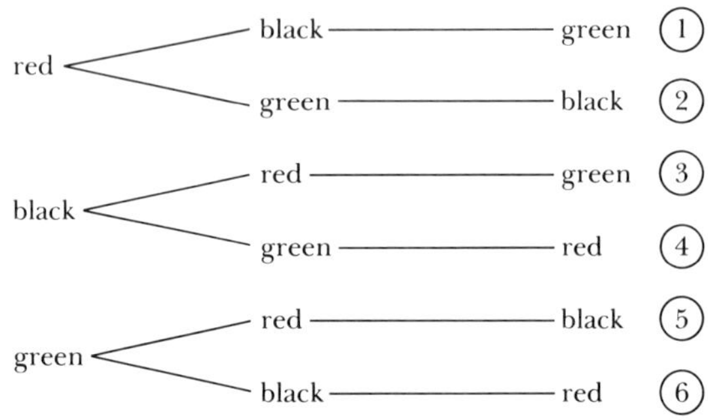

In this case there are six different outcomes in the set.

As a variation of this method, consider the following problem. A man throws a die and then tosses a coin. What are the logically possible outcomes? The following tree diagram shows them:

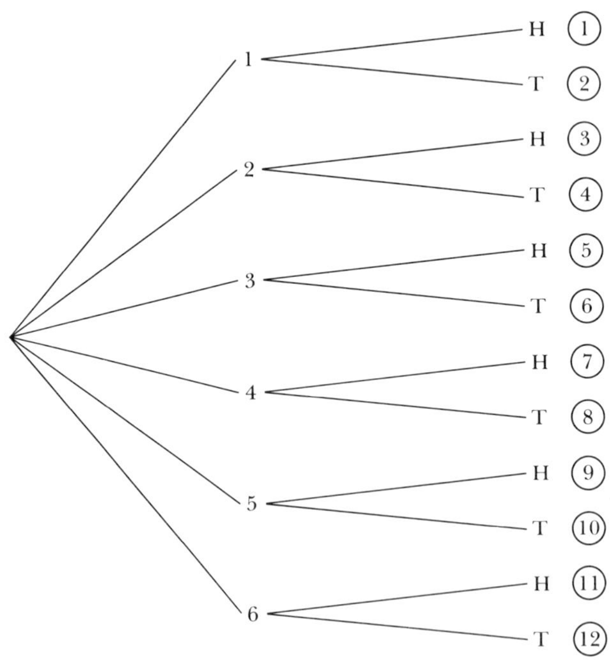

There are 12 outcomes. For these examples the following general rule could be deduced for determining how many logical possibilities there are for a compound experiment made up of a number of simple experiments where the possibilities of each are known. If a compound experiment is made up of a sequence of simple experiments $E_1, E_2, E_3, \ldots, E_n$ (here E_1 is the first experiment, and the . . . indicates some missing E's and n in E_n some unspecified number of experiments) and if each of the simple experiments has a specific number of possible outcomes, say, $O_1, O_2, \ldots, O_n$ (O_1 is the number of logically possible outcomes of E_1, and so on), then the number of logical possibilities for the compound experiment is $O_1 \cdot O_2 \cdot O_3 \cdots O_n$.

For example, the first chip problem was made up of three experiments, with three outcomes each, so that the total number of possibilities was $3 \times 3 \times 3$ or 27. The second chip problem was also made up of three experiments, but although the first had three possible outcomes, the second had only two, and the third had only one. Applying the method, we correctly get $3 \times 2 \times 1 = 6$ possible outcomes. The coin-die problem consisted of two experiments, tossing the die and tossing the coin. The first had six outcomes, and the second had two; thus the compound experiment had $6 \times 2 = 12$ outcomes.

Actually only a few problems can be analyzed by using the methods shown here. The most important concept that should be examined is the definition of a set of logically possible outcomes. Any single experiment can have any number of sets of logically possible outcomes. The only requirement is that each of the members of any set of outcomes is mutually exclusive from all the other members of the same set and that one of these members occurs in any trial of the basic experiment. As a final example, consider the determination of the logically possible truth-value patterns of a compound statement involving three simple statements p, q, and r; that is, how many entries would a truth table for such a compound require? Consider the following tree diagram:

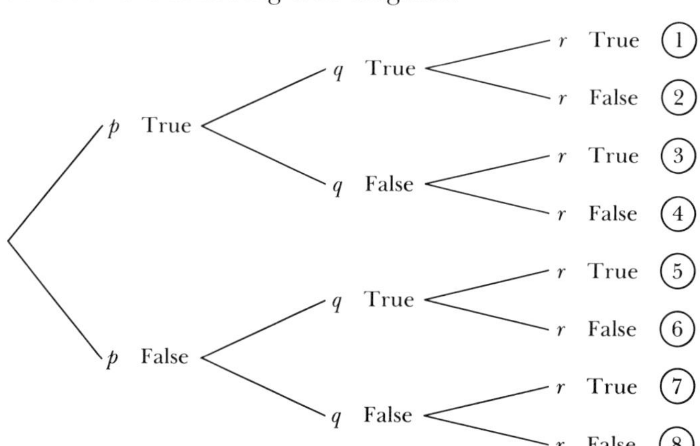

We summarize them in a table:

Case	*p*	*q*	*r*
1	T	T	T
2	T	T	F
3	T	F	T
4	T	F	F
5	F	T	T
6	F	T	F
7	F	F	T
8	F	F	F

This result agrees with the less formal analysis in Chap. 5.

9-3 EXERCISES

1. Using a tree diagram, determine a set of logically possible outcomes for tossing three coins (see Sec. 10-1 for answer).

2. An alternate set of outcomes for tossing three coins may be indicated as follows:

{No heads, one head, two heads, three heads}

In what ways does this set of outcomes differ from that found in Exercise 1?

3. How many entries does a truth table for a compound statement built of six simple statements have?

4. Consider the experiment of tossing a pair of dice, one red and one green. Also consider the following set of outcomes:

{Both dice with an even number; the red die with an even number, and the green die with an odd number; both dice with an odd number}

In what ways does this set of outcomes fail to meet the requirements for a set of logically possible outcomes of the experiment?

5. A tract-home builder designs homes by linking garages to central units and adding bedroom wings. If the builder has 4 different garage designs, 5 different central units, and 7 different bedroom wings, how many different homes can he build?

9-4 PERMUTATIONS AND COMBINATIONS

To this point in the discussion we have been able to calculate only a few simple probabilities, based on a limited number of methods of calculating the number of outcomes of an experiment. It is beyond the scope of the present work to go far into these methods; however, we can, at least, extend our ability to calculate outcomes by adding a few more tools for calculation. Two basic concepts, which are often useful in the calculation of probabilities, are permutations and the closely related concept of combinations.

A permutation of a set of objects is a selection in a specific order of a number of objects in the set; that is, if we have a set of objects and we choose a number of them from the set and indicate that the order of selection is to be considered as well as the objects themselves, we say that we are dealing with a permutation of the set. The example of license plates was a permutation of objects—the set of letters of the alphabet. The selection of AZT as a license plate was considered different from that of ZAT or TZA or ATZ, although only the same three letters A, Z, and T were used. Each of these selections is called a permutation of three letters taken from the set of 26 English letters. Although AZT, ZAT, TZA, and ATZ are different permutations of letters, they are the same combination. This, then, is the difference between a combination and a permutation. In a combination the only interest is the objects; in a permutation the order of the objects is also considered. If one begins with a set of objects and makes a combination of them, the combination can be rearranged into a number of permutations.

Suppose that one wishes to calculate the number of permutations of three objects from a set of seven. If each object can be used only once, then applying the line of reasoning used for determining sets of logical possibilities, there are $7 \times 6 \times 5 = 210$ permutations. The first object can be chosen in seven ways, and the second in six ways for each of the seven choices of the first object. And the first two objects, then, can be chosen in $7 \times 6 = 42$ ways. For each of these 42 ways there are five choices for the third object, making a total of $42 \times 5 = 210$ possibilities or permutations. If, on the other hand, any of the objects can be used as many times as one wishes, there are $7 \times 7 \times 7 = 343$ permutations, because there are seven choices for the third object. Three of the objects may be used once, four may be used twice, and so on.

Because it seems unreasonable to try to develop a general formula that would work in any of these cases, we limit ourselves while attempting to generalize a method of calculation of permutations. Suppose that we attempt only to develop a result if each object is to be used only once. How can we use the patterns we have seen to build a usable formula? First, a general statement of the problem we are facing should be made. Suppose that we have a set of n objects and we wish to make a permutation of x of these n objects, using each object only once. How many permutations are there? The answer is given by the formula

FORMULA 1 $_nP_x = n(n-1)(n-2) \cdot \cdot \cdot (n-x+1)$

where $_nP_x$ is the number of permutations of n objects taken x at a time. Where does this formula come from? We can reason as in the examples. The first object can be any of the n objects in the set. Therefore, it can be chosen in any of n ways. For each of these choices there is one less choice for the second object, or a total of $n-1$ possibilities. The total number of possibilities for the first two objects, then, is $n(n-1)$. There are $n-2$ choices for the third object, this being one less than $n-1$. The total number of possibilities for three objects, then, is $n(n-1)(n-2)$. We also see that the final answer is the product of x numbers, each one smaller than that which precedes it. If we start at n and count backward x numbers, at what number do we arrive? A little experimentation leads to $(n-x+1)$. Although the formula seems reasonable, we should try to use it to check the sample problems. The problem asked to find $_7P_3$, or the number of permutations of seven objects taken three at a time. The answer should have three factors, the first one or n should be seven, and the last one 5 ($x = 3$ or $n - x + 1 = 7 - 3 + 1 = 4 + 1 = 5$). Therefore

$$_7P_3 = 7 \times 6 \times 5 = 210$$

This agrees with the previous answer. We can try a few more examples.

How many permutations of five letters can one make from our alphabet using a letter only once? In terms of permutations we are asking for the value of $_{26}P_5$.

$$_{26}P_5 = 26 \times 25 \times 24 \times 23 \times 22 = 7{,}893{,}600$$

Here $n = 26$; therefore, $n - x + 1$ equals $26 - 5 + 1 = 21 + 1 = 22$. All that is required is to begin with 26 and keep including factors, each one smaller than the previous one, until there are five factors.

A company plans to run a contest among its 20 leading salesmen. If a prize of \$100 is given to the winner and a prize of \$50 to the runner-up, how many ways can the contest turn out? The answer is given by ${}_{20}P_2$:

$${}_{20}P_2 = 20 \times 19 = 380$$

Therefore, the contest can turn out in 380 ways.

On reflection, formula 1 can be used to answer the problem of selecting three objects from the set of seven objects in which three can be used once and four could be used twice. If one notes that, in effect, the problem says that there are eleven objects (counting the three once and the four twice), one is really looking for ${}_{11}P_3 = 11 \times 10 \times 9 = 990$. Therefore 990 permutations are possible.

For a final example, suppose one has five cities to visit on a vacation trip. If each city is to be visited only once, how many ways can the trip be planned? The answer is given by ${}_5P_5$, because it is a question of how many permutations of five objects one can get from a set of five objects:

$${}_5P_5 = 5 \times 4 \times 3 \times 2 \times 1 = 120$$

Note that, when one takes all five objects, one ends up with all the factors from five down to one. Because we encounter this kind of product often, that is, the product of all the whole numbers from a certain number down to one, it is worth our while to set aside a special symbol and name for it. The product $5 \times 4 \times 3 \times 2 \times 1$ is denoted by 5!. This 5! is read 5 factorial. If we wish to denote the product of all the numbers from 12 to 1, we use 12!. This is read 12 factorial and is $12! = 12 \times 11 \times 10 \times 9 \times 8 \times 7 \times 6 \times 5 \times 4 \times 3 \times 2 \times 1$. In general, then we use the following notation:

$$k! = k(k-1)(k-2) \cdot \cdot \cdot 3 \times 2 \times 1$$

Thus

$$2! = 2 \times 1 = 2$$
$$3! = 3 \times 2 \times 1 = 6$$
$$4! = 4 \times 3 \times 2 \times 1 = 24$$
$$5! = 5 \times 4 \times 3 \times 2 \times 1 = 120$$
$$6! = 6 \times 5 \times 4 \times 3 \times 2 \times 1 = 720$$
$$7! = 7 \times 6 \times 5 \times 4 \times 3 \times 2 \times 1 = 1540$$

and so on.

It is also useful to define two special factorials, 0! and 1!. Because the original definition of factorial does not make sense when one attempts to apply it to 0 and 1, we are free to define them in any manner we wish. This is reasonable in the light of previous comments on the nature of symbols. Because symbols are basically arbitrary, we can define any symbol to our advantage. Although these definitions may seem strange at this point, they will be convenient later. 1! is defined as 1, and so is 0!; that is, $1! = 1$ and $0! = 1$.

How can the introduction of the factorial notation help in writing permutations? Consider ${}_7P_3$. How do we write it using factorials? Because ${}_7P_3 = 7 \times 6 \times 5$, to make $7 \times 6 \times 5$ into a factorial one has to multiply it by the additional factor of $4 \times 3 \times 2 \times 1$. But this changes the value of ${}_7P_3$. Therefore, one also has to divide it by this same $4 \times 3 \times 2 \times 1$, so that its value remains unchanged; that is,

$$ {}_7P_3 = \frac{7 \times 6 \times 5 \times 4 \times 3 \times 2 \times 1}{4 \times 3 \times 2 \times 1} $$

This expression reduces to the original ${}_7P_3 = 7 \times 6 \times 5$. If we examine this closely, we note that the numerator of this new expression is 7! and the denominator is 4! Therefore, ${}_7P_3 = 7!/4!$. And so

$$ {}_{26}P_5 = 26 \times 25 \times 24 \times 24 \times 22 = \frac{26!}{21!} $$

$$ {}_{20}P_2 = 20 \times 19 = \frac{20!}{18!} $$

$$ {}_{11}P_3 = 11 \times 10 \times 9 = \frac{11!}{8!} $$

What happens to the general formula? Because ${}_nP_x = n(n-1)(n-2) \cdots (n-x+1)$, what additional factors are required to convert this into a combination of factorials? The number that is one smaller than $(n-x+1)$ is $(n-x)$. Therefore,

$$ \begin{aligned} {}_nP_x &= n(n-1)(n-2) \cdots (n-x+1) \\ &= \frac{n(n-1) \cdots (n-x+1)(n-x)(n-x+1) \cdots 3 \times 2 \times 1}{(n-x)(n-x-1) \cdots 3 \times 2 \times 1} \\ &= \frac{n!}{(n-x)!} \end{aligned} $$

or

FORMULA 2 $$ {}_nP_x = \frac{n!}{(n-x)!} $$

because $(n-x)! = (n-x)(n-x-1)(n-x-2) \cdots 3 \times 2 \times 1$. This result can be checked for the four examples above. In ${}_7P_3$, $n = 7$ and $x = 3$; then $n - x = 7 - 3 = 4$, and the formula yields ${}_7P_3 = 7!/4!$, which agrees with the original answer. In ${}_{26}P_5$, $n = 26$ and $x = 5$; then $n - x = 26 - 5 = 21$, and again there is agreement. The reader can easily verify that ${}_{20}P_2$ and ${}_{11}P_3$ are also given correctly by this factorial form of the permutation formula.

Of the specific examples only one, ${}_5P_5$, has not been written with factorials to this point. Now consider the factorial form used to calculate ${}_5P_5$. Both n and x are five; therefore $n - x = 0$, and then

$$ {}_5P_5 = \frac{5!}{0!} $$

Now is is easy to see why it was useful to define $0! = 1$. The formula is still usable as ${}_5P_5 = 5!/0! = 5!/1 = 5!$. This agrees with the original calculation. We can now conclude that either formula 1 or formula 2 can be used to calculate the number of permutations that can be made when x objects are selected from a group of n, where the objects are distinct and each can be used only once. Before we develop a similar formula for determining the number of combinations in a problem, a simple property of factorials should be noted. If one examines the short list of factorials given above, he notes the following:

$$ 7! = 7 \times 6 \times 5 \times 4 \times 3 \times 2 \times 1 $$
$$ 7! = 7 \times 6! \qquad \text{as} \qquad 6! = 6 \times 5 \times 4 \times 3 \times 2 \times 1 $$
$$ 7! = 7 \times 6 \times 5! \qquad \text{as} \qquad 5! = 5 \times 4 \times 3 \times 2 \times 1 $$
$$ 7! = 7 \times 6 \times 5 \times 4! \qquad \text{as} \qquad 4! = 4 \times 3 \times 2 \times 1 $$

and so on. In general this could be noted as

$$ \begin{aligned} n! &= n(n-1)! \\ &= n(n-1)(n-2)! \end{aligned} $$

and so on. There will be occasions to use this property later.

We now want to turn our attention to questions involving combinations. Rather than attempting inductive reasoning from examples to general rule, let us attempt to develop a general formula for combinations directly.

As before, a statement should be made that limits problems. The formula will be for the number of combinations of x possible objects from a set of n distinct objects. Here the key word is *distinct.* First notice that any combination of x objects can be arranged in ${}_xP_x$ or $x!$ permutations of the x objects. This is always so. If one selects a combination of x objects from a set of n, these x objects can be arranged in $x!$ permutations of x objects from the set of n; that is, each of these combinations of x objects represents $x!$ permutations of the original set of n objects. Furthermore, there is no overlap between the permutations generated from different combinations of objects. Any permutation of x objects from a set of n represents only one combination of x objects and can be generated only from this set. It is also true that each permutation can be generated from at least one of the possible combinations. If the number of distinct combinations of x objects from a set of n is ${}_nC_x$, then ${}_nC_x = {}_nP_x/x!$. One arrives at this conclusion by observing that each combination of x objects can be used to generate $x!$ permutations. Because each possible permutation of n objects is a permutation of a set of x objects from the set of n objects, one arrives at the following chain of conclusions. The total number of permutations of x objects that can be made from a set of n is ${}_nP_x$. This total can also be found by reasoning that it is also the number of combinations of x objects from a set of n multiplied by the number of permutations generated by each such combination. Because the notation for the number of combinations of x possible objects from a set of n objects is ${}_nC_x$ and each of these can be used to generate $x!$ permutations, one arrives at the formula given:

$${}_nC_x \cdot x! = {}_nP_x$$

One can also write this as follows:

$${}_nC_x = \frac{{}_nP_x}{x!}$$

Because ${}_nP_x = n!/(n-x)!$, one concludes

$${}_nC_x = \frac{n!}{x!\,(n-x)!}$$

For a simple example of the use of this formula, consider the following. How many poker hands are possible from an ordinary deck of playing cards? A poker hand is a combination of five cards. The answer is ${}_{52}C_5$. Then $n = 52$, $x = 5$, and therefore $n - x = 52 - 5 = 47$. Thus

$$\begin{aligned} {}_{52}C_5 &= \frac{52!}{5!47!} \\ &= \frac{52 \times 51 \times 50 \times 49 \times 48 \times 47!}{5! \times 47!} \end{aligned}$$

which reduces to

$$\frac{52 \times 51 \times 50 \times 49 \times 48}{5 \times 4 \times 3 \times 2 \times 1} = 2{,}598{,}960$$

This follows from the comments on factorials; that is, $52! = 52 \times 51 \times 50 \times 49 \times 48 \times 47!$. $(52 \times 51 \times 50 \times 49 \times 48)/(5 \times 4 \times 3 \times 2 \times 1)$ reduces to $52 \times 21 \times 5 \times 49 \times 4$, or 2,598,960.

For a second example, consider the calculation of the number of three-man committees one could select from a group of 18 people. Because the order of the selection of members does not matter, the answer is ${}_{18}C_3$:

$${}_{18}C_3 = \frac{18!}{3!15!} = \frac{18 \times 17 \times 16}{3 \times 2 \times 1} = 3 \times 17 \times 16 = 816$$

Another standard notation for ${}_nC_x$ is $\binom{n}{x}$. This notation has arisen because this number occurs in other areas of mathematics. This notation is also useful if one attempts to extend the formula to handle cases where some of the objects in the original set are not distinct. To develop such a formula, one has to use the concept of sets and build it up from notions involving the division of sets into subsets. This notation can also be extended to handle problems where n and x are not whole numbers. To do so, one has to use and define such strange things as "gamma" and "beta" functions. Although they sound strange, they are relatively simple concepts compared with the whole of mathematics.

9-4 EXERCISES

1. Give some examples of permutations of objects.
2. Give some examples of combinations of objects.
3. Calculate ${}_8P_3$ and ${}_{10}P_4$.
4. Calculate ${}_8C_3$ and ${}_{10}C_4$.
5. The lock of a safe has 60 positions, each indicated by a number. If a sequence of 3 is required to open the lock, how many opening sequences are possible? (For answer, see Sec. 9-5.)
6. In the prefix dialing system used in telephones, how many different telephone numbers can be made?
7. In the all-digit dialing system used in telephones, how many different telephone numbers can be made?
8. A bowl contains 100 balls, each one of which has a numeral from 00 to 99 stamped on it. If 10 of these are drawn from the bowl, how many sets of 10 numerals can be drawn from the bowl?
9. How many six-man committees can be formed of United States senators?
10. In many studies of memory, students are asked to memorize three-

letter "words" that consist of a consonant followed by a vowel followed by another consonant. How many such words can be devised?

11. A coin-operated machine takes two coins, both dropped into one slot, but the coins must be of different value. A sign telling which two coins should be used has been lost. If a man wishes to use the machine and has a penny, a nickel, a dime, a quarter, and a half-dollar, how many different ways might he have to try before he finds the right coins? If the coins do not work, they are returned.

12. A group of 18 students is to be divided into two baseball teams of 9 players each. How many ways can this be done?

13. Why do not the formulas given in the present section always work when we want to calculate a permutation or combination?

14. What is wrong with the name *combination lock*?

9-5 PROBABILITIES USING PERMUTATIONS AND COMBINATIONS

With the notation of combinations and permutations we examined in Sec. 9-4, we can attempt more complex problems. For example, suppose that we draw 5 cards from a deck of 52. What is the probability they are all spades? We can answer this without combinations and permutations by reasoning that we have a chain of dependent experiments. The probability of the first spade is $\frac{13}{52}$, the probability of the second, if we have drawn the first, is $\frac{12}{51}$. The probability of the third spade is $\frac{11}{50}$, and the fourth spade $\frac{10}{49}$, and if we have drawn four spades, the probability of drawing the fifth is $\frac{9}{48}$. We can then apply the appropriate probability formula and get $P(5 \text{ spades}) = \frac{13}{52} \times \frac{12}{51} \times \frac{11}{50} \times \frac{10}{49} \times \frac{9}{48}$, which reduces to $(11 \times 3)/(4 \times 17 \times 5 \times 49 \times 4) = \frac{33}{66{,}640}$.

How do we use the permutation and combination formulas to solve this problem? First we ask how many combinations of 5 cards are possible in a deck of 52. The formula for ${}_{52}C_5$ gives us the answer:

$$ {}_{52}C_5 = \frac{52!}{5!47!} = \frac{52 \times 51 \times 50 \times 49 \times 48}{5 \times 4 \times 3 \times 2 \times 1} = 26 \times 51 \times 10 \times 49 \times 4 $$
$$ = 2{,}598{,}960 $$

How many of these consist entirely of spades? We observe that any combination of spades is selected from the 13 in the deck, so that the total number of possible combinations of 5 spades is ${}_{13}C_5$. We then calculate this quantity:

$$_{13}C_5 = \frac{13!}{5!\,8!} = \frac{13 \times 12 \times 11 \times 10 \times 9}{5 \times 4 \times 3 \times 2 \times 1}$$

which reduces to $13 \times 11 \times 9$ or 1287. Because we feel that all the possible combinations of 5 cards are equally likely, and there are $26 \times 51 \times 10 \times 49 \times 4$ such combinations of which $13 \times 11 \times 9$ are spades, the probability of 5 spades is $(13 \times 11 \times 9)/(26 \times 51 \times 10 \times 49 \times 4)$. This reduces to $(11 \times 3)/(2 \times 17 \times 10 \times 49 \times 4) = \frac{33}{66,640}$. We see that this is the same answer we got from the other method.

Suppose that 2 persons are to be chosen at random from a group of 20 married couples. What is the probability that the pair of persons chosen are married to each other? We reason that the total number of possible choices is $_{40}C_2$, and the total number of married pairs is 20. Because $_{40}C_2 = 40!/(2!38!) = (40 \times 39)/(2 \times 1) = 20 \times 39 = 780$, the answer is $\frac{20}{780}$ or $\frac{1}{39}$.

A third problem might involve the safe described in Exercise 5, Sec. 9-4. In this exercise there was a possible choice of $60 \times 60 \times 60$ "combinations." Note that the term *combination* for the sequence of numerals that unlock a safe is incorrect. One should refer to the *permutation* that unlocks the safe. Suppose that a safe cracker breaks into the room with the safe and has enough time to try 10 "combinations" before making his getaway. What are the chances that he hits on the correct one? To make the problem simpler, we assume that he cannot see the dial and does not know what "combinations" he has tried. The probability that any of the 10, as individuals, is the correct sequence is $\frac{1}{60 \times 60 \times 60}$. If we try to calculate his chances of opening the safe, we have to consider the probabilities of 10 distinct outcomes: (1) opening it on the first try, (2) not opening it on the first try but doing so on the second, (3) not opening it on the first two tries but doing so on the third, and so on, the tenth outcome being failure on the first nine tries but success on the tenth. Because these are mutually exclusive events, and since we can calculate the probability of each, we could use the formulas on the combination of probabilities (Sec. 9-2) to determine the final answer. This is a lot of work, and there is a shorter way. Because it is relatively simple to calculate the chances that the safe cracker fails to open the safe, we can subtract this answer from 1 to find the answer to the problem. The probability that he fails to open the safe on one trial is $1 - \frac{1}{60 \times 60 \times 60}$; therefore, the probability that he fails to open the safe on any of the 10 trials is $(1 - \frac{1}{60 \times 60 \times 60})^{10}$; x^{10} means $x \cdot x$

$\cdots x$, ten times. This is because $1-\frac{1}{60}^3$ is the probability of failure on any one trial and the trials are independent, on account of the assumption that he does not know what permutation he is trying. The P (fails on the first trial and fails on the second trial, and so on) is the product of the individual probabilities. Assuming that we carry out this calculation (here a computer would be very handy), we can use this answer to find the answer to the problem. By using a computer, we get the following:

$$P(\text{failure to open in 10 trials}) = (1-\tfrac{1}{60}{}^3)^{10} = 0.99995370466$$

Therefore, the probability that he opens the safe is $1 - P(\text{failure}) = 0.00004629534$.

In these examples, we again illustrate a point we have made before. We do not need to know the nature of each outcome to determine probabilities. Instead, what we need to know is the number of possible outcomes of the experiment, and the combination and permutation formulas help determine them.

Before we go on to discuss more complex and, we hope, more interesting examples of probabilities and their use, we should mention the other measure of chance that is often used. This is, of course, "odds." Odds are not so much a measure of probability as an "equalizer" for betting on unequally probable outcomes. Suppose that you and a friend are going to bet on the rolls of a die. You are to receive \$1 from your friend every time a 6 appears, and you are to pay \$1 every time a 6 does not. In the long run, a 6 appears about once in six rolls. Therefore, you pay \$5 and receive \$1 every six rolls. To equalize this, your sporting friend offers you 1 to 5 odds; that is, every time a 6 does not appear, you pay \$1, and every time it does, you receive \$5. Now, in an average group of six rolls, you pay \$1 five times and receive \$5 once. Thus, by using odds, the amount won or lost by either participant is equalized. A more formal version of equalization involves the idea of an "expected value" of an experiment, which is considered in Chap. 10. The formal definition of odds on the outcome of an experiment is the ratio of favorable to unfavorable outcomes. For example, the odds in favor of throwing a pair of dice and getting a 7 are $\frac{6}{30}$ or $\frac{1}{5}$. This follows from the calculations we made relating to outcomes of throwing a pair of dice in Sec. 9-3. When we examine the table of outcomes for this experiment, we see that there are 6 outcomes with a 7 as the total and 30 outcomes that are not 7. As with the definition of probability, the outcomes must be mutually exclusive and equally likely.

If we say that the odds against a candidate's being elected are $\frac{7}{5}$, we mean that, based on past records, and so on, if the election were held a number of times, it is estimated that the candidate would lose seven times for every five times he won. Naturally, we cannot hold the election again and again to test our estimate. In dealing with probabilities and odds in everyday events, we are much more likely to estimate their value than to have an actual basis of calculation.

The principal advantage of using the probability of an outcome rather than odds is that the mathematical analysis of probability is easier. Because probabilities are always fractions between 0 and 1 and can be compared more easily, they lend themselves to a more formal analysis than odds do. In fact, when mathematicians and statisticians began trying to analyze the outcomes of an event, they found that the probability of an outcome appeared in their analysis. There is, however, a simple relationship between the odds and the probability of an outcome. It is given by the following formulas. If $O(A)$ stands for the odds in favor of outcome A, m stands for the number of favorable outcomes, and n stands for the number of unfavorable outcomes, then:

FORMULA 1 $$O(A) = \frac{m}{n}$$

then

$$P(A) = \frac{m}{n+m}$$

FORMULA 2 $$O(A) = \frac{P(A)}{P(\text{not } A)}$$

These formulas are easy to verify. First, let us examine formula 1. If $O(A) = m/n$, then m is the number of favorable and n the number of unfavorable outcomes. The total number of outcomes is $m + n$. If we then apply the definitions of $P(A)$, we arrive at $P(A) = m/(m+n)$. To verify formula 2 we have to remember that $P(\text{not } A) = 1 - P(A)$. Thus, $P(\text{not } A) = 1 - m/(n+m)$. Recalling that we can add two fractions if they have the same denominator, we have

$$\begin{aligned} P(\text{not } A) &= \frac{n+m}{n+m} - \frac{m}{n+m} \\ &= \frac{n+m-m}{n+m} \\ &= \frac{n}{n+m} \end{aligned}$$

If your algebra is rusty, you can obtain the same result from the basic definition, because the number of favorable outcomes for the event *not A* is given here as n. The basic definition of probability then yields $P(\text{not } A) = n/(n + m)$. At this point, we must either recall or accept the following algebraic manipulation:

$$\frac{P(A)}{P(\text{not } A)} = \frac{m/(m+n)}{n/(m+n)} = \frac{m}{m+n}\,\frac{m+n}{n} = \frac{m}{n}$$

Because m/n is also equal to $O(A)$, we conclude that formula 2 is correct.

Odds are often written $O(A) = m$ to n; that is, the odds in favor of tossing a 7 with a pair of dice, in a single toss, is 1 to 5, or the odds against a certain candidate are 7 to 5.

As with probabilities, there are formulas and rules for combining the odds on parts of a compound experiment to determine the odds on the compound itself. Because these are a dead end mathematically, we do not examine them; that is, we cannot extend the idea of odds in the same way we can probabilities, and we end our discussion of them with their definition.

9-5 EXERCISES

1. Assuming that 1 card has been drawn from a deck of 52 playing cards, what is the probability that 4 more cards of the same suit are drawn?

2. A "flush" is a 5-card poker hand where all the cards are of the same suit. If 5 cards are chosen at random from a deck of 52 playing cards, what is the probability that they form a flush?

3. Two United States senators are chosen at random to be a special committee. What is the probability that both are from the same state?

4. In Exercise 3, what is the probability that both come from one of the 11 Western states?

5. If the odds of an outcome are "a million to one" against its taking place, what is the probability that it does not?

6. An event can take place in 64 mutually exclusive, equally likely ways. If, in 16 of these ways, you win $1000, what is

(*a*) The probability that you win $1000?
(*b*) The odds in favor of your winning $1000?

7. The sum of the probabilities for and against an outcome must add up to 1. Do the sum of the odds in favor of an outcome and against it add up to a particular result? (*Hint:* Try a few examples.)

8. What is the probability that the 10 balls drawn from the bowl in Exercise 8, Sec. 9-4, all have even numbers? (Assume that 00 is even.)

9. A baby is playing with an electric typewriter, touching keys at random. If he strikes four keys in a row, what is the probability that the word *math* is typed?

10. What are the odds in favor of the safe cracker in the example in this section?

9-6 SETS AND PROBABILITY

We have used concepts from the theory of sets in calculating probability. We have even used the set-theory tool of Venn diagrams to visualize and develop the rules for the calculation of probability. In the present section, we mention some of the other parallels that exist between these two theories and indicate some additional effects set theory has had on probability.

The theory of probability is built on the basic concept of sets of outcomes. The computation of probability depends on the number of elements in a specific subset of these sets. Keeping this in mind, we can restate several of the formulas for probability in terms of these sets. For example, if we have two outcomes of an experiment, say, a and b, we can let A be the subset of the total set of equally likely outcomes U that are favorable to outcome a. Here we designate the universal set as the basic set of outcomes. We can say that B is the subset of U favorable to outcome b. To say that a and b are mutually exclusive outcomes can then be interpreted as saying that A and B are disjoint sets. We can also point out that the probability of outcome a depends on the number of elements in set A and that the probability of a not taking place depends on the number of elements in set $\tilde{A}$.

Many of the formulas for probability bear a striking similarity to formulas involving sets. For example, the formula for the number of elements in the union of two disjoint sets parallels that for the probability of two mutually exclusive outcomes of an experiment:

$P(A \text{ or } B) = P(A) + P(B)$ probability
$N(A \cup B) = N(A) + N(B)$ sets

If we modify these to handle outcomes that are not mutually exclusive and sets that are not disjoint, we have

$$P(A \text{ or } B) = P(A) + P(B) - P(A \text{ or } B)$$
$$N(A \cup B) = N(A) + N(B) - N(A \cap B)$$

Although we did not develop these formulas in our discussion of set theory, it is easy to see that they are the correct modification required when sets overlap.

Many of the complex tools developed for the theory of probability come from the theory of sets. For example, many of the problems of probability involve more complex notation about combinations. To handle them, we introduce the concept of the "partition" of a set. A set is partitioned if it is divided into disjoint subsets that are exhaustive; that is, it is divided into subsets so that every element of the original set is a member of one and only one of the subsets forming the partition. These subsets are known as "cells" of the partition. Problems involving permutations and combinations usually involve the number of possible ways of forming a partition of a specified kind. The problem of the number of different possible combinations of x objects from a set of n involves the number of different ways we could partition the original set of n objects into two cells, one containing x objects and the other $n - x$ objects. More complex probability problems involve the partitioning of the original set of experiments into three or more cells of outcomes. As an example, the chances of 4 "perfect" bridge hands of 13 cards, all of the same suit, involve the number of different ways the 52 cards can be divided, or partitioned, into 4 groups of 13. We do not go into the theory necessary to make this calculation but mention only that it is a generalization of the method developed in the section on combinations. The actual number calculated as the number of bridge deals is approximately 53,645,000,000,000,000,000,000,000,000. The list of applications of set theory to probability could go on indefinitely. Many formal studies of probability theory are done entirely in the language of sets. We at least can see that, once again, the theory of sets is an underlying and fundamental concept in mathematics.

9-6 EXERCISES

1. Write an argument for the following formula given in the text:

$$N(A \cup B) = N(A) + N(B) - N(A \cap B)$$

2. The number of possible bridge deals was given in the text. How many deals result in one player's having all the cards of one suit?

3. Identify at least one new parallel between set theory and probability theory.

4. Describe two different partitions of the set $\{0,1,2,3,4,5,6,7,8,9\}$. In one, have two cells, and in the other, three.

5. Find two similarities of the Boolean algebra theorems in Chap. 8 to the probability rules in the present chapter.

9-7 SPECIAL PROBLEMS AND NONINTUITIVE EXAMPLES

Sometimes when we carry out theoretical calculations of the probability of an experiment we get a result that is contrary to our feeling of what should be taking place. Such examples remind us that the major purpose of the construction of a careful analysis of probabilities is to handle problems our intuition cannot. In the present section we examine two examples in which analysis conflicts with feeling.

Suppose that you are shown three cards: one that is red on both sides, one that is white on both sides, and one that is red on one side and white on the other. Each of the three cards is then placed in a separate envelope. There is no way to tell the envelopes apart externally. The three envelopes are then mixed thoroughly and one of them is placed on a table. A card from inside the envelope is slid out without being lifted, and you observe that, as it lies on the table, a red side is exposed. What is the probability that the other side of the card is also red? One feels that there are two possible outcomes: the other side of the card is either red or white. Therefore, the probability of a favorable outcome is $\frac{1}{2}$.

There is a flaw in this. To use the basic probability, each member of the set of outcomes must be equally likely. This is not so here. The experiment is not selecting a color from two colors but selecting one of three envelopes. In the selection there are three equally likely outcomes. When we ask what is the probability that we have a red side down, we are really asking what are the chances that we have selected a card with two sides of the same color. Because two outcomes of the experiment of selecting a card are favorable to this outcome, the actual probability is $\frac{2}{3}$.

A standard reaction to this is that we do not have three outcomes, because the white-white card has been eliminated. We can arrive at the answer of $\frac{2}{3}$ another way. Suppose that we modify the problem by imagining that we placed two cards in each envelope. One side of each card was blank. In the first envelope, we placed two cards with a red face; in the second, two cards with a white face; and in the third, one card with a red face and one with a white face.

In Fig. 9-7, we identify the three envelopes as E_1, E_2, and E_3, and the cards as W_1, W_2, and W_3, and R_1, R_2, and R_3, to be able to analyze the problem. If we select an envelope and draw a red card from it, we know that we have selected either E_2 or E_3. The red cards we can draw are R_1, R_2, or R_3.

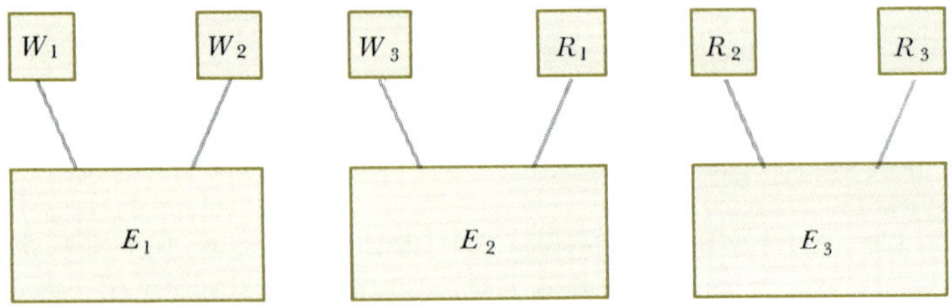

FIG. 9-7 ENVELOPE EXPERIMENT.

What possible outcomes are there for the remaining cards? If we have drawn R_1, then the card remaining in the envelope is W_3. If we have drawn R_2, then the remaining card is R_3. Finally, if we have drawn R_3, the remaining card is R_2. There are three equally likely possibilities for the card remaining in the envelope. They are R_2, R_3, or W_3. So, once again, we see that the probability that it is a red-faced card is $\frac{2}{3}$.

A simple experiment should convince doubting readers. Get three envelopes and six 3- by 5-in. cards. Place an X on three of the cards and leave the others blank. Mix the cards and place two in each envelope. Mix the envelopes and draw one and remove one card from it. If you get a blank card, begin again; but if you get a card with an X, record whether the other card also has an X or not. Tabulate your results like this:

Number of draws:	𝍸 𝍸 𝍸 𝍸 𝍸 𝍸
Second card X:	𝍸 𝍸 𝍸 𝍷𝍷𝍷𝍷
Second card blank:	𝍸 𝍸 𝍷

Try the experiment about seventy-five times. If our analysis is correct, you should get about 50 X cards to 25 blank cards. If it is incorrect, you should get about 37 to 37. Do not expect to get exactly 50 X cards, for as we have pointed out, probability represents the average outcome and not necessarily the exact one.

For the second example, let us consider the following problem. Suppose that you enter a room and find someone you have not before met. What is the probability that both of you have the same birthday (month and day, not year)? To make the problem simple, let us assume in what follows that no one is ever born on February 29. If your birthday is, say, May 3, then the probability that the other person has the same birthday is

$\frac{1}{365}$. Here we assume that being born on a particular day is as likely as being born on any other day. So far, so good. Now let us make the problem more complex. Suppose that there are two persons in the room that you enter. What is the probability that at least two of you have the same birthday?

A direct computation is relatively complex, but it is simple to calculate the probability that none of the three persons in the room has the same birthday. The probability that the second person does not have the same birthday as the first is $\frac{364}{365}$, and the probability that the third person does not have the same birthday as the first two is $\frac{363}{365}$. Because these two elements are independent, we reason that P(none shares a birthday) is $\frac{364}{365} \times \frac{363}{365}$. This fraction is approximately equal to 0.991796. This makes P(at least two share a birthday) $= 1 - 0.991796 = 0.008204$ or about eight times in 1000 tries. We could carry out a similar calculation with four persons in a room and arrive at these figures:

$$\begin{aligned} P(\text{none shares a birthday}) &= \tfrac{364}{365} \times \tfrac{363}{365} \times \tfrac{362}{365} \\ &= 0.98364 \\ P(\text{at least two share a birthday}) &= 0.01636 = 1 - P(\text{none}) \end{aligned}$$

These examples establish a pattern we can use to calculate the probability of two people sharing a birthday, given a certain number of people in the room. Table 9-1 shows these results calculated with the help of a computer. Some of these results are indeed startling. If we recall the fraction-of-the-time interpretation of probabilities, we see that, in a group of only 23 people, more than half the time there are 2 sharing a birthday. Ninety-seven times out of 100 there are 2 sharing a birthday in a group of 50 people. If there are 58 people in a group, the probability of 2 sharing a birthday is about the same as 2 not sharing a birthday if there are only 3 in a group. In a group of 116 or more, the odds are "a million to one" that there are 2 sharing a birthday.

An interesting example of having the same birthday is the first 36 Presidents of the United States. Of this group, 2 share a birthday, Warren G. Harding and James K. Polk. Both were born on November 2. The same calculation also gives probabilities of dates of death. Of the 33 Presidents who are dead at the time of this writing, 2, Millard Fillmore and William Howard Taft, died on the same day of the year, March 8. Even stranger is the fact that 3 other Presidents died on the Fourth of July. These were John Adams and Thomas Jefferson, who died in 1826, and James Monroe, who died in 1831.

TABLE 9-1 PROBABILITIES OF TWO PEOPLE SHARING A BIRTHDAY

Number of people	*P(none shares a birthday)*	*P(at least two share a birthday)*
1	1.000	0.000
2	.997	.002
3	.991	.008
4	.983	.016
5	.972	.027
6	.959	.040
7	.943	.056
8	.925	.074
9	.905	.094
10	.883	.116
11	.858	.141
12	.832	.167
13	.805	.194
14	.776	.223
15	.747	.252
16	.716	.283
17	.684	.315
18	.653	.346
19	.620	.379
20	.588	.411
21	.556	.443
22	.524	.475
23	.492	.507
24	.461	.538
25	.431	.568
26	.401	.598
27	.373	.626
28	.345	.654
29	0.319	0.680
30	.293	.706
31	.269	.730
32	.246	.753
33	.225	.774
34	.204	.795
35	.185	.814
36	.167	.832
37	.151	.848
38	.135	.864
39	.121	.878
40	.108	.891
41	.096	.903
42	.085	.914
43	.076	.923
44	.067	.932
45	.059	.940
46	.051	.948
47	.045	.954
48	.039	.960
49	.034	.965
50	.029	.970
51	.025	.974
52	.021	.978
53	.018	.981
54	.016	.983
55	.013	.986
56	.011	.988
57	.009	.990
58	.008	.991

Although these results seem to indicate that our calculations are correct, a word of caution is necessary. Although it may be a "million to one" in favor of a result, we may be looking at the "one" rather than "one of the million." With 120 people in a group, there may be none sharing a birthday. We also know that we could toss an "honest" coin a thousand times and get a thousand heads. With caution in mind, we end our discussion of the calculation of probabilities. In Chap. 10 we examine how we can use these concepts to our advantage in discussing and analyzing events.

9-7 EXERCISES

1. We said that it was possible to toss an "honest" coin a thousand times and get 1000 heads. Because 1000 is a large number of calculations, suppose that we toss a coin twenty times. We have indicated that the average outcome is 10 heads. Calculate the following directly. Each refers to an exact number of heads.

(*a*) P(9 heads)
(*b*) P(10 heads)
(*c*) P(11 heads)
(*d*) P(20 heads)

2. Using Table 9-1, calculate:

(*a*) The probability that two baseball players on a field have the same birthday.
(*b*) The probability that two junior members of the United States Senate have the same birthday.

3. What is the minimum number of pairs of shared birthdays in the United States House of Representatives?

4. In the example of red and white cards above, what is the probability that, in 10 draws, we get no red-red or white-white results?

5. If, in the experiment of 75 draws described above, how do you interpret a result of 75 X second cards? What is the most likely cause of such a result?

6. If you tossed a coin twenty times, as described in Exercise 1, and got 20 heads, how would you bet on the next toss of the coin? Why?

REVIEW EXERCISES

1. Why did we choose a definition for probability so that the total probabilities of an experiment added up to 1?

2. What is the probability that none of the next seven people you meet was born on the same day of the week as you?

3. Why is the concept of mutually exclusive outcomes needed in defining probability?

4. What is the probability on the single draw of a card from a deck that:

(*a*) It is red?
(*b*) It is a picture card?
(*c*) It is neither red nor a picture card?

5. What assumptions about drawing a card from a deck are needed to solve Exercise 4?

6. A bowl contains three red, three green, and three white chips. If two chips are drawn from the bowl, without being replaced, what is the probability that:

    (*a*) Both are of the same color?

    (*b*) Neither is green?

    (*c*) They are of different colors?

7. What assumptions are needed to solve Exercise 6?

8. Make a table for two sets of logically possible outcomes for the event of tossing a pair of white dice.

9. The following table gives the time blocks for television programs for a Wednesday evening: (numbers indicate starting times; last program over at 11:00 PM.)

<table>
<tr><th></th><th>ABC</th><th>CBS</th><th>NBC</th><th>NET</th></tr>
<tr><td>7:30</td><td>A_1</td><td rowspan="2">C_1</td><td rowspan="3">N_1</td><td rowspan="2">E_1</td></tr>
<tr><td>8:00</td><td>A_2</td></tr>
<tr><td>8:30</td><td rowspan="2">A_3</td><td>C_2</td><td rowspan="3">E_2</td></tr>
<tr><td>9:00</td><td>C_3</td><td rowspan="2">N_2</td></tr>
<tr><td>9:30</td><td rowspan="3">A_4</td><td>C_4</td></tr>
<tr><td>10:00</td><td rowspan="2">C_5</td><td rowspan="2">N_3</td><td>E_3</td></tr>
<tr><td>10:30</td><td>E_4</td></tr>
</table>

How many "programs" of television watching can be made for a viewer who wishes to watch from 7:30 to 11:00 PM?

10. A student is taking three classes. How many permutations of grades (A, B, C, D, F) can he receive?

11. Find the following quantities and arrange the results in the same pattern as shown below:

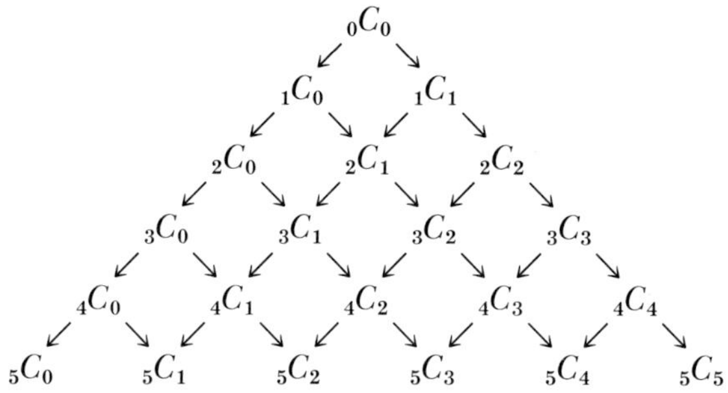

that is, replace the C's in the diagram with their numerical value. Is there any meaning to the arrows? Do they indicate any pattern?

12. List three permutations and three distinct combinations of the set of English letters.

13. (*a*) How many 4-card cribbage hands are possible from an ordinary deck of 52 cards.

(*b*) What is the probability that all 4 cards are jacks?

14. A dice player in a London casino has been watching the play for a while and has observed that no 12s have come up in the last 72 rolls of the dice. He reasons that, because the probability of a 12 is $\frac{1}{36}$, he can expect two 12s in the next few rolls. What, if anything, is wrong with his reasoning?

15. In the discussion on sets and models we mentioned partitions. A useful variation of this concept is considered here. Let $A = \{a,b,c,d,e,f,g,h,i,j,k\}$. Two partitions of A are $\{\{a,b,c,d,e\},\{f,g,h,i,j,k\}\}$ and $\{\{a,b,c\},\{d,e,j,k\},\{f,g,h,i\}\}$. Let $A_1 = \{a,b,c,d,e\}$ and $A_2 = \{f,g,h,i,j,k\}$, so that the first partition is $\{A_1,A_2\}$. In a similar manner, let $B_1 = \{a,b,c\}$, $B_2 = \{d,e,j,k\}$, and $B_3 = \{f,g,h,i\}$, so that the second partition is $\{B_1,B_2,B_3\}$. Find all the members of the set.

$$\{A_1 \cap B_1,\ A_1 \cap B_2,\ A_1 \cap B_3,\ A_2 \cap B_1,\ A_2 \cap B_2,\ A_2 \cap B_3\}$$

Explain why this set of sets is a partition of A? When two partitions of a set are used to form a new partition in this manner, the new partition is called a "cross-partition" formed from the original partitions.

16. In what manner is the word *taxonomy* related to Exercise 15?

For selected references see Chap. 10.

Model of a joint probability distribution (one in which two variables are examined simultaneously). Each row of the model shows the distribution of one variable for a particular value of the other.
FROM M. DWASS, "FIRST STEPS IN PROBABILITY," MCGRAW-HILL, 1967.

chapter ten

THE INTERPRETATION AND USE OF THE LAWS OF CHANCE

The ideas and methods in Chap. 9 are interesting, not because they permit us to analyze poker hands or the roll of a pair of dice, but because they can be applied to everyday situations so as to gain a real advantage over impulse and intuition. To do so, it is necessary to examine experiments as a whole rather than only some of the possible outcomes. In the present chapter we examine ideas related to how this is done. In addition some of the ways probability and statistics fit into everyday processes are examined.

10-1 SAMPLE SPACES, PROBABILITY DISTRIBUTION, AND AREA REPRESENTATION OF PROBABILITY

We mentioned that a set of outcomes of an experiment is useful when considered as a whole, for with it we can get a complete picture of the experiment. We call such a set of mutually exclusive, exhaustive outcomes a "sample space." From a different point of view a sample space is any set of logically possible outcomes of an experiment. The additional condition is that every outcome of the experiment that may take place can be identified with one and only one member of the set. For example, we mentioned two sample spaces in the experiment of tossing a pair of dice. One was the set of 36 outcomes given in Sec. 9-3, and the other the set of 11 sums (outcomes) for the two dice given by {2,3,4,5,6,7,8,9,10,11,12}. Both of these sets meet the requirements of the definition of a sample space. Any result of the toss of a pair of dice can be identified with one and only one of the 36 outcomes and can also be classified as one and only one sum of the spots.

Consider the experiment of tossing three coins. Several sample spaces are easily determined. In one we list the eight permutations of head and tail that may take place:

Outcome	*Coin 1*	*Coin 2*	*Coin 3*
1	Head	Head	Head
2	Head	Head	Tail
3	Head	Tail	Head
4	Head	Tail	Tail
5	Tail	Head	Head
6	Tail	Head	Tail
7	Tail	Tail	Head
8	Tail	Tail	Tail

In another we consider only the total number of heads observed:

Outcome	*Result*
1	No heads
2	One head
3	Two heads
4	Three heads

Although each of the eight outcomes in the first sample space is equally likely, this is not so in the second. Each outcome in the first sample space has a probability of $\frac{1}{8}$. It is simple to calculate the probabilities for the outcomes in the second sample space. This calculation was discussed in

Exercise 2, Sec. 9-3, and a result is given below:

Outcome	*Result*	*Probability*
1	No heads	$\frac{1}{8}$
2	One head	$\frac{3}{8}$
3	Two heads	$\frac{3}{8}$
4	Three heads	$\frac{1}{8}$

The list of probabilities associated with the members of a sample space is called a "probability distribution" for the sample space. The probability distribution for tossing a pair of dice for the sum is given below (see for these probabilities Exercise 6, Sec. 9-1):

Outcome	*Result*	*Probability*
1	Sum 2	$\frac{1}{36}$
2	Sum 3	$\frac{2}{36} = \frac{1}{18}$
3	Sum 4	$\frac{3}{36} = \frac{1}{12}$
4	Sum 5	$\frac{4}{36} = \frac{1}{9}$
5	Sum 6	$\frac{5}{36}$
6	Sum 7	$\frac{6}{36} = \frac{1}{6}$
7	Sum 8	$\frac{5}{36}$
8	Sum 9	$\frac{4}{36} = \frac{4}{9}$
9	Sum 10	$\frac{3}{36} = \frac{1}{12}$
10	Sum 11	$\frac{2}{36} = \frac{1}{18}$
11	Sum 12	$\frac{1}{36}$

Because the outcomes in a sample space are mutually exclusive and exhaustive, the sum of the probabilities of a probability distribution is always 1. This follows from the fraction-of-the-time concept of probability. In effect, the probability distribution shows the fraction of the time each outcome occurs. Because these outcomes are nonoverlapping, in the sense

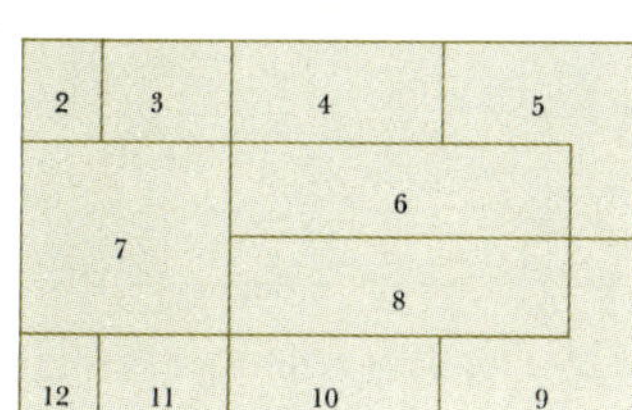

FIG. 10-1 PROBABILITY DIAGRAM OF TOSSING DICE.

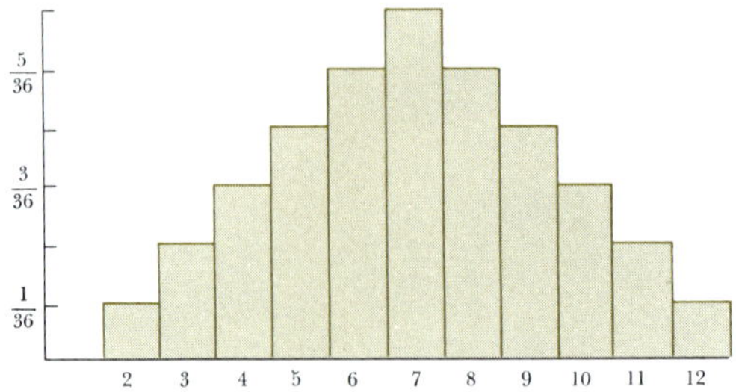

FIG. 10-2 A HISTOGRAM—AN ALTERNATIVE DIAGRAM OF TOSSING DICE.

that, if one of them occurs, all the others are excluded, and because one of them must take place, the sum of the fractions each takes place is the total time, or 1.

The same thing can be said using the language of sets. Thus, a sample space of an experiment is a partition of the set of outcomes of the experiment. Each cell of the partition represents a fraction of the whole set, and the sum of the fractions must add up to the whole experiment. We can make a modification of a Venn diagram to illustrate this further (Fig. 10-1). This Venn diagram is divided into 36 units of equal area. Because a sum of 2 has a probability of $\frac{1}{36}$, it is represented by 1 square unit of area. A sum of 3 is represented by 2 square units or $\frac{1}{36}$ of the whole, and so on. In Fig. 10-1 each possible sum has an area of appropriate size. We can get a better picture of what is taking place if we cut apart the Venn diagram and rearrange the areas, keeping the total area the same (Fig. 10-2). The only difference between this figure and the rectangular one is that we place a column of appropriate area above each sum and place the sums in order. Such diagrams are known as "histograms." We can do the same thing for the probability distribution for the tossing of three coins (Figs. 10-3 and 10-4). These figures are interpreted in the same

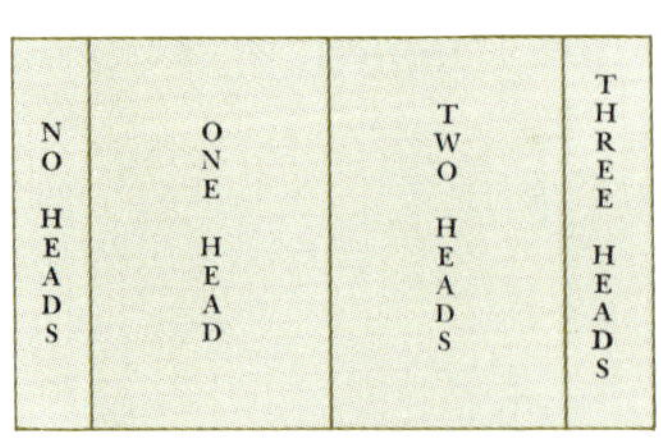

FIG. 10-3 PROBABILITY DIAGRAM OF TOSSING THREE COINS—VENN FORM.

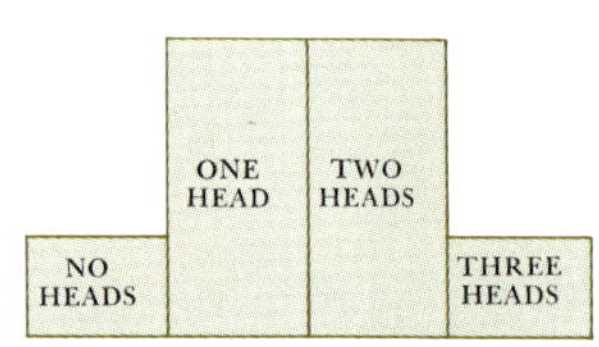

FIG. 10-4 PROBABILITY DIAGRAM OF TOSSING THREE COINS—HISTOGRAM FORM.

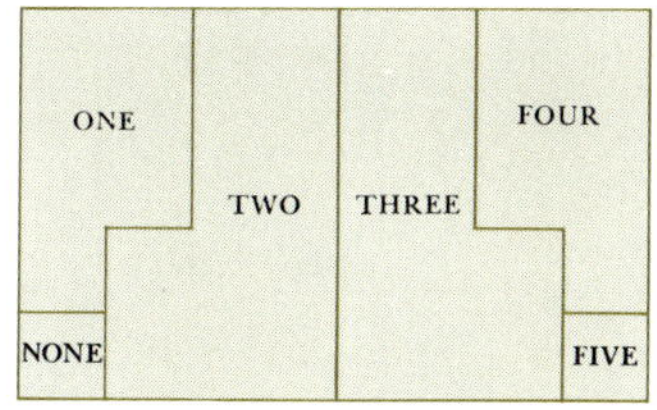

FIG. 10-5 PROBABILITY DIAGRAM OF TOSSING FIVE COINS.

manner as Figs. 10-1 and 10-2; that is, the total area in Fig. 10-3 is divided into $\frac{1}{8}$, $\frac{3}{8}$, $\frac{3}{8}$, and $\frac{1}{8}$ sections. These are the same fractions of the total represented by the columns in Fig. 10-4.

Let us examine the probability distributions and histograms of some other experiments. If 5 coins are tossed, the outcomes and probabilities are as follows:

Outcome	*Probability*
No heads	$\frac{1}{32}$
1 head	$\frac{5}{32}$
2 heads	$\frac{10}{32}$
3 heads	$\frac{10}{32}$
4 heads	$\frac{5}{32}$
5 heads	$\frac{1}{32}$

The Venn diagram for this event is given in Fig. 10-5, and the histogram in Fig. 10-6. If we toss 11 coins, the number of heads (outcome) and probabilities are as follows:

Outcome	*Probability*
0	$\frac{1}{2048}$
1	$\frac{11}{2048}$
2	$\frac{55}{2048}$
3	$\frac{165}{2048}$
4	$\frac{330}{2048}$
5	$\frac{462}{2048}$
6	$\frac{462}{2048}$
7	$\frac{330}{2048}$
8	$\frac{165}{2048}$
9	$\frac{55}{2048}$
10	$\frac{11}{2048}$
11	$\frac{1}{2048}$

The histogram for this event is given in Fig. 10-7.

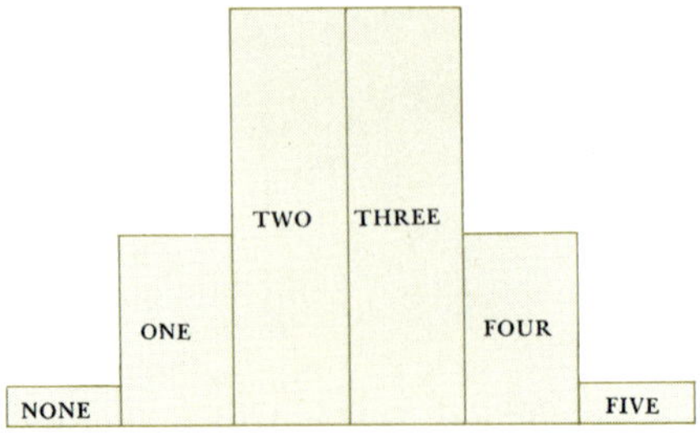

FIG. 10-6 HISTOGRAM OF TOSSING FIVE COINS.

We have made the bases of these column diagrams all the same width so that we can compare results. Let us try to visualize what takes place if we toss 1000 coins. We can calculate the probabilities, but the diagram looks more like smooth lines than columns. For 1000 coins, there are 1000 columns, each of such a height that the area of the column represents the appropriate fraction of the whole (Fig. 10-8). The whole histogram looks like a smooth curve but is not.

In actual situations, mathematicians assume that these curves, such as in Fig. 10-8, are smooth rather than steps and, on this basis, calculate probabilities and work out advanced theories of probability. When students are "graded on a curve," the usual reference is to a smoothed curve for a theoretical probability distribution developed by the German mathematician Karl Friedrich Gauss. Gauss' curve looks like that in Fig. 10-9. Because of its shape, this probability distribution is often called the "bell curve," and because it describes so many naturally occurring events, it is often called the "normal curve." Its algebraic equation (remember the algebraic models of geometry in Chap. 7) is very complex:

$$y = ce^{(x-a)^2/2b}$$

In this equation, a, b, and c are fixed numbers for any specific problem and are chosen so that the total area between the curve and the base line is equal

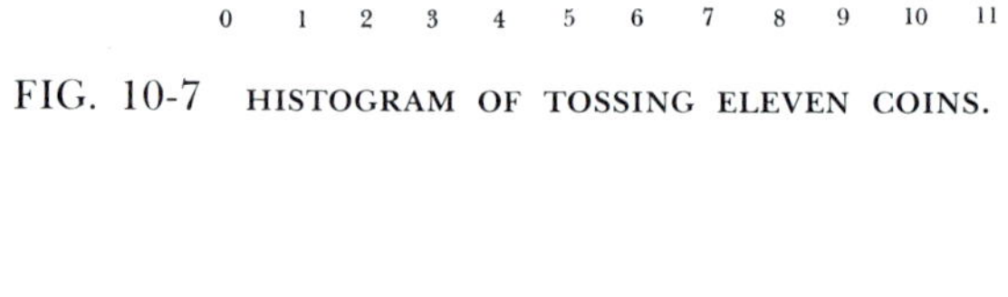

FIG. 10-7 HISTOGRAM OF TOSSING ELEVEN COINS.

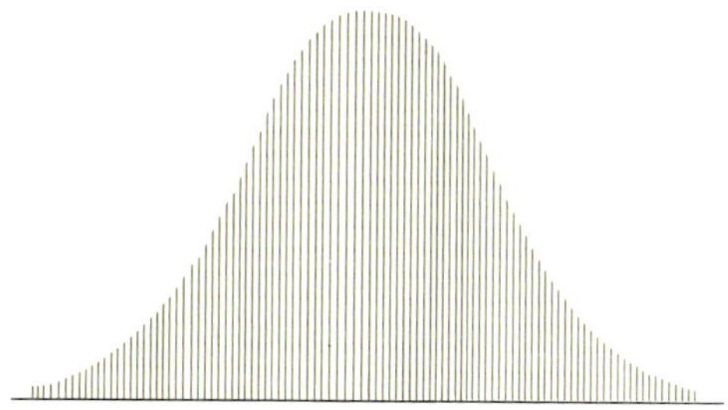

FIG. 10-8 HISTOGRAM OF THE DISTRIBUTION OF HEADS IN ONE TOSS OF 1000 COINS.

to 1. And e is a symbol for a special number that, like π in problems involving a circle, has no exact fractional or whole-number value; it is approximately equal to 2.718281828459.

There are many other smoothed curves of probability. For example, if we consider the probability that the next person you meet is of a particular age, we get a smoothed curve like that in Fig. 10-10. This curve does not have the symmetry of the normal curve. Many other theoretical distributions of probability exist. They have names like the "binomial distribution" and the "Poisson distribution" and they are used and analyzed by mathematicians and statisticians when working with probabilities. Although we cannot use these concepts in our discussion directly, we can examine some of the ways we can usefully apply the theories of probability to actual problems. Naturally, we have to limit our considerations to simple problems for which we have tools to handle the calculations. The tools, such as theoretical distributions, available to the professional give him many methods for applying probability.

The calculation of probabilities using theoretical-distribution curves is made relatively easy by the use of tables. Table 10-1 is that for the gaus-

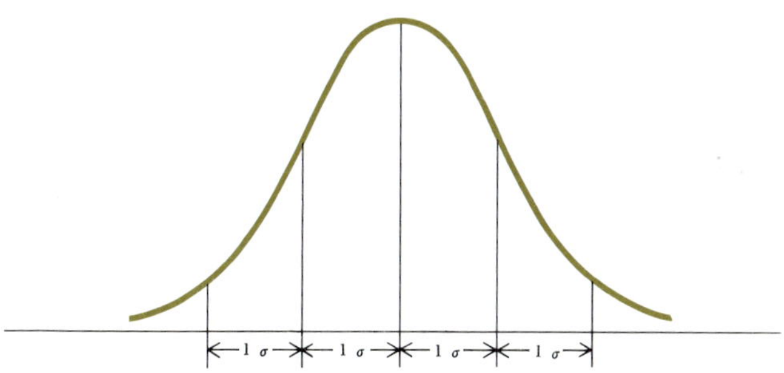

FIG. 10-9 GAUSSIAN OR NORMAL DISTRIBUTION OF PROBABILITIES

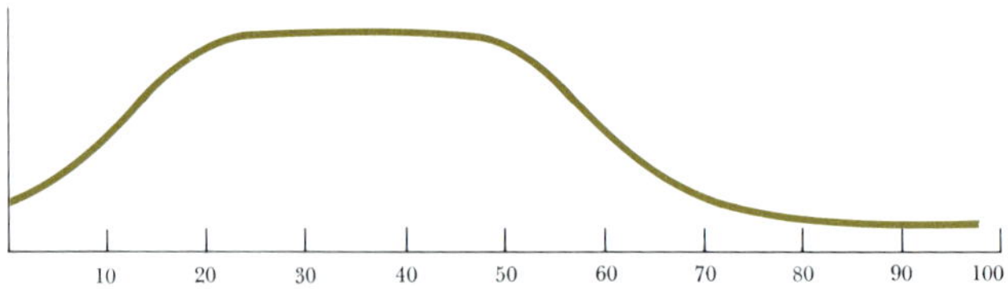

FIG. 10-10 POSSIBLE DISTRIBUTION OF AGES (FICTIONAL).

TABLE 10-1 GAUSSIAN OR NORMAL-CURVE AREAS

z	*.00*	*.01*	*.02*	*.03*	*.04*	*.05*	*.06*	*.07*	*.08*	*.09*
0.0	.0000	.0040	.0080	.0120	.0160	.0199	.0239	.0279	.0319	.0359
0.1	.0398	.0438	.0478	.0517	.0557	.0596	.0636	.0675	.0714	.0753
0.2	.0793	.0832	.0871	.0910	.0948	.0987	.1026	.1064	.1103	.1141
0.3	.1179	.1217	.1255	.1293	.1331	.1368	.1406	.1443	.1480	.1517
0.4	.1554	.1591	.1628	.1664	.1700	.1736	.1772	.1808	.1844	.1879
0.5	.1915	.1950	.1985	.2019	.2054	.2088	.2123	.2157	.2190	.2224
0.6	.2257	.2291	.2324	.2357	.2389	.2422	.2454	.2486	.2517	.2549
0.7	.2580	.2611	.2642	.2673	.2704	.2734	.2764	.2794	.2823	.2852
0.8	.2881	.2910	.2939	.2967	.2995	.3023	.3051	.3078	.3106	.3133
0.9	.3159	.3186	.3212	.3238	.3264	.3289	.3315	.3340	.3365	.3389
1.0	.3413	.3438	.3461	.3485	.3508	.3531	.3554	.3577	.3599	.3621
1.1	.3643	.3665	.3686	.3708	.3729	.3749	.3770	.3790	.3810	.3830
1.2	.3849	.3869	.3888	.3907	.3925	.3944	.3962	.3980	.3997	.4015
1.3	.4032	.4049	.4066	.4082	.4099	.4115	.4131	.4147	.4162	.4177
1.4	.4192	.4207	.4222	.4236	.4251	.4265	.4279	.4292	.4306	.4319
1.5	.4332	.4345	.4357	.4370	.4382	.4394	.4406	.4418	.4429	.4441
1.6	.4452	.4463	.4474	.4484	.4495	.4505	.4515	.4525	.4535	.4545
1.7	.4554	.4564	.4573	.4582	.4591	.4599	.4608	.4616	.4625	.4633
1.8	.4641	.4649	.4656	.4664	.4671	.4678	.4686	.4693	.4699	.4706
1.9	.4713	.4719	.4726	.4732	.4738	.4744	.4750	.4756	.4761	.4767
2.0	.4772	.4778	.4783	.4788	.4793	.4798	.4803	.4808	.4812	.4817
2.1	.4821	.4826	.4830	.4834	.4838	.4842	.4846	.4850	.4854	.4857
2.2	.4861	.4864	.4868	.4871	.4875	.4878	.4881	.4884	.4887	.4890
2.3	.4893	.4896	.4898	.4901	.4904	.4906	.4909	.4911	.4913	.4916
2.4	.4918	.4920	.4922	.4925	.4927	.4929	.4931	.4932	.4934	.4936
2.5	.4938	.4940	.4941	.4943	.4945	.4946	.4948	.4949	.4951	.4952
2.6	.4953	.4955	.4956	.4957	.4959	.4960	.4961	.4962	.4963	.4964
2.7	.4965	.4966	.4967	.4968	.4969	.4970	.4971	.4972	.4973	.4974
2.8	.4974	.4975	.4976	.4977	.4977	.4978	.4979	.4979	.4980	.4981
2.9	.4981	.4982	.4982	.4983	.4984	.4984	.4985	.4985	.4986	.4986
3.0	.4987	.4987	.4987	.4988	.4988	.4989	.4989	.4989	.4990	.4990

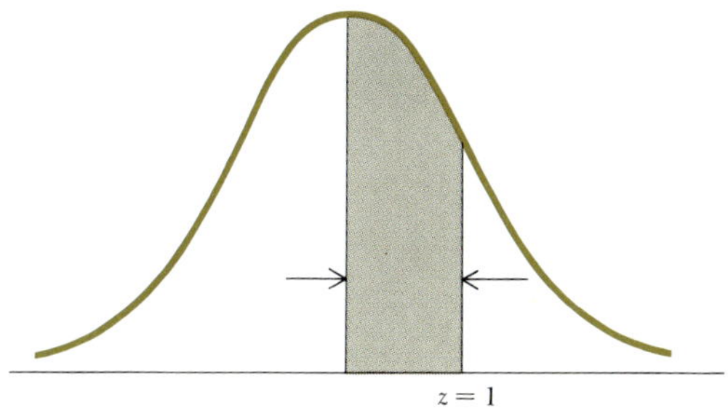

FIG. 10-11 NORMAL-CURVE AREA IF $z = 1$.

sian or normal curve. This table gives the fraction of the total area, which has a value of 1 square unit, between the central axis and a line z standard units from the axis. The numerical value of z is usually referred to as the "z score." For example, suppose that one wanted to determine the area between the axis and a z score of 1, or a line 1 standard unit from the axis (Fig. 10-11). From Table 10-1 the entry for $z = 1$ is 0.3413. This means that 34.13 percent of the total area below a normal curve can be found between the central axis and a line 1 standard unit from it. In Fig. 10-11 a z score of $+1$ is shown, meaning that values to the right of the central axis are being considered. If the z score were -1, the same table entry would be used, but the negative sign would indicate that the line bounding the area was to the left of the central axis (Fig. 10-12). Table 10-1 gives areas for positive values of z only. The symmetry of the normal curve ensures that negative z scores have corresponding areas.

Table 10-1 can also be used to find areas between two z scores or on the side of a z line opposite the central axis. The examples below illustrate this.

EXAMPLE 1 Area between $z = 1.20$ and $z = 2.43$
The area is shown in Fig. 10-13. Table 10-1 indicates that the area be-

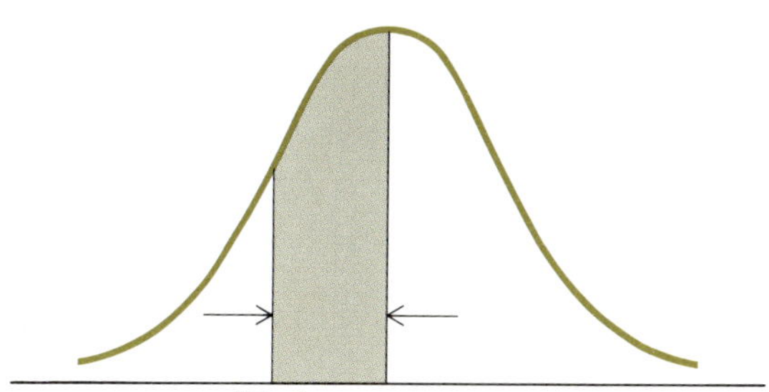

FIG. 10-12 NORMAL-CURVE AREA OF $z = -1$.

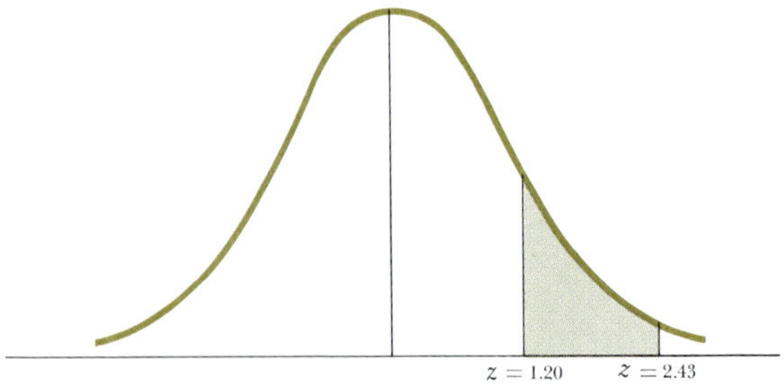

FIG. 10-13 NORMAL-CURVE AREA BETWEEN $z = 1.20$ AND $z = 2.43$.

tween the central axis and the line at $z = 1.20$ is 0.3849. The area between the central axis and the line at $z = 2.43$ is 0.4925. The area between the two z lines is the difference between the two areas. Thus the actual area is $0.4925 - 0.3849 = 0.1076$, or 10.76 percent of the total area below the curve.

A similar line of reasoning is used in determining the area to the right or left of a z value.

EXAMPLE 2 Area to the right and left of $z = 2.43$
The central axis separates the total area into two parts each containing 0.5 of the whole. Because the area between the central axis and the line at $z = 2.43$ is given in Table 10-1 as 0.4925, the area to the right of the line is $0.5000 - 0.4925 = 0.0075$ square unit, and that to the left of it $0.5000 + 0.4925 = 0.9925$ square unit.

EXAMPLE 3 Area between $z = -1.33$ and $z = 0.53$
The area is shown in Fig. 10-14. From Table 10-1 the area between the central axis and the line at $z = -1.33$ is 0.4082, and that between the cen-

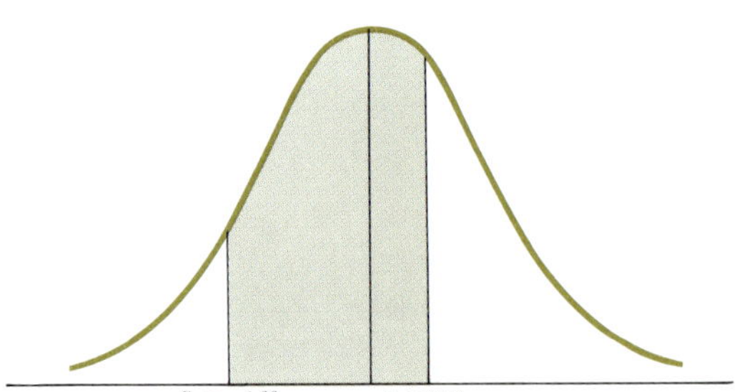

FIG. 10-14 NORMAL-CURVE AREA BETWEEN $z = -1.33$ AND $z = 0.53$.

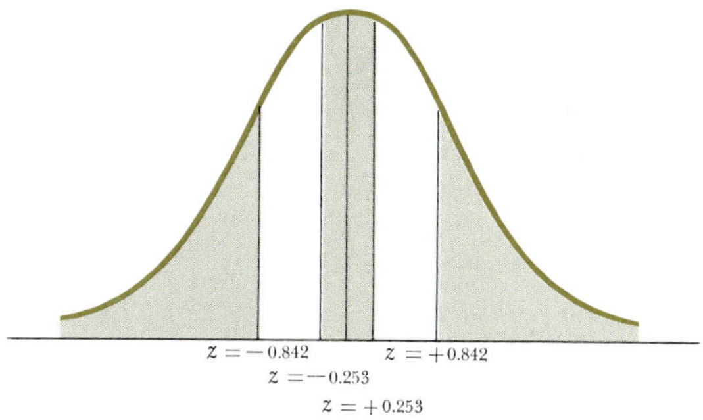

FIG. 10-15 z SCORES OF FIVE EQUAL AREAS.

tral axis and $z = 0.53$ is 0.2019. The total area between the two lines is the sum of the two areas or $0.2019 + 0.4082 = 0.6101$. Therefore there is 61.01 percent of the whole area between the two values.

These examples illustrate how any part of the total area can be calculated. Table 10-1 can also be used in reverse; that is, from a percentage of the total area the corresponding z scores can be calculated.

EXAMPLE 4 z scores that divide the total area into 5 equal parts
The values of z in Fig. 10-15 are determined by using Table 10-1 "backwards." To divide the total area into 5 equal parts, z scores for 10 and 30 percent have to be found. These then yield the desired result, as indicated in Fig. 10-15. To find these values, Table 10-1 has to be searched for the values of 0.1000 and 0.3000. Because there are only 310 entries in Table 10-1 and there are 10,000 possible four-digit decimals, it seems unlikely that any actual value, such as 0.1000 or 0.3000, would be found in the table. For example, 0.1000 does not appear in the table. The value in the table nearest 0.1000 is 0.0987, which corresponds to $z = 0.25$, and 0.1026, which corresponds to $z = 0.26$. The actual z score that corresponds to 0.1000 then lies between $z = 0.25$ and $z = 0.26$. Because 0.0987 is closer to 0.1000, one can estimate the value as $z = 0.253$. The table values nearest 0.3000 are 0.2995, which corresponds to $z = 0.84$, and 0.3023, which corresponds to $z = 0.85$. The actual z value can be estimated as $z = 0.842$. Symmetry indicates that $z = -0.253$ and $z = -0.842$ are the other two values.

In many cases, when dealing with quantities that are normally distributed, the central value and the number of actual units needed to make a standard unit can be calculated. Once this has been done, probabilities can be calculated by converting actual quantities to standard units relative to the central value. The central value is known as the "mean," and the number of actual units needed to make a standard unit is known as the "standard deviation." The method by which these quantities are calculated and the way they can be used are examined in Sec. 10-4.

10-1 EXERCISES

1. Find a sample space, and calculate the probability distribution for tossing a single die. Draw the histogram.
2. A single die is tossed twice, and the average number of spots showing is noted. Make a histogram or probability diagram for this event.
3. Why do we assume that the total area of a diagram is 1 square unit?
4. What must one assume if he wants to say that the theoretical probability distribution given by a smooth curve is the real probability distribution of an event?
5. Why do you suppose that the gaussian or bell theoretical distribution is often called the normal distribution?
6. A box contains three white, two red, and one green chip. List three different sample spaces for the event of drawing one chip from the box.
7. Why is the nonoverlapping nature of the cells of a sample space important?

10-2 PROBABILITY IN USE

In Sec. 9-1 we indicated that one of the main purposes for studying probability was to quantify, to some extent, the processes of decision making. The last few years have seen a tremendous growth in the application of mathematical statistics to these processes. In the present section, we show two kinds of processes with which decisions can be made on a formal, mathematical basis rather than intuition.

EXPECTATION

Let us begin with a problem facing Mr. A. S. (Analytical Store) Owner. Mr. Owner is thinking about having a special one-day sale on Friday of this week. Today being Wednesday, he telephoned the Weather Bureau and found that there was a 70 percent chance of rain on Friday. From experience, he knows that, if it rains on the day of a sale, he makes approximately \$700 less than if he simply carries on with business as usual. This is because of the additional help he must hire, the advertising, and so on.

If it does not rain on the day of the sale, he can expect to make \$1500 more than usual, because he has not held this kind of sale for several months. Should he hold the sale?

Mr. Owner, having read Chap. 9 of this book and being a clever man, reasons that a 70 percent chance of rain means that 70 percent of the time it will rain and 30 percent of the time it will not. He reasons further that 30 percent of the time he will make \$1500 and 70 percent of the time he will lose \$700. Although he knows that these conditions may never occur again, he imagines what would happen if they did, say, 100 times. He thinks that 30 times, that is, 30 percent of 100 times, he will make \$1500 or a total of $30 \times \$1500 = \$45{,}000$ and 70 percent of the time he will lose \$700. This is 70 percent of 100 or 70 times—a total loss of $70 \times 700 = \$49{,}000$. While looking at these figures, he sees that, if the sale took place 100 times, he would lose \$4000. He thinks of this in terms of an average, and it occurs to him that he would, in effect, lose an average of $\$4000/100 = \$40$ every time he held a sale. Being a good businessman, he cancels his plan for a one-day sale.

Mr. Owner has an even greater insight. It occurs to him that it was not necessary to assume that the conditions took place 100 times. He reasons that 70 percent of $\$700 = \$490$ would be his average loss no matter how often he had a sale. And 30 percent of $\$1500 = \$450$ would be his average gain for each sale he held. The number of trials, or times, the conditions are met does not matter when one considers an average. He confirms this by noting that $\$490 - \$450 = \$40$, or his calculated average loss.

Let us examine the key elements of Mr. Owner's method of making his decision. He has an experiment, weather on Friday, and he knows how much he will make, or win, if the outcome of no rain takes place; he also knows how much he will lose if the outcome of rain takes place. The basic elements of this problem appear to be an experiment and several outcomes. For each outcome a probability and a measure of gain or loss were given. The final figure was the sum of the products of amounts gained (or lost) and the probabilities of these gains (or losses). Wins were treated as credits, losses as debits. If the sum had been positive (a credit), he would have made money; the negative sum he got (a debit) indicated a loss.

The quantity described in the last few paragraphs is known as the mathematical "expectation" of the experiment. More generally, we can define it as follows. If an experiment has a number of mutually exclusive outcomes $A, B, C,$ and so on, and each outcome o has a probability $P(o)$ and a measure of gain or loss $D(o)$ (if $D(o)$ is positive it is a gain, if negative it is a loss), then the mathematical expectation of the experiment is

$$ME = D(A)P(A) + D(B)P(B) + \cdot\cdot\cdot + D(D)P(D)$$

There is one product or term for each possible outcome of the experiment.

For a second example of mathematical expectation, we are running the roulette wheel at the casino at Monte Carlo. The wheel has 37 pockets, numbered 0 to 36. For the game, 18 of the pockets are red, and 18 are black. The pocket labeled 0 is green. A ball is spun around the top of the wheel and permitted to drop into one of the pockets. The winning number is the number of this pocket. Among other bets a player may make with us (remember, we are the "house") is "even money" on red or black. Let us say that a player bets 100 francs on black. The ball is spun. If it drops into a black pocket, he wins 100 francs (and keeps his own 100 francs). If it drops into a red or green pocket, he loses his 100 francs. *Even money* means that the winner is paid what he bets. We can consider three outcomes: red, black, and green. The following probabilities then occur:

$$P(\text{red}) = \tfrac{18}{37} \qquad P(\text{black}) = \tfrac{18}{37} \qquad P(\text{green}) = \tfrac{1}{37}$$

Because we pay only if the ball drops into a red pocket, the mathematical expectation is

$$\begin{aligned} ME &= ME(\text{red}) + ME(\text{green}) + ME(\text{black}) \\ &= \tfrac{18}{37} \times 100 + \tfrac{1}{37} \times 100 + \tfrac{18}{37} \times (-100) \\ &= \tfrac{100}{37} \\ &= 2.7 \text{ francs} \end{aligned}$$

As you can see, this wager is excellent from our point of view, because on the average, we win 2.7 francs every time the player makes his bet. Of course, we never actually win 2.7 francs; sometimes we win 100 francs, sometimes we lose 100 francs, but we average more wins than losses (19 to 18 overall) and we average 2.7 francs per spin of the wheel.

There are other possible bets at roulette. For example, one can bet on 1 of the 37 numbers. If it comes up, he wins thirty-five-times his original bet. If any of the other 36 numbers comes up, he loses his bet. This time, suppose that we are the player. What is the mathematical expectation for this bet? The reader can verify the following calculations for a bet of 1 franc:

$$ME = \tfrac{1}{37} \times 35 + \tfrac{36}{37} \times (-1) = -\tfrac{1}{37} = -0.027$$

The negative value indicates a loss, on the average, of 0.027 franc per spin of the wheel. This figure also gives some insight into people who say that they have a "system." How can any system get around an average loss per spin?

HYPOTHESIS TESTING

Consider the drug problem mentioned in Sec. 9-1. In this problem, a disease that had a probability of $\frac{8}{10}$ or 0.8 of being fatal was described. When a new drug was administered to 20 victims of the disease, 15 recovered. We then asked whether the drug was any good? Suppose that the drug has no effect. If we make this assumption, we have a definite probability, namely, $P(\text{victim lives}) = 0.2$. What is the probability that the observed result occurred by chance; that is, what is the probability that, although the drug was not effective, simply by chance 15 out of 20 survived? We should decide in advance how we are going to interpret the answer. In other words, we should state the criteria we are going to use before finding the results. We decide arbitrarily that, if the chances of the observed result's happening by coincidence are less than $\frac{5}{100}$, we shall reject the idea that the drug does not help. However, if the observed result can happen 6 or more times out of 100 by coincidence, we agree that the drug has no effect.

Instead of computing the probability of this observed result, we determine a range of results that includes 95 percent of all possible outcomes. We reason that we have a certain distribution of probability, such as indicated in Fig. 10-16, which is the theoretical distribution of the percentage of people who recover from the disease. The figure is based on generalizations of the methods we examined in Chap. 9 and experience with the disease. As with previous histograms, we assume that the total area of all the columns is 1. The height of a column is such that its area compares with the total area of all the columns in the same manner that the probability of the outcome compares with 1. The numerals under each column represent the number who live out of a group of 20. For example, the column labeled 0 has an area of 1.15 percent of the total (the first number above the column indicates the area of the column, the second the accumulated total area of columns from left to right). This figure tells us that, out of a group of 20 victims, there is a 0.0115 chance that all 20 succumb to the disease. The column labeled 4 indicates that there is a

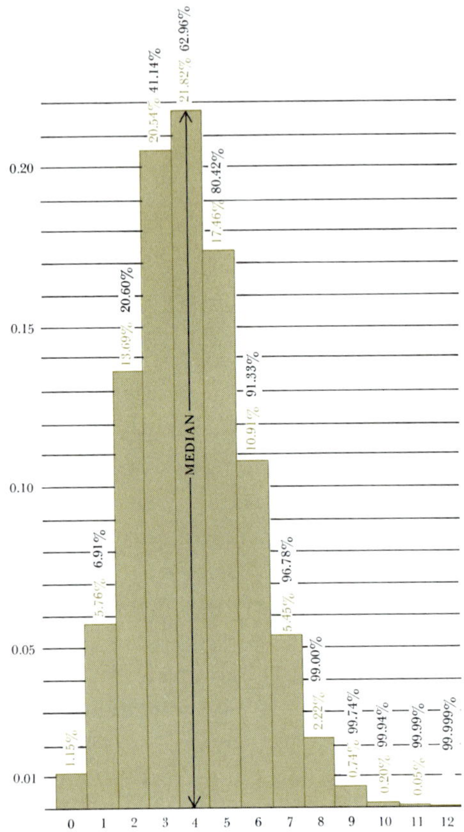

FIG. 10-16 DISTRIBUTION OF PROBABILITIES IN THE DRUG PROBLEM.

0.2182 chance that, in a group of 20 victims, 4 live. The 41.14 percent indicates that there is a 0.4114 chance that, out of a group of 20 victims, 4 or fewer survive.

What about the drug? If we examine the chart, we see that the chance that 15 out of a group of 20 victims survive is almost vanishingly small. We should certainly reject the idea that 15 people got well simply by chance. We should say that the drug was effective. The chart also indicates that, if we were willing to accept the drug if it was an improvement on 95 percent of the outcomes that take place, we should have been willing to accept it if as few as 8 people recovered. This follows from the indicated 96.78 percent total of the column labeled 7. This figure tells us that 0.9678 of the time, 7 or fewer out of a group of 20 victims recover.

This is an example of a general technique that mathematicians and statisticians call "hypothesis testing." The name comes from the hypothesis or conjecture we set out to check and use in the calculation of probabilities. In this example the hypothesis was that the drug was not effective and that the observed result occurred simply by chance. This method is not foolproof. If 8 victims had survived, we might have accepted the drug's effectiveness, although there was more than a 3 percent probability that the test cases formed one of the groups out of which 8 survived without assistance. However, suppose that only 6 had survived. We should have said that the drug was not effective. But perhaps it was. Once we make a hypothesis, there are at least two errors we can make. We can reject the hypothesis, although it is true; and we can accept it, although it is false. But the probabilities of making these errors are small.

Our examples of the use of probabilities are very simple. However, they illustrate how decisions are quantified. Naturally, most applications of probability require techniques that are more complex than those examined.

10-2 EXERCISES

1. A promotional contest held by a store to stimulate business has a prize of \$1000, and the chance of winning is $\frac{1}{15,000}$. Assume that to take part in this contest requires about 45 min of a customer's time, valued at \$1.40 per hr. What is the mathematical expectation for a person playing this contest?

2. Suppose that the following payoffs occur, based on the roll of a single die: for 1 or 2 spots, the player wins \$10; for 3 spots, the player loses \$20; for 4, 5, or 6 spots, no money changes hands. What is the mathematical expectation for the player? What is the relation of his expectation to that of the house?

3. An event is "fair" if its mathematical expectation is zero. Is a roulette bet on a single number fair? Explain.

4. Give two examples of fair bets.

5. In the example of hypotheses testing we described two errors for the specific example and two general classes of error for this kind of analysis. Which specific error is a case of which general class?

6. Why was it necessary to make assumptions as to what is taking place to test our conjecture about the effectiveness of the new drug?

10-3 STATISTICS

Some words are almost automatically associated with others. Ham and eggs, night and day, black and white, probability and statistics. We now know something about probability, but what are statistics, or what is a statistic? Generally, we think of statistics as special numbers: the number of people in San Francisco who are over 35 years old, or the average income of adults in New York State. Sometimes a statistic is something unpleasant, such as the number of traffic deaths on a Christmas weekend. Generally, a statistic is any numerical quantity used to describe or characterize an entity. Averages, largest values, smallest size, total, and so on, are examples.

When a mathematician or statistician looks for a statistic, he hopes to find a quantity to help him calculate probabilities and make decisions based on them. In the past the term *statistics* was used only for a collection of descriptive numerical quantities. Now it is considered the study and structure of decision making when chances are involved. Although we have some understanding of the universe, physicists are not sure that the ultimate answers to questions concerning structure exist. When measurements are made, there are always errors. It seems to be impossible to measure different characteristics of atomic particles at the same time. For example, if we measure the location of a particle, we cannot measure its momentum at that instant. If we measure its momentum, we cannot measure its location. These and other observations suggest that there is a basic uncertainty in the universe. In other words, chance seems a basic aspect of life. Many things are not completely determined in advance but are decided by a toss of Nature's coin. Conditions apparently as similar as man's ability to measure can make them, produce different outcomes with no apparent reason. However, known patterns of chance or uncertainty are far more useful than unknown patterns, and many theorists believe that the laws of chance are our greatest hope of understanding the universe.

What kinds of numbers are we looking for, as scientific statistics? The answer, of course, depends on the kind of problem. Two interesting statistics are those which determine the shape of a specific bell-shaped curve (Fig. 10-17). The first of these is the "mean," or arithmetic average of the outcomes, assuming that numerical quantity is being measured. The mean locates the position of the curve. In Fig. 10-17 there are three probability distributions, all with the same shape but different means.

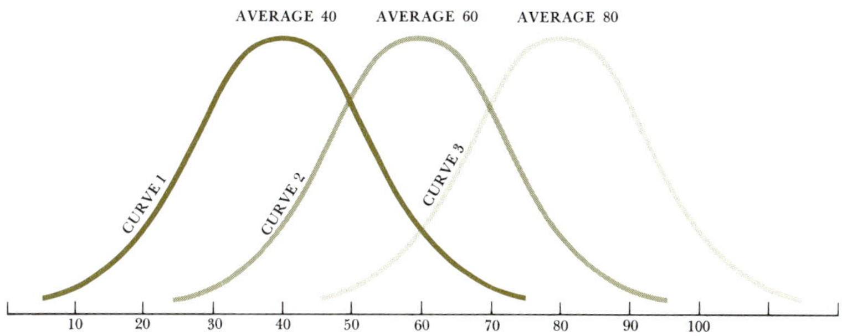

FIG. 10-17 SIMILAR NORMAL CURVES WITH DIFFERENT CENTRAL AXES.

The mean for a normal distribution is the value at the "axis," or the line about which the curve is symmetrical. Suppose that the values along the bottom of the curve have a particular meaning, say, the grades on a test given to various groups of students. The left curve has an average of 40, the middle curve an average of 60, and the right curve an average of 80. The mean, then, is a measure of location. The reason for the name is obvious. It locates the center and thus the position of a probability distribution. For a normal distribution of outcomes, the mean is the only measure of location needed. Other probability distributions, both theoretical and those compiled from observed data, require other measures of location. One of these is the "median."

If the outcomes of an event are listed in descending order, or ascending order, the median is the outcome such that half the outcomes are larger than it and half are smaller. Take, for example, the distribution of outcomes 1, 1, 3, 5, 6, 6, 7. The median is 5. Note that the average of this simple distribution is 4. For symmetrical distributions of outcomes, such as the normal distribution, the mean and the median are the same.

A third measure of location is the "mode," or the value that occurs most often in the distribution. In the distribution in Fig. 10-16, the mode is 4, because the most frequent number of survivors is 4, this outcome occurring 21.8 percent of the time. These three quantities—the mean, the median, and the mode—are the most common measures of the location, although there are others. Consider the three probability distributions in Fig. 10-18. These three curves are all normal distributations; furthermore, they all have the same mean. But they are certainly different. Each is so constructed that it encloses an area of 1 square unit, but the "spread," or variation of values, is different in each. To work with these differences, one must find a way of describing them. A numerical quantity describing

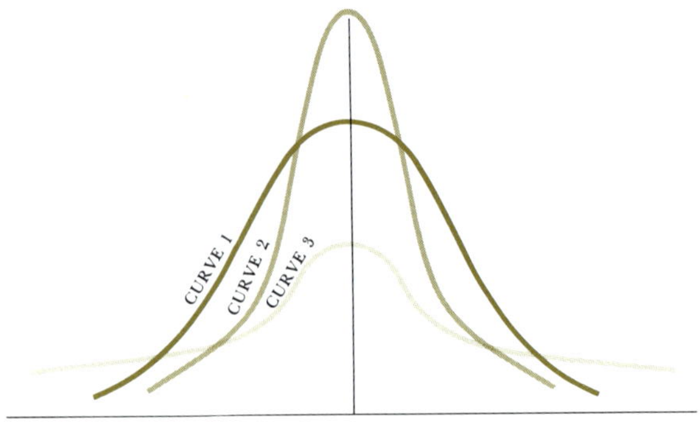

FIG. 10-18 NORMAL CURVES WITH THE SAME CENTRAL AXIS BUT DIFFERENT STANDARD DEVIATIONS.

this variation is known as the "standard deviation" or "root-mean-square deviation" of a distribution. This quantity tells how far from the right or left of the mean one must go to enclose a specified area or part of the whole. One standard deviation to the right or left encloses about 34 percent of the total area.

For other distributions of outcomes, other measures of variation are used. One of these is the "range" of variation or difference in value between the largest and smallest outcomes. Another is the "average deviation," or average difference. This is the average value of the difference between individual outcomes and the mean. The latter quantity is zero for a symmetrical distribution. The list does not end here. There are many other measures of spread or variation. The standard deviation is the most commonly used, because, coupled with the mean, it describes a normal probability distribution completely.

We can illustrate the mean and standard deviation by calculating them for a selected set of data. Consider the following set of numbers, which we assume represent measurements of a quantity: 20, 20, 25, 25, 25, 30, 30, 30, 35, 40, 40, 40, 45, 45, 45, 90, 95. Perhaps they are the numbers of pounds of apples yielded by the trees in an orchard, or the scores on an examination. The source of the data is not a factor in the calculations. The mean, or average value, is defined as the sum of the observations divided by the number of observations. In this case, the sum is 680, and there are 17 observations; the average is thus $\frac{680}{17} = 40$, and the mean is 40. We can examine the data more easily if we arrange them in column form, as in Table 10-2.

TABLE 10-2 SAMPLE DATA FOR CALCULATION OF MEAN AND STANDARD DEVIATION

	Observed result	*Deviation or difference between observations and mean*	*Square of the entries of deviations*
	20	−20	$20 \times 20 = 400$
	20	−20	$20 \times 20 = 400$
	25	−15	$15 \times 15 = 225$
	25	−15	$15 \times 15 = 225$
	25	−15	$15 \times 15 = 225$
	30	−10	$10 \times 10 = 100$
	30	−10	$10 \times 10 = 100$
	30	−10	$10 \times 10 = 100$
	35	−5	$5 \times 5 = 25$
	40	0	$0 \times 0 = 0$
	40	0	$0 \times 0 = 0$
	40	0	$0 \times 0 = 0$
	45	+5	$5 \times 5 = 25$
	45	+5	$5 \times 5 = 25$
	45	+5	$5 \times 5 = 25$
	90	+45	$45 \times 45 = 2025$
	95	+50	$50 \times 50 = 2500$
Total	680	−10	6400

We can easily find the median for this set of data. The median was defined as the value such that half the results are larger and half smaller. Because there are 17 observations, the median is the ninth value from the top or bottom. By inspection the median is 35.

The standard deviation is the square root of the average value of the squared deviations. The sum of the squared deviations is 6400. Because there are 17 observations, the average squared deviation is $\frac{6400}{17}$ or approximately 376.47. The square root of 376.47 is approximately 19.4. This number, 19.4, is the standard deviation of these observations. Thus about 34 percent of the scores are numbers between 40 (the mean) and 59 (the mean plus one standard deviation). The area encompassed by the middle four standard deviations (two on each side of the mean) contains more than 95 percent of all scores (see Fig. 10-9). In this example, two standard deviations are about $19 + 19 = 38$. Thus 95 percent of the scores should lie in the range of 2 to 78. We see that that is about right, only two scores (90 and 95) being outside this range. These values are calculated using the normal-curve methods in Sec. 10-1.

We have calculated the mean and standard deviation directly from their definitions. When actually solving a problem, there are many shortcut formulas and methods that can be used. If large numbers of data are being considered, then we should use a computer, which could shorten even the shortcuts.

Because the mean and standard deviation completely describe a normal distribution as well as a number of other theoretical probability distributions, they in turn can be calculated from theoretical considerations directly; that is, it often happens that knowledge of the probability distribution is used to arrive at the mean and standard deviation of the set of quantities rather than calculating these quantities from raw numerical data. For example, in Sec. 10-1 a distribution for the number of heads expected in tossing 11 coins was given. The values were calculated by finding a set of equally likely, mutually exclusive outcomes for the experiment of tossing 11 coins. This set of outcomes contained 2048 individual outcomes. Using methods related to the concepts of permutation and combination, the number of outcomes favorable to 0, 1, 2, 3 heads, and so on, can then be determined. Once this was done, the table of probabilities was established.

On theoretical grounds, related to the nature of the experiment, the mean number of heads to be expected when a number of coins are tossed is

$$\mu = n \times \tfrac{1}{2}$$

where μ is the mean and n the number of coins. The standard deviation is

$$\sigma = \sqrt{n \times \tfrac{1}{2} \times \tfrac{1}{2}}$$

where σ is the standard deviation and n again the number of coins. In this example, these values are

$$\mu = 11 \times \tfrac{1}{2} = 5.5$$

and

$$\sigma = \sqrt{11 \times \tfrac{1}{2} \times \tfrac{1}{2}} = \sqrt{2.75} = 1.66$$

Section 10-4 illustrates how these results can in turn be used to calculate probabilities.

10-3 EXERCISES

1. Give three examples of statistics used descriptively.

2. In what ways does a statistician make use of statistics?

3. Descriptive statistics seek to characterize two important aspects of the distribution of data. What are they?
4. Calculate the mean and standard deviation for the following set of numbers: 70, 72, 72, 74, 74, 74, 76, 76, 76, 76, 78, 78, 78, 78, 78, 80, 80, 80, 80, 80, 80, 82, 82, 82, 82, 82, 84, 84, 84, 84, 86, 86, 86, 88, 88, 90. Use the method shown in the example in the text.
5. Estimate the median for the data in Exercise 4. Why is this difficult?
6. An experiment of tossing 20 coins is to be performed. Use the formulas in the text to calculate the mean and standard deviation of this experiment.
7. Suppose that the experiment described in Exercise 6 is repeated a number of times, say, 100. If the number of heads observed in each toss is averaged, must this average agree with the result of Exercise 6? Explain.
8. Assume that a group of 11 coins is tossed 2048 times, the number of heads observed in each toss being tallied. Further assume that when all the tosses are complete, the tally follows the theoretical probability distribution given in Sec. 10-1; that is, of the 2048 tosses, 1 had no head, 11 had 1 head, 55 had 2 heads, 165 had 3 heads, 330 had 4 heads, and so on. With this information as raw data, calculate the mean number of heads for the 2048 tosses and the standard deviation of the heads using the methods outlined in the present section.
9. Interpret the expected value described in Sec. 10-2 as an average calculated from a theoretical distribution of probabilities.
10. If an outcome of an experiment has a probability p of occurring when the experiment is repeated n times, the average number of times the specified outcome takes place is

    $$\mu = n \cdot p$$

    The standard deviation of the number of times the outcome is observed is

    $$\sigma = \sqrt{n \cdot p \cdot (1 - p)}$$

    Tossing a 5 with a pair of dice is such an outcome, with $p = \frac{5}{36}$. If $n = 50$, calculate μ and σ.
11. What is the average number of times "snakes eyes," or 2, can be expected if a pair of dice is tossed 100 times?
12. Must the answer to Exercise 11 be real, in the sense that it must be a number that could be observed if the dice were actually tossed? Explain.

10-4 STATISTICS IN USE

We are all familiar with the use of statistics as descriptive quantities. To demonstrate how they can be useful in calculation, consider the following comparison between an actual set of probabilities as given in Sec. 10-1 and a corresponding set calculated by assuming that the distribution of probabilities follows a normal distribution with mean and standard deviation calculated as in Sec. 10-3.

The calculation of the probabilities in the experiment of tossing 11 coins and counting the number of heads were carried out in Sec. 10-1. In Sec. 10-3 it was determined that the mean number of heads in this experiment was 5.5, and the standard deviation 1.66. If a normal curve of distribution with mean 5.5 and standard deviation 1.66 is superimposed over the histogram in Fig. 10-7, the result is Fig. 10-19.

It should be remembered that both the area below the smooth normal curve and the total of the columns of the histogram are to equal 1 square unit. From Fig. 10-19 it seems that the area in column 7 can be estimated by the area bounded by the normal curve and the lines at 6.5 and 7.5. This area can be calculated by using the methods in Sec. 10-1. The line at 6.5 is 1 actual unit or $\frac{1}{1.66} = 0.6$ standard unit from the central axis at 5.5. The line at 7.5 is 2 actual units or $\frac{2}{1.66} = 1.21$ standard units from the central axis. Using Table 10-1, a line 0.6 ($z = 0.6$) standard unit from the central axis has 0.2257 of the total area between itself and the central axis, and a line 1.21 ($z = 1.21$) standard units from the central axis has 0.3869 square unit between itself and the central axis. The area between the

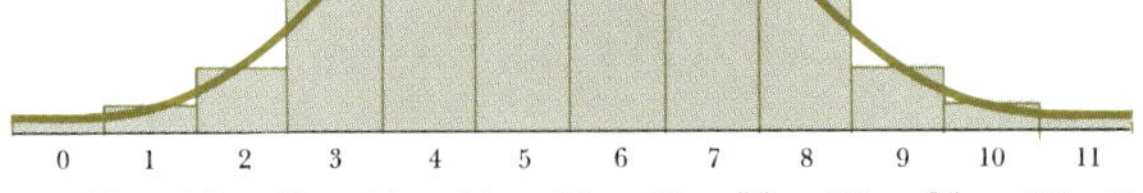

FIG. 10-19 HISTOGRAM OF THE DISTRIBUTION OF HEADS IN 11 TOSSES OF A COIN AND THE NORMAL-CURVE ESTIMATE.

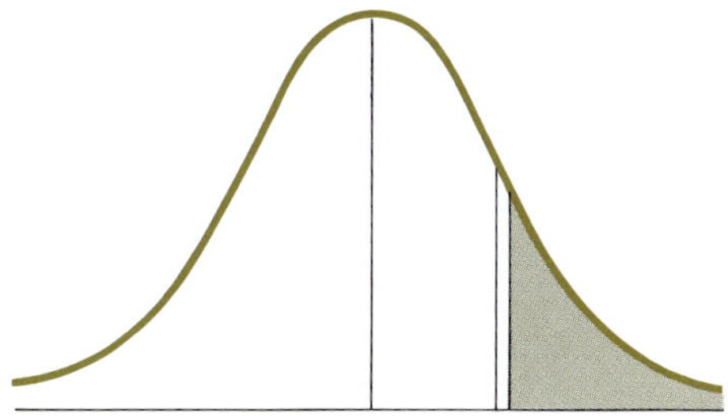

FIG. 10-20 NORMAL-CURVE ESTIMATE OF MORE THAN 210 HEADS IN 400 TOSSES OF AN HONEST COIN.

two lines is then $0.3869 - 0.2257 = 0.1612$ square unit. The area in the column above the 7 is $\frac{330}{2048} = 0.16113$ square unit. Thus the normal-curve area differs from the actual area by only $0.1612 - 0.16113 = 0.00007$. If one calculates the probability using the normal-curve estimate rather than the actual "long" way, he makes an error of less than 0.01 percent.

The same method can be used to calculate other estimates of the probabilities related to the experiment of tossing 11 coins.

As another example, suppose that an experiment of tossing a coin 400 times is to be conducted. Further suppose that you, as one of the players involved, feel that your opponent is not the most honest person in the world and that he will use a coin weighted toward heads. You are able to get the following rule included in the experiment. If there are more than 210 heads when the coin is tossed, the bet is off, and a new coin will be used for a second trial. What are the chances that an "honest" coin will be thrown out?

To calculate this probability using the actual data would be difficult; however, the normal-curve estimate is relatively easy. The average number of heads is $400 \times \frac{1}{2} = 200$, and the standard deviation $\sqrt{400 \times \frac{1}{2} \times \frac{1}{2}} = \sqrt{100} = 10$. Figure 10-20 shows the situation, using the smooth normal curve rather than a series of 400 individual columns. The answer is the fraction of the total area to the right of the line identified as 210.5, which is 10.5 actual units or $\frac{10.5}{10}$ standard units from the central axis of 200. Table 10-1 indicates that a line 1.05 units from the central axis has 0.3531 square unit between itself and the central axis. Therefore there is $0.5000 - 0.3531 = 0.1469$ square unit to the right of the 210.5 line. This means that 14.69 percent of the time an honest coin produces more than 210 heads when tossed 400 times.

In summary, if a set of outcomes has a normal distribution, the mean and standard deviation identify the particular shape of the distribution.

Other theoretical distributions require other statistics to complete their mathematical description.

Statistics, then, is the study and use of numerical statistics as objects. At one time the only use for statistics was to fill pages of reports. The sole function of a statistician was to collect and tabulate these numeral quantities. The only area in which there was any use of these quantities, in a modern sense, involved gambling and its sophisticated cousin, insurance. Today, with ideas such as those discussed, it is one of the important tools in the quantification of decision making.

Naturally, most applications require methods and concepts beyond those we have examined. We can, however, examine in a general descriptive manner some examples of the advanced application of probability and statistics.

One of the exercises in Chap. 7 described the test marketing of a new product offered in three boxes of different color: red, green, and blue. How would one decide that one color was better than another? In practice, the manufacturer should have consulted with statisticians before making his survey. The design of a statistical experiment is a key factor in statistical analysis. A properly designed experiment provides many useful results from a very small sample of a whole. To illustrate this point, consider briefly a problem faced by the state of California every year.

In the complexities of finance, California faces the problem of the distribution of state funds to the 58 counties of the state. The counties range in population from Los Angeles with millions to Modoc with 8000, Mono with 2000, and Alpine county with 400 inhabitants (half this figure when the native Indian population is not counted, as the census sometimes does not). The distribution of funds depends on the value of the land in a county, and the problem is to estimate the worth of all the property in a county. This is carried out by the state board of equalization. How does one determine how much the property in an area is worth? One way is to send appraisers into the area and evaluate each parcel of land. With the millions of pieces of land in California and with the rapidly changing value of land in this fast-growing state, this is impossible. What would be the next best method? Probably taking a sample of the properties in a county and, on the basis of this sample, estimating the worth of the total property in the county. How big a sample must be used? Suppose that one tried a 10 percent sample of the property in a county. In Alpine county, with approximately 400 pieces of property, this would amount to 40 appraisals; in Los Angeles, with approximately 4,000,000 pieces of property, this would amount to 40,000 appraisals. Even this solution would not work. In a small county, such as Alpine, some single pieces of property represent 5 percent of the total value of the county. Using advanced theories of sampling and probability, the board of equalization's statisticians were able to work out a method of sampling properties that required a sample of about 60 properties in Alpine county, and only 400 in Los Angeles county. The overall results were excellent.

Size is not the only factor that controls the usefulness or reliability of a sample. The nature of the set to be sampled and the probabilities of various outcomes of the sampling procedure can affect how good the sample will be. To determine the color of 10,000 beads in a box, one has only to look at one bead if they are all the same color; but one must look at all 10,000 if there are 10,000 different colors. All too often, a sample is worthless when taken without forethought. Often the desired result could be found at an expenditure of half the cost and effort.

A second kind of advanced application involving probability can be illustrated with a standard approach used by many branches of the Weather Bureau across the country. It is not uncommon to turn on a radio and hear, "The probability of rain is 60 percent today, 80 percent tonight, and 30 percent tomorrow." The use of probability in forecasting is a very reasonable concept. Not only weather, but many predictions of physical events are best handled in terms of probability. Many problems in physics and related areas can be carried out only in terms of probabilities. Sometimes it seems that atomic particles and huge weather systems behave more like a pair of dice than as predetermined events following immutable laws of nature. Physical scientists of today sometimes take the point of view that Nature plays dice with the world. They feel that the future is not ordained by the events of the past. Until recently, many felt that, if we had detailed knowledge of the past and sufficient knowledge of the laws of nature, the future could be foretold completely. Modern theorists, however, think of the universe operating, to some degree, by pure chance. What more natural way to work with forecasts, of tomorrow's weather or of a nuclear reaction, than with the laws and methods of probability?

Actually, we are considering two different ideas. In one, we use probability theory to analyze an event, and, in the other, we use results that, rather than definite conclusions, are probabilities. The probability distribution for tossing a pair of dice formed a complete picture of the event. Given an event, such as tossing a pair of dice, the probability distribution is an excellent total picture of what is taking place.

A third major use of probability in advanced applications involves the search for relationships among results. For a simple example, suppose that we wish to determine whether any relationship exists between cigarette smoking and lung cancer. We have the records of deaths due to this disease, and we also have records of how much the victims smoked. What are the chances that there is a relationship? Could an apparent relationship be the result of coincidence? Without explaining the details of the analysis, we now know that it has been proved that such a relationship does exist beyond pure chance. Similar questions include the following. Is there any relationship between abilities in mathematics and English? Is there a relationship between economic status and IQ? Is there a relationship between soil moisture and crop yield? And so on, with a hundred more questions, each lending itself to a form of statistical analysis.

Advanced methods, such as multiple nonlinear regression-analysis correlation, analysis of variance, and stratified random sampling procedures, may not have any meaning to most of us, but their use has an effect on the cash in our pockets, the television entertainment we are provided, and the color and kind of new cars we find in showrooms. Of more significance, they help determine the usefulness of a new drug; they assure us that the airliner we fly is safe from equipment failure; that wastes dumped into a river are safe and will not pollute the water; that a food product is free from dangerous bacteria. These are only a few examples of the use of statistical methods to answer questions of vital interest to us all. Using carefully constructed probability models of actuality, we make decisions on the basis of rational criteria rather than feelings. Certainly, these are powerful tools we can use as we carry out the process of making and acting on decisions.

FIGURES DON'T LIE BUT . . .

We cannot end our discussion without some remarks on the misuse, rather than the use, of probability. When we looked at relatively complex examples, we saw that our intuitive feeling was not in agreement with the facts of the laws of chance. How many of us would have had a correct feeling about the birthday probability? We now examine some events for which probabilities correctly calculated lead to incorrect results. In short, as with the other concepts we have examined, the methods of probability and statistics are not all-powerful and do have their limitations.

Statistically, tomatoes are 95 percent fatal. "But, but, but," you say. Read the claim again, "Statistically, tomatoes are 95 percent fatal." What are we claiming here? If our understanding of probability is sound, we are saying that 95 percent of the time a person who has eaten a tomato has died, and this is a true statement. How? Simple. We estimate that, since modern man arrived on the earth, about 60 billion persons have been born or lived. Of these 60 billion, approximately 3 billion are still alive, or perhaps it would be better to say that 3 billion are alive today. This leaves 57 billion, or about 95 percent of the total, who have died. Therefore, our claim that 95 percent of the people who have eaten tomatoes are now dead is true. (For that matter birth is fatal 95 percent of the time, because 95 percent of the people who have been born have died.) The fallacy, of course, lies in relating, by statistics, two things which are unrelated in a statistical sense.

We could use these same facts about the world's population to disprove statistically the old saw that nothing is certain but death and taxes. Without comment on taxes, we state that there is a 5 percent probability of not dying, for 1 out of every 20 persons who have been born has not died. Well, this has been a trick, a special case in which the probabilities are correct but have been taken out of context. Or has it been?

Suppose that you are an investor in the stock market and want to invest in the transportation industry. Because you are concerned with cost and efficiency, you want to choose a company or area of transportation that is strong in both of these factors. You know that there are methods of measuring them. A usual measure of cost is the cost of transporting 1 ton of cargo 1 mile, or the cost per ton per mile. As to efficiency, one standard method is the load factor, or the percentage of available seating or freight space in use. After some searching, you stumble across precisely the method of transportation you are looking for. It costs only pennies per ton-mile and has had an average load factor of 100 percent without turning down requests for shipment. You say that you have never heard of such a marvelous mode of transportation, but you have. You have seen and examined every detail of it, for we are talking about space capsules. Using the measures of ton-mile and load factor that we have described, this method of travel is by far the best.

One more example. If the odds on an event were 250,000,000 to 1 against its taking place, would you be willing to bet that it would take place? Of course not. No one in his right mind would be willing to bet on such an outcome. The chances of such an event's taking place are even less than the chances that the random motion of air molecules will suddenly cause all the air in a room to rush to one corner. It is our good luck that our 250,000,000 to 1 chance does take place. Medical experts estimate that only one sperm cell in 250,000,000 actually reaches and fertilizes an ovum. The event we were considering is the birth of a human being.

Probabilities and statistics have to be taken in a specified context. In the examples considered above we have, of course, chosen cases where the events and probabilities are atypical, but the same flaws and problems arise. In more realistic cases, probabilities, taken at face value, may be correct but may lead, through carelessness, to major errors.

10-4 EXERCISES

1. Calculate the normal-curve estimate for the probability of tossing 4 heads in 11 tosses of a coin.

2. Calculate the actual probability of tossing 4 heads in 11 tosses of a coin.

3. Why was the number 210.5 used in the second example in the present section rather than the actual number of heads being considered, namely, 210?

4. Use the method in the present section to calculate the probability that in 11 tosses of a coin at least 3 heads are observed.

5. In problems asking "at least" or "at most" questions, what advantages does a normal-curve estimate have over direct calculations?

6. Find a range of values around the mean of 200 such that 95 percent of the time 400 coins are tossed the number of heads observed falls in this range.

7. In what ways do the examples in the present section reflect the idea of symbolic model building?

8. What word is usually used to describe the conditions indicated in the text example of a box with 10,000 beads all of the same color and another box with 10,000 beads of 10,000 different colors?

9. How may weather be a "dice-tossing" event?

10. Suppose that a probability model of the weather had been constructed in 1900. Why would such a model be useless without modern technology?

11. How do the applications mentioned in the present section reinforce the idea that technology must advance along a front rather than by long probes into the unknown?

12. If one divided the sky into 10 billion equal areas, and these areas were searched by a photoelectric device sensitive to the visual limits of the human eye, only once in 1,250,000 tries would there be any response due to the presence of stars. Why could we then say that statistically stars do not exist?

REVIEW EXERCISES

1. There is a method for solving probability problems known as the "Monte Carlo" method. What do you think is its underlying principal?
2. Explain the significance of an event having a positive mathematical expectation and of one having a negative mathematical expectation.
3. Why must the outcomes of a sample space be nonoverlapping?
4. The sum of the probabilities of a sample space must add up to 1, and the total area of a histogram is 1 square unit. Explain.
5. In the text, it was mentioned that, if the mathematical expectation of a bet was negative, no system for placing a sequence of such bets could result in a net gain for a player. Explain this in terms of long-run averages.
6. If a bet has a negative mathematical expectation, does this mean that every player who makes the bet loses?
7. The Chamber of Commerce of Marvelous Vacation Area claims that 3 out of 4 scuba divers prefer air that has been bottled in their area. What additional facts might be needed to attach any real meaning to this bit of information?
8. A major change has taken place in the work of statisticians during the past several years. What is it?
9. Explain insurance as a bet. Could an insurance company afford to have a "fair" bet with a client?
10. Explain in terms of expectation, and so on, why a person would be less likely to lose money at a friendly neighborhood poker game than in Las Vegas.
11. Find the mode of the data in Exercise 4, Sec. 10-3.

SELECTED REFERENCES

Fujii, John N.: "An Introduction to the Elements of Mathematics," John Wiley & Sons, Inc., New York, 1961.

Thorough discussion of set counting and abstract elementary probability.

Freund, John: "Modern Elementary Statistics," 2d ed., Prentice-Hall, Inc., Englewood Cliffs, N.J., 1960.

Standard introductory text in normal-curve areas and statistical methods.

Kemeny, John, J. L. Snell, and G. L. Thompson: "Introduction to Finite Mathematics," Prentice-Hall, Inc., Englewood Cliffs, N.J.

Probability introduced from set theory. Very elementary.

Meserve, Bruce E., and Max A. Sobel: "Introduction to Mathematics," Prentice-Hall, Inc., Englewood Cliffs, N.J., 1964.

Probability introduced from a computational point of view. Very elementary.

ANSWERS TO ODD-NUMBERED EXERCISES

SECTION 1-2

1. (*a*) Table
(*b*) Chair (almost any word will do)

3. When they can be paired, with no member of either group without a mate.

5. Make a stack of 10 dimes; then stack the others in piles of the same height.

7. The same number of objects.

SECTION 1-4

1. (*a*) 65
(*b*) 1010
(*c*) 200,087

3. (*a*) 75
(*b*) 809
(*c*) 3421
(*d*) 9060

5. (*a*) 328 (*b*) 291
(*c*) 1501 (*d*) 2149
(*e*) 2448

7. None is needed. If one wishes to exclude multiples of 10, for example, no heel bones are drawn.

9. If a particular order is used, it is possible to approximate numbers at a glance.

11. (*a*) *aa* 9
(*b*) *bbb aaa*
bbbb aaaa 7
(*c*) *cccc bb* 9
(*d*) *dddd ccc*
dddd cccc aaa 5

SECTION 1-5

1. 1080

3. 226

5. 51,499

7. 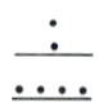

9. 

11. 

13. 

15. (*a*) Base 4
(*b*) say-say
say-say-oh
say-say-ah
say-say-so
(*c*) say-say-say-oh
(*d*) No

SECTION 1-6

1. 1202

3. 47,536,328

5. 31,752,966,603

7. 

9. 

11. 

13. 60 seconds = 1 minute, 60 minutes =1 degree.

SECTION 1-8

1. 0, 1, 10, 11, 100, 101, 110, 111, 1000, 1001, 1010

3. 42

5. 23110_4

7. $13e_{12}$

9. 55_{10}

11. *a.* 66_{12}
b. 900_{12}
c. $e74_{12}$
d. 5003_{12}
e. $e0t0_{12}$
f. $teet_{12}$

13. You must invent a single symbol for 19, such as *N*, for the same reason that *e* is 11 in base 12.

15. (*a*) *zywc*
(*b*) *azayawc*
(*c*) *aaoac*

REVIEW EXERCISES FOR CHAPTER 1

1. Numerals are the names of numbers. A number is one of the things the wheels on a car have in common with the legs of a horse; this number is called four.

3. 200,362

5. 145,823

7. 2449

9. 0, 1, 2, 10, 11, 12, 20, 21, 22, 100, 101, 102, 110, 111, 112, 120

SECTION 2-1A

1. 

3. 

5. The traditional Chinese-Japanese. It is base-10, and although it is not a positional system, its addition combination and "carries" are the same as the Hindu-Arabic.

7. 

9.
```
 9452
 9984
  699
18925
 2013
  0
```

11.

carry

	1(100) 1(10)	
498 =	4(100) + 9(10) + 8(1)	
752 =	7(100) + 5(10) + 2(1)	
	1(1000) + 2(100) + 5(10) + 0(1)	
	= 1250	

SECTION 2-1B

1. Regrouping in terms of base, base times base, and so on.

3. 1112_5

5. 2312_5

7. 130_8

9. 14505_5

11. 10153_8

13. (*a*) Excess 4
 (*b*) Excess 1
 (*c*) Excess 1
 (*d*) Excess 6
 (*e*) Excess 3

15. 1001000_2

17. 10001011_2

19. $e9_{12}$

21. $159e_{12}$

23. $1e507_{12}$

25. (*a*) The excess of fives is 1.
 (*b*) The excess of nineteens is 14 ($1496 \div 19 = 78$ with a remainder of 14).

27. All addition seems to check, because base 2 has digits 1 and 0. If 1 is cast out, 0 remains no matter what the problem.

SECTION 2-2

1.
```
 1      32
 2      64
 4     128
 8 ✓   256
16 ✓   512
32 ✓  1024
```
$56 \times 32 = 256 + 512 + 1024 = 1792$

3. (*a*)
```
   4
  32
1268   (1428)
  42
  34
```
(*b*)
```
  79
 634
5624   (6794)
  86
  79
```

5.

x	0	1	2	3	4	5	6	7
0	0	0	0	0	0	0	0	0
1	0	1	2	3	4	5	6	7
2	0	2	4	6	10	12	14	16
3	0	3	6	11	14	17	22	25
4	0	4	10	14	20	24	30	34
5	0	5	12	17	24	31	36	43
6	0	6	14	22	30	36	44	52
7	0	7	16	25	34	43	52	61

7. (*a*) 4210_8
 (*b*) 310661_8
 (*c*) 4530_8
 (*d*) 1576115_8

9. Because the order of multiplication is irrelevant.

SECTION 2-3

1. 110_5

3. 13_5

5.

$$\begin{array}{rcl} & \textit{Check} & \\ 303_5 & 3+0+3=11_5 & \textit{Excess of fours} \\ \underline{232_5} & \underline{2+3+2=\ 3} & 11_5-3=3 \\ 21_5 & 2+1\ =\ 3 & \end{array}$$

Note that if the excess of 303_5 is called 2, then we must subtract $2-3$. Therefore instead of 2 we used 11_5.

7. 4224_{12}

9. $65t7_{12}$

11. $10t2_{12}$

13. $701t5_{12}$

15. 10

17. 1011_2

19.
$$\begin{array}{r|l} & 4\,0\,0\,0 \\ & 9\,2\,6\,7 \\ \hline 1 & 3\,2\,6\,7 \\ & 1 \\ \hline & 3\,2\,6\,8 \end{array}$$

21.
$$\begin{array}{r|l} & 3\,5\,2\,0\,4 \\ & 9\,0\,0\,8\,8 \\ \hline 1 & 2\,5\,2\,9\,2 \\ & 1 \\ \hline & 2\,5\,2\,9\,3 \end{array}$$

23.
$$\begin{array}{r|l} & 1\,1\,1 \\ & 1\,0\,0 \\ \hline 1 & 0\,1\,1 \\ & 1 \\ \hline & 1\,0\,0 \end{array}$$

25.
$$\begin{array}{r|l} & 1\,0\,0\,0 \\ & 1\,0\,0\,0 \\ \hline 1 & 0\,0\,0\,0 \\ & 1 \\ \hline & 1 \end{array}$$

27.
$$\begin{array}{r|l} & 9\,e\,4\,9 \\ & 8\,6\,7\,t \\ \hline 1 & 6\,6\,0\,7 \\ & 1 \\ \hline & 6\,6\,0\,8 \end{array}$$

29.
$$\begin{array}{r|l} & 4\,0\,0\,0 \\ & 3\,1\,2\,0 \\ \hline 1 & 2\,1\,2\,0 \\ & 1 \\ \hline & 2\,1\,2\,1 \end{array}$$

31. (*a*) If one removes 20 apples from a group of 100 apples, the rest numbers 80.

(*b*) From \$100 I shall keep \$75 and give you the rest.

SECTION 2-4

1.
$$\begin{array}{rrl} 1 & 7 & 88 \div 7 = 8 + 4 = 12 \\ 2 & 14 & \\ 4 & 28\checkmark & \\ 8 & 56\checkmark & R = 88 - 84 = 4 \end{array}$$

3.
$$\begin{array}{rrr} 1 & 23 & 7004 \div 23 = 256 + 32 \\ 2 & 46 & + 16 = 304 \\ 4 & 92 & R = 7004 - 6992 \\ 8 & 184 & = 12 \\ 16 & 368\checkmark & \\ 32 & 736\checkmark & \\ 64 & 1472 & \\ 128 & 2944 & \\ 256 & 5888\checkmark & \end{array}$$

5.
$$\begin{array}{r|l} 5 & \\ 6\,1 & \\ 5\,1\,5\,5 & \\ 1\,4\,3\,5\,7 & 1\,5\,6 \\ 9\,2\,2\,2 & \\ 9\,9 & \end{array}$$

7. Multiply the excess of nines from the divisor and quotient, and to this product add the excess from the remainder. If the division is correct, this sum has the same excess of nines as the dividend.

9. 302 $R = 95$

11. 131_5 $R = 23_5$

13. $16t_{12}$ $R = 5$

15. 1011_2

17. 1001001_2

SECTION 2-5

1. 23110_4

3. $13e_{12}$

5. xte_{13}, where $x = 12$

7. 324_5

9. 75_{12}

11. 20101_5

13. $8t4_{12}$

SECTION 2-6

1. Incorrectly used (to name a few): Heading 2-5, page 53, and lines 18 and 22, page 28.
Correctly used: lines 1 and 2, page 58.

3. No

5. $10 - 3 \neq 3 - 10$

7. *Operation* Go from city A to city B.
Possible processes By train, plane, or bus.

Operation To have a good time.
Possible processes Shows, parties, or water skiing.

9. One case is sufficient to show that division is not associative, but to prove that addition is associative for all numbers would be an unending task.

REVIEW EXERCISES FOR CHAPTER 2

1. 8

3.

x	0	1	2	3
0	0	0	0	0
1	0	1	2	3
2	0	2	10	12
3	0	3	12	21

5.
```
 2642
  573
   86
 ----
 2191
 320
  3
```

7. 132031_5

SECTION 3-1

1. Two examples are:
(*a*) *Objective* Driving a car from point A to point B. The workings of the engine, and so on, are not important to the driver, only how to arrive at his destination.
(*b*) *Objective* Adjust the picture of a color television to good color. The electronics of the tint and color controls are not of interest, nor is the basic principle of such transmissions, only the observed picture quality.

3. The natural examples for a mathematics text would be any of the arithmetic operations. Others include the construction of any object on an assembly line; the manner in which almost all men shave, by habit rather than thinking it out each time.

5. (*a*) Make a stack of 10 dimes.
(*b*) Place the rest of the dimes in a number of stacks that are the same height

as the first stack; the last stack may be shorter than the rest.

(*c*) Count the number of full stacks.

(*d*) Count the number of dimes in the last stack, and multiply the number by 10.

(*e*) Step *c* gives the number of dollars, step *d* the cents.

SECTION 3-2

1.

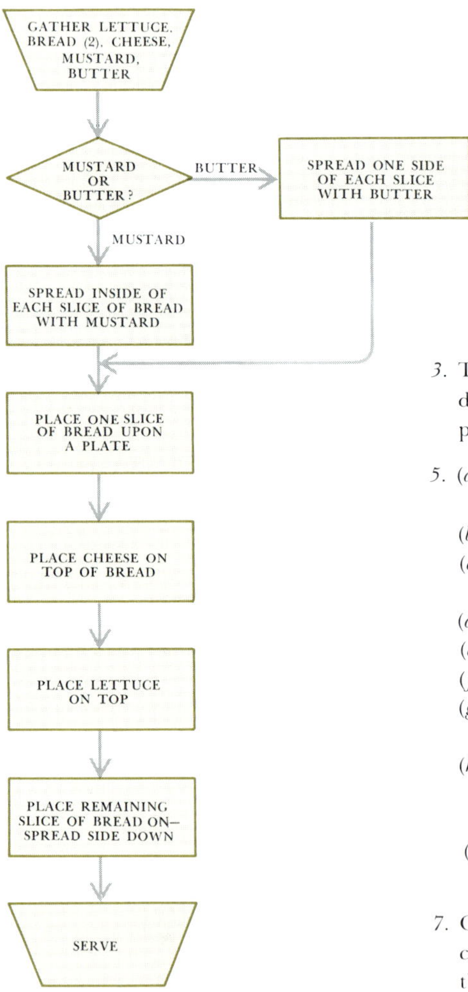

3. The actual method used by the machine does not matter, only the input and output methods.

5. (*a*) Sit behind steering wheel and buckle seat belt.
(*b*) Put key in ignition lock.
(*c*) Set gearshift lever to N (neutral) or P (park).
(*d*) Push gas pedal to floor and release.
(*e*) Turn key to start position.
(*f*) Release key when engine starts.
(*g*) Check ignition dials for normal readings.
(*h*) If readings are normal, move gearshift lever to D (drive) and slowly depress accelerator.
(*i*) If dials are not normal, turn engine off and call for help.

7. One must speculate on exactly how the car starts. Not all cars are started with the key in the ignition lock.

9.

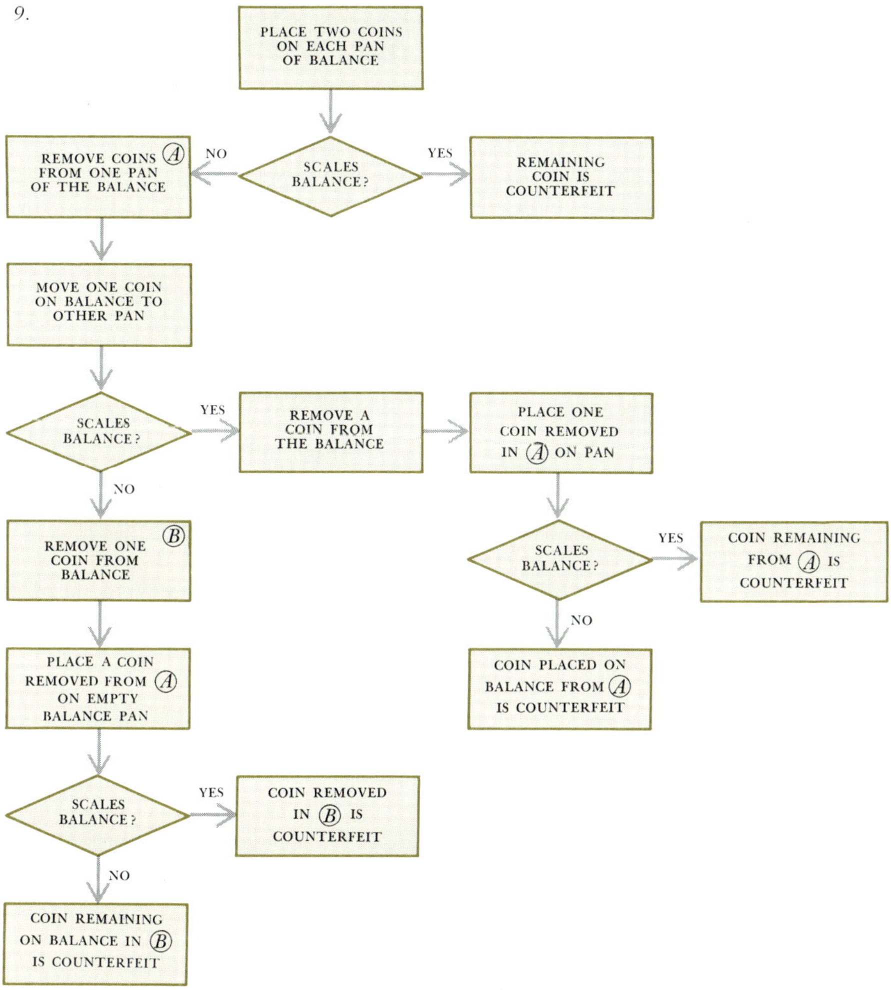

SECTION 3-3

1. $137 = 254_7$

3. (*a*) Yes
(*b*) No
(*c*) Yes

5. (*a*) Steps 5, 6, and 7.
(*b*) Write the remainders in the reverse order of their appearance.

7. The answer is both yes and no. Some of the trivial steps can be changed in a limited manner; the important steps cannot.

SECTION 3-4

1. *Input* Television waves received by set from the television station.
Output Picture itself.
Storage "Glow" that remains as electron beam hits screen.

Control Circuits that cause electron beam to "scan" tube.

3. *Thermostat* Control temperature in a room.
Dial numbers on television Select a particular group of television waves to translate into a picture.

5. (*a*) *Input* Steps 1, 2, and 3.
(*b*) *Output* Step 11.
(*c*) *Memory* Steps 5, 6, 7, and 9.
(*d*) *Arithmetic* Steps 4, 5, 6, and 7.
(*e*) *Control* Steps 8 and 10 (all steps).

7. "Each is a part of the main."

REVIEW EXERCISES FOR CHAPTER 3

1. A systematic method of calculation.

3. (*a*) Compute percentage for each team.
(*b*) Compare percentage of first and second team in list.
(*c*) Save team with larger percentage.
(*d*) If any teams remain to be checked go to step *e*; otherwise go to step *g*.
(*e*) Compare next team in list with result of step *c*.
(*f*) Go to step *c*.
(*g*) Team saved in step *c* is winner.

5. Less chance for error if a large number of numbers is considered.

7. A very simple version:

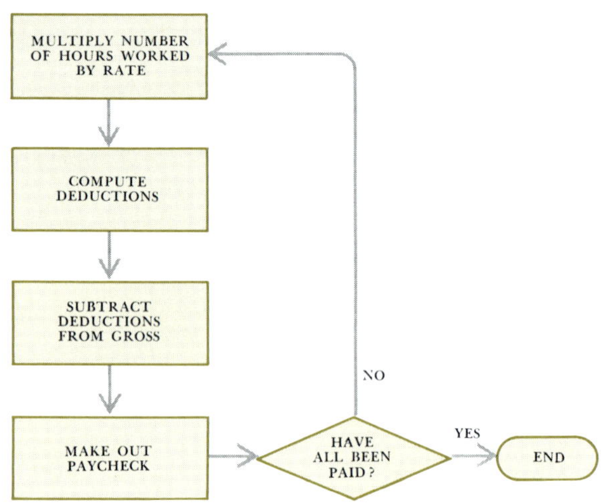

SECTION 4-1

1. Two devices with two states that lend themselves to functioning as binary storage devices are the light bulb and the electric switch. Other things that could perform a similar function are faucets, which permit a fluid to flow or shut it off; doors, which are either open or

closed; candles, which could store a one if they were lit and a zero if they were not.

3. A base-8 digit could always be stored in three binary storage devices.

5. 04154160

7. 05052077
03077052
05077113
03113052
05113154
03154052

9. In Fortran an excutable statement is one that is to be carried out in the actual solution of the problem involved; nonexecutable statements, such as FORMAT statements, tell the translator program something about the nature of the program as it is being converted from the Fortran language to the numerical or absolute language that the computer will actually execute.

11.
```
   TYPE 11, NUM
11 FORMAT (I5)
```

13.
```
2 READ 1, N
1 FORMAT (I4)
  IF (N-5000)3, 2, 2
3 TYPE 1, N
  GO TO 2
  END
```

15. There are 12 one-punch characters:

$\frac{12 \times 11}{2 \times 1} = 66$ two-punch characters (see Chaps. 9 and 10)

$\frac{12 \times 11 \times 10}{3 \times 2 \times 1} = 220$ three-punch characters

making a total of $12 + 66 + 220 = 298$.

SECTION 4-2

1. The program checks for zero to find when it has read all the data cards. Unless this is changed, a zero or negative input value will cause the program to jump to the output section of the program without reading in any remaining data values.

3. Program statement 3 (step 5) could be changed from 3 Z = A − B to 3 Z = B − A. Under these new conditions the smaller of values of A and B will be saved.

5.

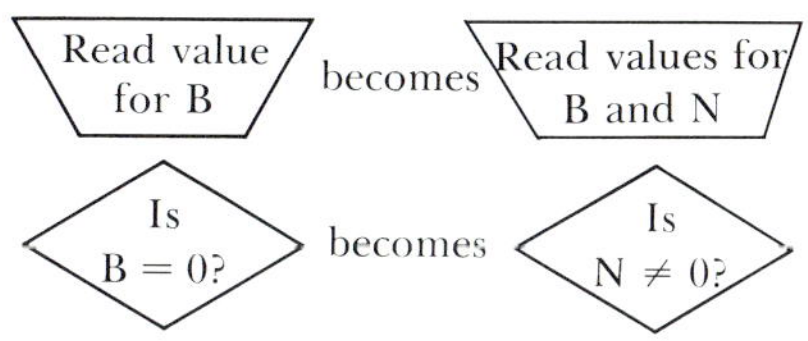

These are the only changes required.

7. Approximately $201.64.

9. The FORMAT statements would have to be changed to the I form.

11. Because another part of the program had to refer to them.

13. Between steps 1 and 3 insert
C = A
Change step 6 to
IF (Z) 5,4,8
Between steps 8 and 9 insert
```
  Y = C − B
  IF (Y) 4,4,7
7 C = B
  GO TO 4
```
Between steps 9 and 10 insert
TYPE 101,C

SECTION 4-3

1. Basically except for the addition of outside power:
(*a*) Power drill or saw
(*b*) Golf-ball-driving machine
(*c*) Lighter (flint and steel)
(*d*) Mixer

3. No value of T would be available.

5. Set up the program to find the average value of a second set of data.

7. A base-10 digit requires up to 4 cores to store. Therefore 10,000,000 cores could store 10,000,000 ÷ 4 = 2,500,000 digits.

9. 1200 × 150 = 180,000 characters per min. 800,000,000 binary characters would require 800,000,000 ÷ 180,000 = 4444.4 min.

11. Between steps 1 and 2 insert

```
B = O
```

Between steps 7 and 8 insert

```
  Z = B - A
  IF (Z) 3,62,62
3 B = A
```

Between steps 10 and 11 insert

```
TYPE 307,B
```

SECTION 4-4

1. 7 × 287 = 2009

3. 3 × 671 = 2013

5. Modify steps 14, 15, and 16 to read

```
4 TYPE 2, M
2 FORMAT (9HNON-PRIME I6)
  Z = 1
  GO TO 3
12 STOP
```

Between steps 9 and 10 insert

```
   IF (Z) 10,11,10
10 IF (K) 7,7,12
```

Modify step 10 to read

```
11 IF (K) 7,7,8
```

7. Ratio to computer:

$$\frac{\text{Calculation speed}}{\text{Output speed}} = \frac{\text{man's rate}}{\text{output rate}}$$

or

$$\frac{1/300{,}000}{32/180{,}000} = \frac{3}{x}$$

or

$$\frac{3}{160} = \frac{3}{x} \Rightarrow x = 160 \text{ sec } (2\tfrac{1}{2} \text{ min})$$

9. Modify step 14 to read

```
14 IF (SENSE SWITCH 1) 9,10
```

Then insert

```
 9 TYPE 2
   STOP
10 TYPE 6,M
 6 FORMAT (I6)
   STOP
```

SECTION 4-5

1. Modify step 6 to read

```
NQ = NC/8
```

Modify step 7 to read

```
NR = NC - NQ*8
```

Change 7 to 8 in step 15.

3. No. All calculations are in base 10.

5. 12

REVIEW EXERCISES FOR CHAPTER 4

1. The ability to execute a sequence of operations automatically.

3. The fact that all phases of operation are carried out electronically, in other words, the internally stored program.

5. Formula translation

7. An executable statement is one that causes the machine to do something such as add, subtract, and so on.

9. Cogo and Cobol are two other programming languages. No one language is really suitable to all the potential uses of a computer.

11. No. A programmer could use the absolute language of the machine itself.

SECTION 5-2

1. Not a statement of logic.

3. Statement of logic.

5. Statement of logic.

7. $p \vee q$

9. $\sim p \wedge \sim q$

11. $\sim(p \wedge q)$

13. It is raining, and the streets are slippery.

15. It is not raining, or the streets are not slippery.

17. It is not the case that it is raining and the streets are slippery.

19. It is not true that it is not raining and the streets are not slippery.

21. True

23. False

25. True

SECTION 5-4

1. TFFT

3. FFFT

5. FTFT

7. FFFT, $\sim p \wedge \sim q$

9. $\sim(p \wedge q)$

11. FT

13. It is the truth table for $\sim p$.

SECTION 5-5

1. FF

3. TTFT

5. TTTFTTTT

7. FTFTFFFT

9. F

11. T

13. T

15. T

17. TFTT

19. It is the truth table for $p \rightarrow q$.

21. TTTF, $p \vee q$

23. TFTT, $p \rightarrow q$

SECTION 5-6

3. If you are lovely, then you have many dates.

   It is not true that you are lovely and you do not have many dates.

   If you do not have many dates, then you are not lovely.

   You have many dates, or you are not lovely.

5. $\sim(\sim p \wedge \sim q)$

7. Both have truth table FFFTTTTT.

SECTION 5-7

1. $p \rightarrow q$

3. $q \rightarrow p$

5. $q \rightarrow p$

7. $p \leftrightarrow q$

9. If you do not like Valkyries, then you do not like Wagner.

11. The biconditional

13. $r \rightarrow s$

15. $p \rightarrow q$

REVIEW EXERCISES FOR CHAPTER 5

1. $p \to \sim q$

3. $\sim p \leftrightarrow q$

5. If France is not smaller than Texas, then the United States is larger than Canada.

7. FTTTTTTT

SECTION 6-1

1. Valid

3. Invalid

5. Invalid

7. Valid

9. Invalid

11. Invalid

SECTION 6-2

13. Valid

SECTION 6-3

In each of the following only the most likely argument forms are used.

1. $p \to r$ hypothetical syllogism

3. $\sim\sim q$ or q *modus tollens*

5. $p \mid p$ disjunctive syllogism

7. Motherhood is not good (*modus tollens*).

9. If he is a good husband, then he comes home for tea (hypothetical syllogism).

SECTION 6-4

1. They are the same thing.

3. Harold demonstrates invalid methods of reasoning.

7. Nothing

SECTION 6-5

1. (*a*) The processes a child goes through to decide that he will be punished after breaking his mother's favorite lamp.
(*b*) Kepler proposed his laws of planetary motion by observing planets and by reading of the observations of Tycho Brahe.

3. It uses general principles but does not create them.

5. *Primitive terms* Golf ball, stroke, golf club, tee.
Technical terms Putter, wood, iron, hole-in-one.
Axioms There are golf courses. Golf balls exist.
Theorems The best score possible is 18. The first stroke is from a tee.

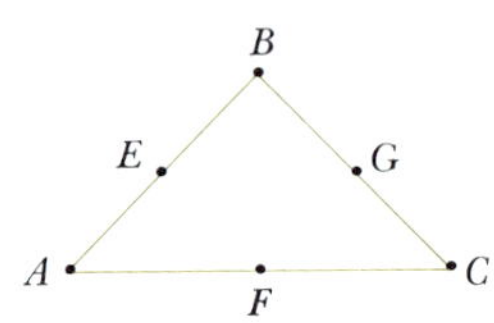

7. Using the figure from Exercise 6 (above),

(*a*) Form lines *EG*, *GF*, and *FE* (axiom 3).
(*b*) Add points *H*, *I*, and *J* (axiom 2).
(*c*) The figure formed by points *A*, *E*, *B*, *G*, *C*, and *F* and lines *AEB*, *BGC*, and *CFA* forms a squiggle, as do points *E*, *I*, *G*, *J*, *F*, and *H* and lines *EIG*, *GJF*, and *FHE* (each is composed of six points and three lines).

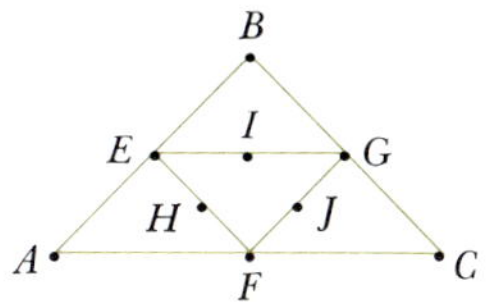

9. (*a*) Points A and B and the line connecting them exist (axiom *d*).
   (*b*) A and B are each on another line (axiom *b*).
   (*c*) These lines each contain another point, called C and D (axiom *a*).
   (*d*) C and D are each on another line (axiom *b*).

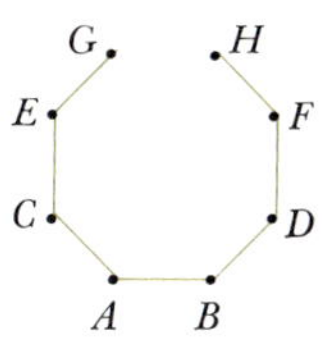

   (*e*) These lines each contain another point, called E and F (axiom *a*).
   (*f*) In the same manner, we create points G and H, which are the seventh and eighth points.

REVIEW EXERCISES FOR CHAPTER 6

1. Valid

3. Valid

5. Because $\sim A \vee \sim B \equiv \sim(A \wedge B)$, it is an example of *modus tollens*.

SECTION 7-1

1. (*a*) Examples of sets:

   Collection of American males over 6 ft $\frac{1}{2}$ in. tall.

   Collection of plays written by T. S. Eliot.

   Collection of theories describing the origin of the earth.

   Collection of copper pennies minted in 1943.

   (*b*) Collections that are not sets:

   Collection of short novels. (When is a novel "short"?)

   Collection of great Presidents of the United States. (Which ones were "great"?)

3. (*a*) The set of positive even numbers less than 9.
   (*b*) The set of the first three letters of the English alphabet.
   (*c*) The set of men who were President of the United States in the first half of the twentieth century.
   (*d*) The set of the fiftieth state to be admitted to the Union.
   (*e*) The set containing the empty set.

5. (*a*) z is an element of set B, or z is contained in set B.
   (*b*) x is an element of set P, but it is not an element of set Q.
   (*c*) A is a subset of set B, and B is a proper subset of set C.
   (*d*) A is a proper subset of set B, but B is an element of set C.
   (*e*) The empty set is a subset of set A.
   (*f*) A is a subset of set B, but B is not a proper subset of set A.
   (*g*) A is a subset of set B, and B is a subset of set A; also A equals B.
   (*h*) A is a proper subset of set B, and B is a proper subset of set A. This, of course, is not possible.

7. We cannot make a list of its elements, because it has an infinite number of members.

SECTION 7-2

1. (*a*) $A \cup B = \{1,2,3,4,5,6,7,8\}$
   (*b*) $A \cap C = \{1,3\}$
   (*c*) $\tilde{D} = \{0,1,2,3,4\} = C$
   (*d*) $E \cup F = \{2,3,5,6,8,9\}$
   (*e*) $C - B = \{0,1,3\}$

3. (*a*) $E \cup (B \cap E) = \{3,6,9\} \cup \{6\}$
   $= \{3,6,9\}$
   $(E \cup B) \cap (E \cup E) = \{2,3,4,6,8,9\}$
   $\cap \{3,6,9\}$
   $= \{3,6,9\}$

   (*b*) $\widetilde{A \cap C} = \{\widetilde{1,3,5,7\} \cap \{0,1,2,3,4}\}$
   $= \{\widetilde{1,3}\} = \{0,2,4,5,6,7,8,9\}$
   $\tilde{A} \cap \tilde{C} = \{0,2,4,6,8\} \cup \{5,6,7,8,9\}$
   $= \{0,2,4,5,6,7,8,9\}$
   (*c*) $(\tilde{\tilde{B}}) = \{\widetilde{0,1,3,5,7,9}\} = \{2,4,6,8\} = B$

5.

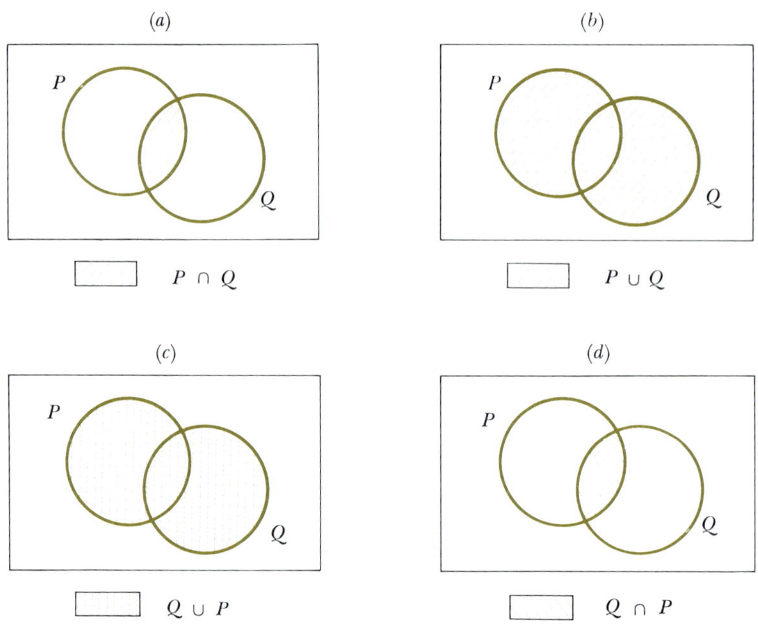

7.

(a)

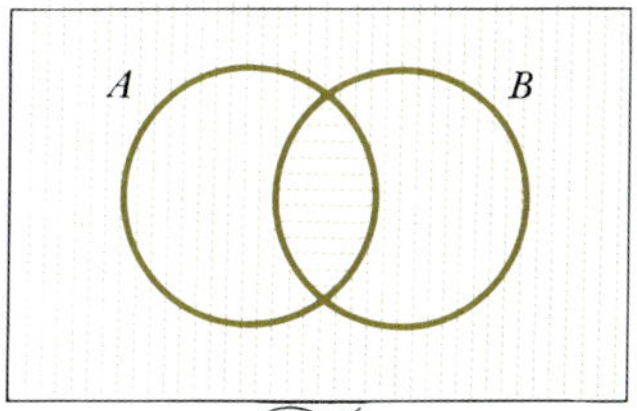

$\widetilde{A \cap B}$

AREA MARKED

=

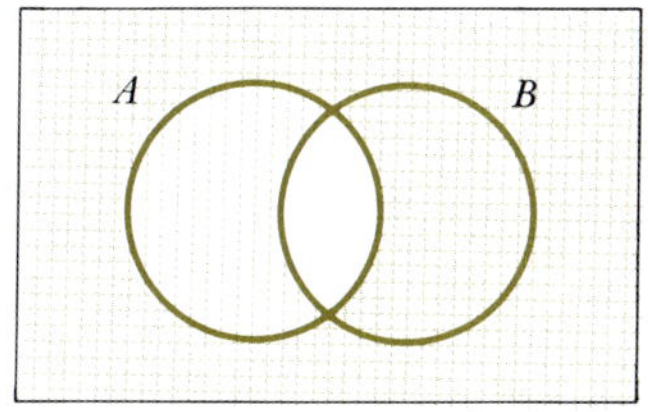

$\widetilde{A} \cup \widetilde{B}$

AREA MARKED BY

AND AND

(b)

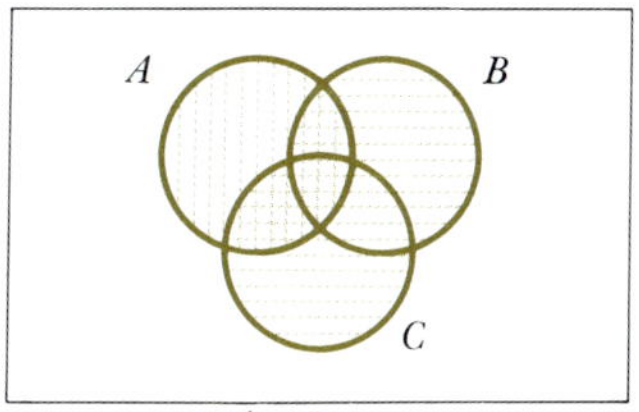

$A \cap (B \cup C)$

AREA MARKED

=

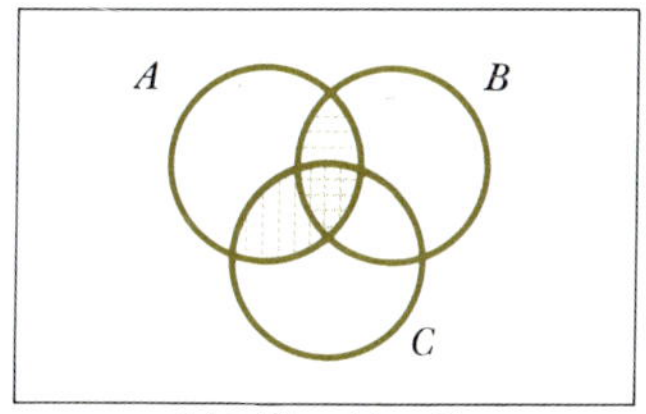

$(A \cap B) \cup (A \cap C)$

AREA MARKED

AND AND

(c)

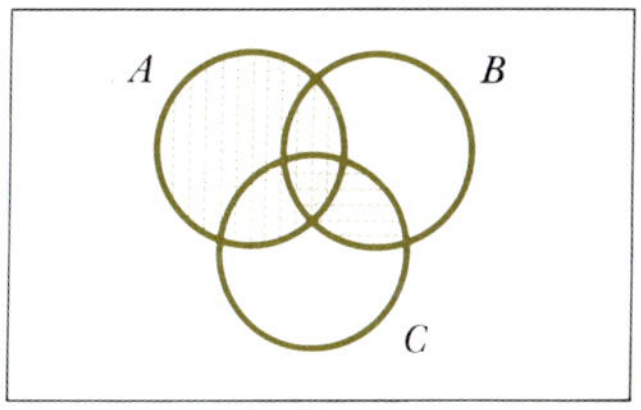

$A \cup (B \cap C)$

AREA MARKED

AND AND

=

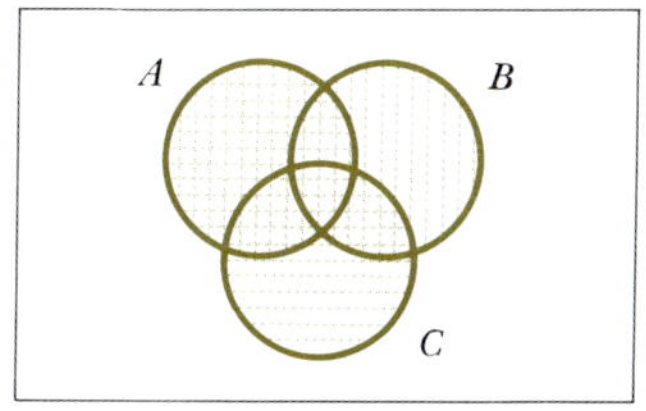

$(A \cup B) \cap (A \cup C)$

AREA MARKED

(d)

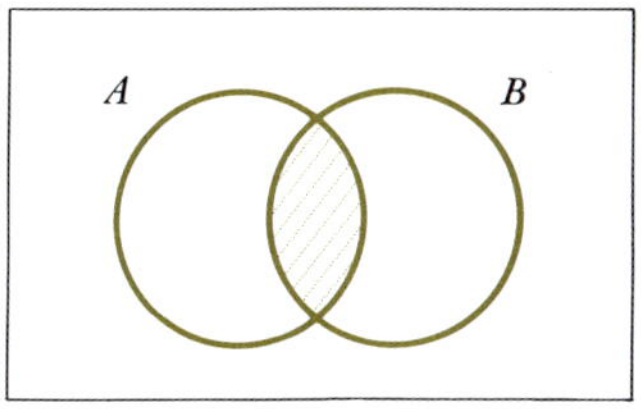

$A \cap B \subseteq A$

9. (a) 9

(b) 28

SECTION 7-3

1. (*a*) Power set of $C = \{\emptyset,\{4\},\{5\},\{6\},\{4,5\},\{4,6\},\{5,6\},\{4,5,6\}\}$.
   (*b*) No, but $\{5\}$ is.

3. $A \cup E = \{a,b,6,7\}$, and thus it has 4 elements. Therefore its power set has 16 elements.

5. The power set of a set with 5 elements contains 32 elements.

7. (*a*) $A \times C$ contains 9 or 3×3 elements.
   (*b*) $B \times E$ contains 6 or 3×2 elements.
   (*c*) $E \times B$ contains 6 or 2×3 elements.

9. Probably the best notation is (a,b,c), where $a \in$ first set, $b \in$ second set, and $c \in$ third set.

SECTION 7-4

1. A and D are equivalent.
   B and E are equivalent.
   A and C are equivalent.
   C and D are equivalent.

3. $N(A) = 5$
   $N(D) = 5$
   $A \cup D = \{1,3,5,7,9,☆,0,□,\triangle,*\}$
   Therefore
   $N(A \cup D) = 10$

5. $N(A) + N(B) \neq N(A \cup B)$

7. Yes, for X products the sets do not have to be disjoint.

9. If $A \subseteq B$, then we can define $N(B) - N(A)$ as $N(B - A)$ (relative complement).

11. $N(R \times S) = N(R) \cdot N(S)$

13. Let $S = \{T,B,F\}$. The number of different round trips is $N(S \times S)$.

15. $N((A \cap B) \times C) = 0 \times 2 = 0$
    $(A \cap B) \times C = \emptyset$

SECTION 7-5

1.

(*a*)

$p \wedge \sim q$

(*b*)

$\sim p \wedge q$

(*c*)

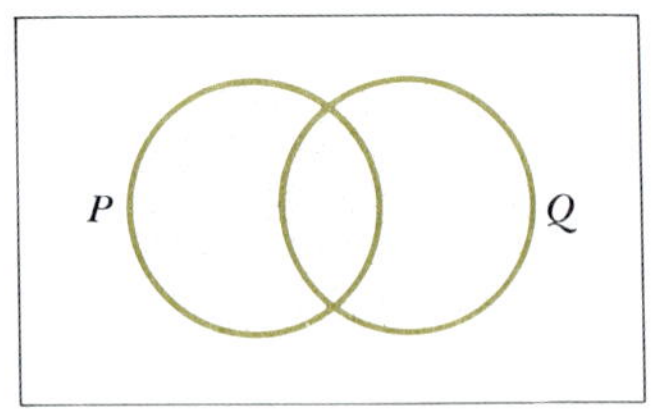

$\sim p \rightarrow q$

(*d*)

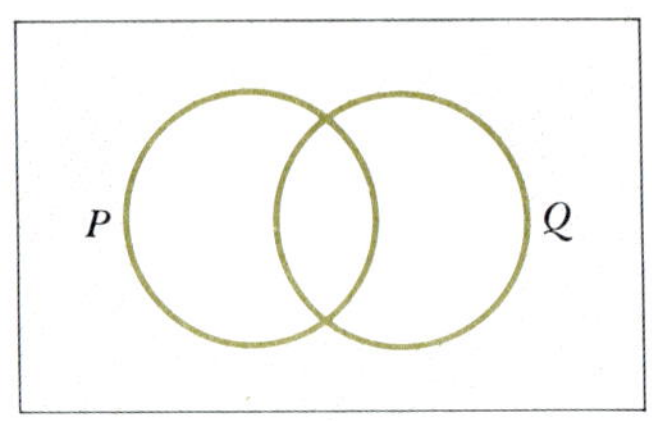

$\sim p \wedge \sim q$

3. If *a* and *c* have the same Venn diagram, $p \rightarrow q$ and $\sim q \rightarrow \sim p$ are equivalent statements. The same result for $q \rightarrow p$ and $\sim p \rightarrow \sim q$ follows from *b* and *d*.

5. The main advantage is in studying one system with its model in the other system. In this way one can obtain results that are not easily seen in the original system.

7. Rather than shade areas of the Venn diagrams, we have numbered them and worked with the numbers to find the "final" area.

   (*a*) $\sim q \wedge (p \rightarrow q) \equiv \sim q \wedge (\sim p \vee q)$

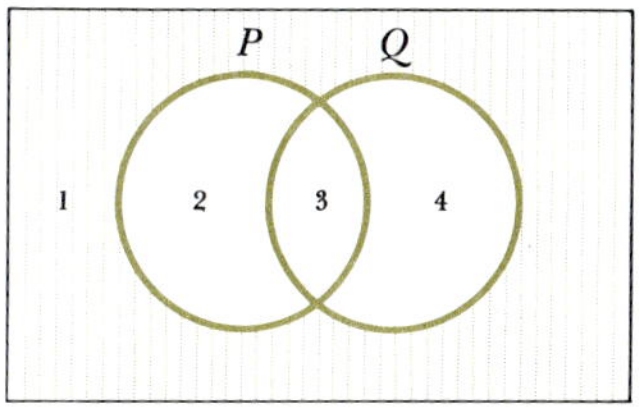

$\tilde{Q} \cap (\tilde{P} \cup Q)$
$\{1,2\} \cap \{1,4,3\}$
$\tilde{Q} \cap (\tilde{P} \cup Q) \subseteq \tilde{P}$

Therefore the argument is valid.

(*b*)

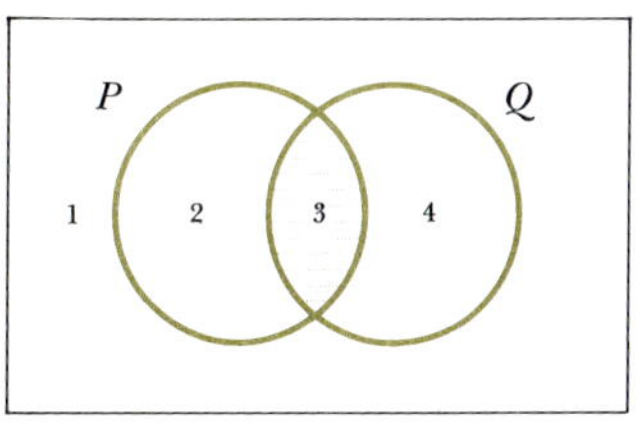

$(p \wedge q) \wedge (\sim p \rightarrow q) \equiv (p \wedge q) \wedge (p \wedge q)$
$(P \cap Q) \cap (P \cup Q)$
$\{3\} \cap \{2,3,4\}$
$(P \cap Q) \cap (P \cup Q) \not\subseteq \tilde{Q}$

Therefore the argument is invalid.

(*c*) $(p \rightarrow q) \wedge (\sim r \rightarrow \sim q)$
$\equiv (\sim p \vee q) \wedge (r \vee \sim q)$
$\sim r \rightarrow \sim p \equiv r \vee \sim p$

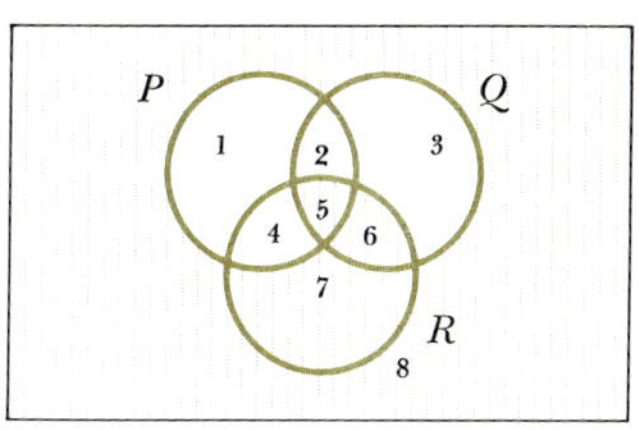

$(\tilde{P} \cup Q) \cap (R \cup \tilde{Q})$
$\{3,6,7,8,2,5\} \cap \{4,5,6,7,1,8\}$
$\{5,6,7,8\}$

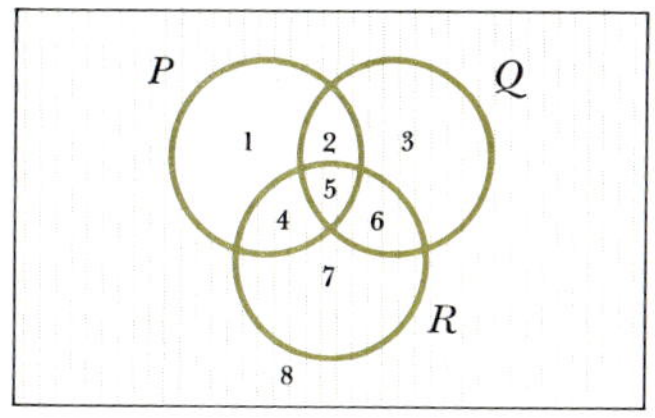

$R \cup \tilde{P}$
$\{4,5,6,7,3,8\}$
$(\tilde{P} \cup Q) \cap (R \cup \tilde{Q}) \subseteq R \cup \tilde{P}$

Therefore the argument is valid.

(*d*) $p \leftrightarrow q \equiv (p \rightarrow q) \wedge (q \rightarrow p)$

So the conjunction of the premises is equivalent to

$(\sim p \vee q) \wedge (\sim q \vee p)$
$\wedge (q \vee r) \wedge \sim r$

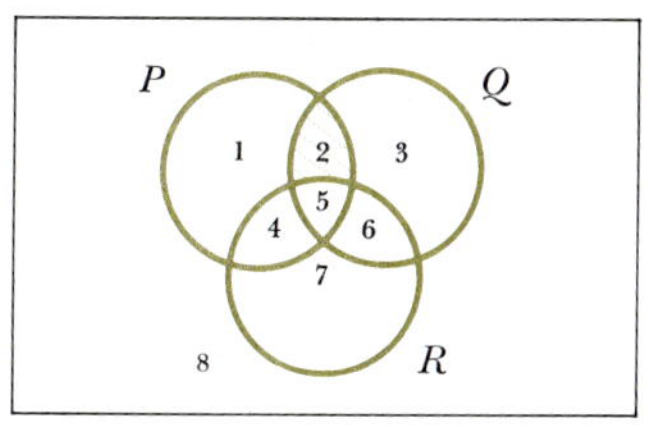

$(\tilde{P} \cup Q) \cap (\tilde{Q} \cup P) \cap (Q \cup R) \cap \tilde{R}$
$\{3,6,7,8,2,5\} \cap \{1,4,7,8,2,5\}$
$\cap \{2,3,4,5,6,7\} \cap \{1,2,3,8\}$
$\{2\}$
$(\tilde{P} \cup Q) \cap (\tilde{Q} \cup P) \cap (Q \cup R) \cap \tilde{R} \not\subseteq \tilde{P}$

Therefore the argument is invalid.

SECTION 7-6

1. (*a*) To reduce development costs, as in an aircraft model built prior to the construction of the aircraft itself.

   (*b*) To study a whole system when it is too large to view or examine, such as a model of the solar system in a science classroom.

(*c*) To examine the workings of an object normally too small to work with practically, such as a model of the human eye.

3. (*a*) A model of the solar system.
   (*b*) A model of an aircraft design.

5. Because they are similar to ordinary maps, which are models of points on the earth.

REVIEW EXERCISES FOR CHAPTER 7

1. We know exactly which objects belong in a well-defined collection, but in one not well-defined we cannot be sure which objects do and which do not belong.

3. Because sets can also be elements of other sets, we often have objects that are both sets and elements.

5. Because the universal set remains unchanged for all complements.

7. (*a*) $A \cap (B \cup C) = \{t,h\}$
   $(A \cap B) \cup (A \cap C) = \{t,h\} \cup \emptyset$
   $= \{t,h\}$
   (*b*) $A \cup (B \cap C) = \{m,a,t,h\} \cup \{e,i\}$
   $= \{m,a,t,h,e,i\}$
   $(A \cup B) \cap (A \cup C)$
   $= \{m,a,t,h,e,i\} \cap \{m,a,t,h,e,i,c,s\}$
   $= \{m,a,t,h,e,i\}$

9. (*a*)

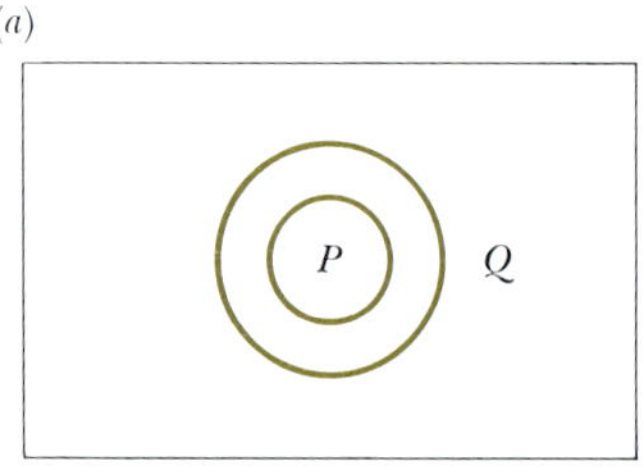

$P \subset Q$

(*b*)

$\widetilde{P} \cap Q = \emptyset$

(*c*)

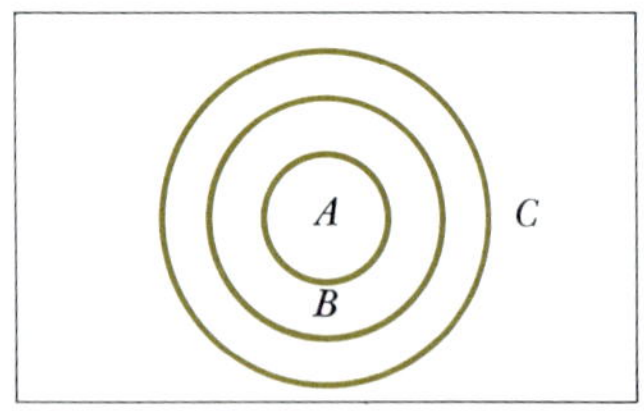

$A \subset B \subset C$

11. $A \times B = \{(0,1),(0,2),(1,1),(1,2)\}$
    $B \times A = \{(1,0),(2,0),(1,1),(2,1)\}$

13. One major reason is that many of the properties of numbers can be studied in terms of sets.

15. Partial list:

Sets	*Logic*
$\cap$	$\wedge$
$\cup$	$\vee$
$=$	$\equiv$
$\subseteq$	$\Rightarrow$

17. Both answers could be supported.

19. Basically, understand its limits.

SECTION 8-1

1. $a \bigcirc b = b \bigcirc a$

3. $a \bigcirc a = a$

5. $(a \bigcirc b)' = a' \ \text{☆}\ b'$

7. $a \bigcirc i_{\bigcirc} = a$

11. The additive and multiplicative identities are 0 and 1 respectively.

SECTION 8-2

1. $A \bigcirc (B \star C)' = A \bigcirc (B' \bigcirc C')$
De Morgan
$= (A \bigcirc B') \bigcirc C'$
associative law

3. $A \star (A' \bigcirc B) = (A \star A') \bigcirc (A \star B)$
distributive law
$= i_{\bigcirc} \bigcirc (A \star B)$
inverse
$= (A \star B) \bigcirc i_{\bigcirc}$
commutative law
$= A \star B$
identity

5. $(A \star B) \bigcirc (A \star B') = A \star (B \bigcirc B')$
distributive law
$= A \star i_{\star}$
inverse
$= A$
identity

7. The statement $i_{\bigcirc}' = i_{\star}$ is the dual of Exercise 6.

9. $a = a \star i_{\star}$
identity
$= a \star (a' \bigcirc (a')')$
inverse
$= (a \star a') \bigcirc (a \star (a')')$
distributive law
$= i_{\bigcirc} \bigcirc (a \star (a')')$
inverse
$= (a' \star (a')') \bigcirc (a \star (a')')$
inverse
$= ((a')' \star a') \bigcirc ((a')' \star a)$
commutative law
$= (a')' \star (a' \bigcirc a)$
distributive law
$= (a')' \star (a \bigcirc a')$
commutative law
$= (a')' \star i_{\star}$
inverse
$= (a')'$
identity

SECTION 8-3

1. Any theorem of sets remains true if the operation $\cup$ and the identity element $\emptyset$ are replaced by $\cap$ and U throughout the theorem.

3. $P \cup P = P$
$P \cap P = P$

5. $P \cup \emptyset = P$
$P \cap U = P$

7. $P \cup \tilde{P} = U$
$P \cap \tilde{P} = \emptyset$

9. $p \vee p \equiv p$
$p \wedge p \equiv p$

11. $p \wedge \mathrm{T} \equiv p$
$p \vee \mathrm{F} \equiv p$

13. $p \wedge \sim p \equiv \mathrm{F}$
$p \vee \sim p \equiv \mathrm{T}$

15.

p	q	r	$p \vee (q \wedge r)$		$(p \vee q) \wedge (p \vee r)$		
T	T	T	T	T	T	T	T
T	T	F	T	F	T	T	T
T	F	T	T	F	T	T	T
T	F	F	T	F	T	T	T
F	T	T	T	T	T	T	T
F	T	F	F	F	T	F	F
F	F	T	F	F	F	F	T
F	F	F	F	F	F	F	F

17. (a) $P \cap \widetilde{(P \cap Q)} = P \cap (\tilde{P} \cup \tilde{Q})$
De Morgan
$= (P \cap \tilde{P}) \cup (P \cap \tilde{Q})$
distributive law
$= \emptyset \cup (P \cap \tilde{Q})$
inverse
$= P \cap \tilde{Q}$
identity

(*b*)

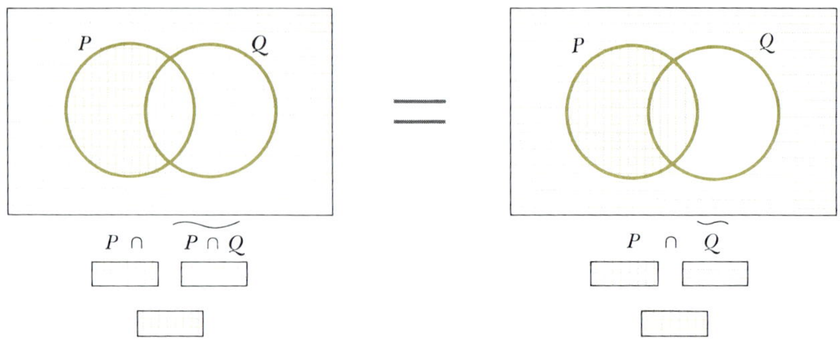

19. $p \wedge (\sim p \vee q) \equiv (p \wedge \sim p) \vee (p \wedge q)$ distributive law

$\equiv \mathrm{F} \vee (p \wedge q)$ inverse

$\equiv p \wedge q$ identity

21. (*a*) $(p \wedge q) \wedge [p \wedge (q \vee r)]$

$\equiv [(p \wedge q) \wedge p] \wedge (q \vee r)$ associative law

$\equiv [p \wedge (p \wedge q)] \wedge (q \vee r)$ commutative law

$\equiv [(p \wedge p) \wedge q] \wedge (q \vee r)$ associative law

$\equiv (p \wedge q) \wedge (q \vee r)$ idempotent law

$\equiv p \wedge [q \wedge (q \vee r)]$ associative law

$\equiv p \wedge q$ absorption

(*b*)

p q r	$(p \wedge q) \wedge [p \wedge (q \vee r)]$	$p \wedge q$
T T T	T T T T T	T
T T F	T T T T T	T
T F T	F F T T T	F
T F F	F F T F F	F
F T T	F F F F T	F
F T F	F F F F T	F
F F T	F F F F T	F
F F F	F F F F F	F

SECTION 8-4

1. (*a*) *Reflexive property* 3 is less than or equal to 3, as are 4 and 5.

(*b*) *Antisymmetry* a is less than or equal to b and $b \leq a$ if and only if a and b are both replaced by 3 or both by 4 or both by 5.

(*c*) *Transitivity* If 3 is less than or equal to 4 and 4 is less than or equal to 5 then 3 is less than or equal to 5.

Because the properties *a, b,* and *c* hold, the relation *less than or equal to* partially orders {3,4,5}.

3. No. Although $0 < a$ for any positive whole number, it is not true that $a < 1$ for all a.

5. (*a*) *Reflexive property*

$a \star a = a$ idempotent law

a ξ— a definition of ξ—

(*b*) *Antisymmetry* If a ξ— b and b ξ— a, then

$a \star b = b$ and $b \star a = a$ definition of ξ—

But $b \star a = a \star b$ commutative law

$a = b$ substitution

(*c*) *Transitivity* If a ξ— b and b ξ— c, then a ξ— c:

$a \star b = b$ and $b \star c = c$ definition of ξ—

$a \star c = a \star (b \star c)$ substitution

$a \star c = (a \star b) \star c$ associative law

$a \star c = b \star c$ substitution

$a \star c = c$ substitution

a ⊱— c definition of ⊱—

7. $a \star i_{\bigcirc} = i_{\bigcirc}$ theorem 3

a ⊱— $i_{\bigcirc}$ definition of ⊱—

9. $x \bigcirc a = a$ and $x \bigcirc b = b$ definition of —⊰

$(x \cup a) \star (x \bigcirc b) = a \star b$ substitution

$x \bigcirc (a \star b) = a \star b$ distributive law

x —⊰ $a \star b$ definition of —⊰

SECTION 8-5

1. $B \cup \emptyset = B$ identity

$B \cup (A \cap \tilde{B}) = B$ substitution

$(B \cup A) \cap (B \cup \tilde{B}) = B$ distributive law

$(B \cup A) \cap U = B$ inverse

$B \cup A = B$ identity

$A \cup B = B$ commutative law

$A \subseteq B$ definition of $\subseteq$

3. $A \cup B = B$ definition of $\subseteq$

$\tilde{A} \cup \tilde{B} = \tilde{A} \cup \tilde{B}$ reflexive property of =

$\tilde{A} \cup \tilde{B} = \tilde{A} \cup (\widetilde{A \cup B})$ substitution

$\tilde{A} \cup \tilde{B} = \tilde{A} \cup (\tilde{A} \cap \tilde{B})$ De Morgan

$\tilde{A} \cup \tilde{B} = \tilde{A}$ absorption

$\tilde{B} \cup \tilde{A} = \tilde{A}$ commutative law

$\tilde{B} \subseteq \tilde{A}$ definition of $\subseteq$

5. $p \vee q \equiv q$ and $p \vee r \equiv r$ definition of $\Rightarrow$

$(p \vee q) \wedge (p \vee r) \equiv q \wedge r$ substitution

$p \vee (q \wedge r) \equiv q \wedge r$ distributive law

$p \Rightarrow (q \wedge r)$ definition of $\Rightarrow$

7. They are identical Boolean algebra statements, one in sets and the other in logic, A, B, C of set notation being replaced by p, q, r of logic.

9. $p \rightarrow q \equiv \sim p \vee q$ the last law in Chap. 6

$[\sim q \wedge (p \rightarrow q)] \vee \sim p$

$\equiv [\sim q \wedge (\sim p \vee q)] \vee \sim p$ substitution

$\equiv [(\sim q \wedge \sim p) \vee (\sim q \wedge q)] \vee \sim p$ distributive law

$\equiv [(\sim q \wedge \sim p) \vee \text{F}] \vee \sim p$ inverse

$\equiv (\sim q \wedge \sim p) \vee \sim p$ identity

$\equiv \sim p \vee (\sim q \wedge \sim p)$ commutative law

$\equiv \sim p \vee (\sim p \wedge \sim q)$ commutative law

$\equiv \sim p$ absorption

$\sim q \wedge (p \rightarrow q) \Rightarrow \sim p$ definition of $\Rightarrow$

The argument is valid.

SECTION 8-6

1.

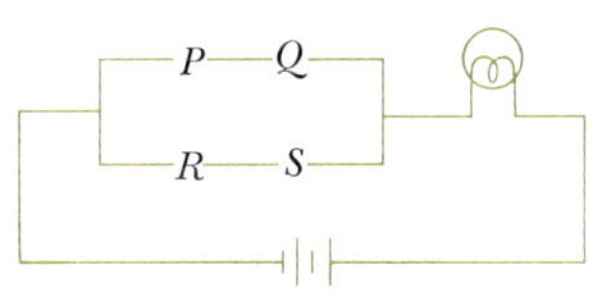

3.

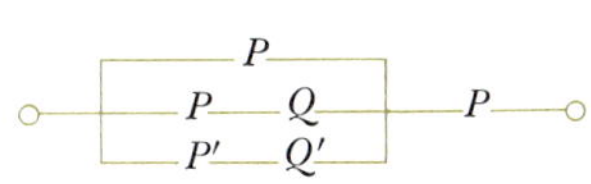

5. $a \vee (\sim b \wedge \sim c)$

7. $(a \wedge b) \vee [(a \vee \sim b \vee c \vee d) \wedge a]$

9.

11.

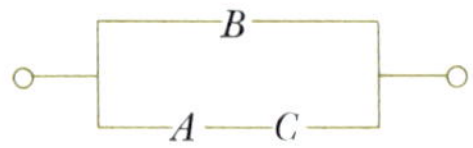

13.

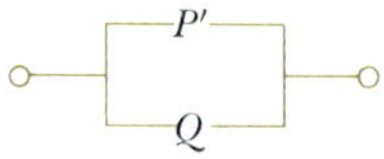

15.

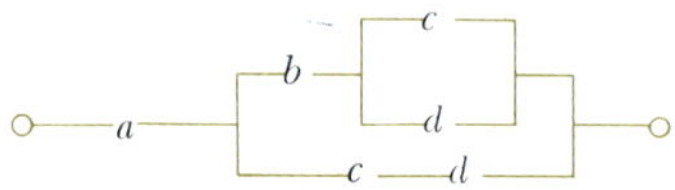

REVIEW EXERCISES FOR CHAPTER 8

1. $A \star (A' \star B)' = A \star (A'' \bigcirc B')$ De Morgan

$= A \star (A \bigcirc B')$ double negation

$= A$ absorption

3. $A' \bigcirc i_{\bigcirc} = A'$ identity

$A' \bigcirc (A \star B) = A'$ substitution

$(A' \bigcirc A) \star (A' \bigcirc B) = A'$ distributive law

$i_{\star} \star (A' \bigcirc B) = A'$ inverse

$A' \bigcirc B = A'$ identity

$B \bigcirc A' = A'$ commutative law

$B \longrightarrow A'$ definition of $\longrightarrow$

5. $\tilde{A} \cup \emptyset = \tilde{A}$ identity

$\tilde{A} \cup (A \cap B) = \tilde{A}$ substitution

$(\tilde{A} \cup A) \cap (\tilde{A} \cap B) = \tilde{A}$ distributive law

$U \cap (\tilde{A} \cap B) = \tilde{A}$ inverse

$\tilde{A} \cap B = \tilde{A}$ identity

$B \cap \tilde{A} = \tilde{A}$ commutative law

$B \subseteq \tilde{A}$ definition of $\subseteq$

7.

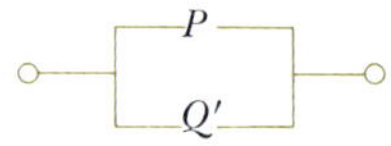

SECTION 9-1

1. One advantage is that numerical probabilities can be compared.

3. (*a*) $2b, 2d, 2e, 2f$
(*b*) $2a, 2b, 2e, 2f$
(*c*) $2b, 2e, 2f$
(*d*) $2d$
(*e*) $2a$
(*f*) $2c$

5. (*a*) $4a, b, d, f, g$
(*b*) Drawing two cards from a deck without replacing the first card.

7. P (sum 2) $= \frac{1}{36}$
P (sum 3) $= \frac{2}{36} = \frac{1}{18}$
P (sum 4) $= \frac{3}{36} = \frac{1}{12}$
P (sum 5) $= \frac{4}{36} = \frac{1}{9}$
P (sum 6) $= \frac{5}{36}$
P (sum 7) $= \frac{6}{36} = \frac{1}{6}$
P (sum 8) $= \frac{5}{36}$
P (sum 9) $= \frac{4}{36} = \frac{1}{9}$
P (sum 10) $= \frac{3}{36} = \frac{1}{12}$
P (sum 11) $= \frac{2}{36} = \frac{1}{18}$
P (sum 12) $= \frac{1}{36}$

9. One common example is the chances of a driver, male, under 25, having a traffic accident. Another is weather forecasting using probability.

11. That there have not been and will not be any changes in the factors affecting life expectancy.

SECTION 9-2

1. (*a*) The probability of Johnny winning the contest and Bill being taller than the next person he meets.

(*b*) The probability of Johnny winning the contest or Bill being taller than the next person he meets.

(*c*) The probability that Bill is taller than the next person he meets, assuming that Johnny has won the contest.

(*d*) The probability that Johnny wins the contest, assuming that Bill is taller than the next person he meets.

(*e*) The probability that Johnny wins the contest and Bill is not taller than the next person he meets.

(*f*) The probability that Johnny does not win the contest.

3. (*a*) $P(\text{both } Q) = P(Q_{\text{I}}) \cdot P(Q_{\text{II}})$

$= \frac{1}{13} \times \frac{1}{13} = \frac{1}{169}$

(*b*) $P(H_{\text{I}} \text{ or } K_{\text{I}}) = P(H_{\text{I}}) + P(K_{\text{I}}) - P(\text{both})$

$= \frac{13}{52} + \frac{4}{52} - \frac{1}{52} = \frac{16}{52} = \frac{4}{13}$

(*c*) $P(\text{not } R_{\text{I}} \text{ and not } R_{\text{I}})$

$= P(\text{not } R_{\text{I}}) \cdot P(\text{not } R_{\text{I}})$

$= \frac{26}{52} \times \frac{26}{52} = \frac{1}{4}$

(*d*) $P(R_{\text{I}} \text{ and } B_{\text{II}}) = P(R_{\text{I}}) \cdot P(B_{\text{II}})$

$= \frac{1}{2} \times \frac{1}{2} = \frac{1}{4}$

(*e*) $P(\text{both same color})$

$= \frac{1}{2}$(first draw only picks color)

(*f*) $P(P_{\text{I}} \text{ or } B_{\text{I}}) = P(P_{\text{I}}) + P(B_{\text{I}}) - P(\text{both})$

$= \frac{12}{52} + \frac{26}{52} - \frac{6}{52} = \frac{32}{52} = \frac{8}{13}$

5. (*a*) $P(\text{no } H) = \frac{1}{8}$

(*b*) $P(\text{one } H) = \frac{3}{8}$

(*c*) $P(\text{two } H) = \frac{3}{8}$

(*d*) $P(\text{three } H) = \frac{1}{8}$

SECTION 9-3

1. Answer in Sec. 10-1.

3. $2 \times 2 \times 2 \times 2 \times 2 \times 2 = 64$ outcomes.

5. $4 \times 5 \times 7 = 140$ different homes.

SECTION 9-4

1. Words are permutations of letters; multidigit numerals are permutations of {0,1,2,3,4,5,6,7,8,9}.

3. ${}_8P_3 = 8 \times 7 \times 6 = 336$

${}_{10}P_4 = 10 \times 9 \times 8 \times 7 = 5040$

5. Answer in Sec. 9-5.

7.

Area code first digit	8 choices
Area code second digit	10 choices
Area code third digit	10 choices
Local number first digit	8 choices
Local number second digit	10 choices
Local number third digit	10 choices
Local number fourth digit	10 choices
Local number fifth digit	10 choices
Local number sixth digit	10 choices
Local number seventh digit	10 choices

Therefore the total number of permutations is

$$8 \times 10 \times 10 \times 8 \times 10 \times 10 \times 10 \times 10 \times 10 \times 10 = 6{,}400{,}000{,}000.$$

9. $${}_{100}C_6 = \frac{100 \times 99 \times 98 \times 97 \times 96 \times 95}{6 \times 5 \times 4 \times 3 \times 2 \times 1}$$

$$= 100 \times 33 \times 49 \times 97 \times 4 \times 19$$

11. $${}_5C_2 = \frac{5 \times 4}{2 \times 1} = 10$$

13. Because they work only when we select distinct items without replacement.

SECTION 9-5

1. $$P(4 \text{ more}) = \frac{{}_{12}C_4}{{}_{51}C_4}$$

$$= \frac{(12 \times 11 \times 10 \times 9)/(4 \times 3 \times 2 \times 1)}{(51 \times 50 \times 49 \times 48)/(4 \times 3 \times 2 \times 1)}$$

$$= \frac{12 \times 11 \times 10 \times 9}{51 \times 50 \times 49 \times 48}$$

$$= \frac{3 \times 11 \times 3}{13 \times 17 \times 5 \times 49} = \frac{33}{16{,}660}$$

3. $P(\text{both from same state}) = \dfrac{50}{_{100}C_2}$

$$= \frac{50}{(100 \times 99)/(2 \times 1)} = \frac{1}{99}$$

5. $\dfrac{1{,}000{,}000}{1{,}000{,}001}$

7. No. $\frac{1}{3} + \frac{3}{1} = 3\frac{1}{3}$, not 1; and 3:1 are the odds against Exercise 6, part *b*.

9. Assuming 44 keys, $\frac{1}{44} \times \frac{1}{44} \times \frac{1}{44} \times \frac{1}{44}$

SECTION 9-6

1. The number of elements in the union of two sets is the sum of the number in the individual set, with the number in the overlap, thus counted twice, subtracted.

3. Mutually exclusive outcomes and disjoint subsets.

SECTION 9-7

1. There are 64 possible outcomes ranging from HHHHHH to TTTTTT. The student should make a table of these outcomes and verify the following:

$$P(\text{no heads}) = {}_6C_0 \cdot \tfrac{1}{64} = \tfrac{1}{64}$$
$$P(1\text{ head}) = {}_6C_1 \cdot \tfrac{1}{64} = \tfrac{6}{64}$$
$$P(2\text{ heads}) = {}_6C_2 \cdot \tfrac{1}{64} = \tfrac{15}{64}$$
$$P(3\text{ heads}) = {}_6C_3 \cdot \tfrac{1}{64} = \tfrac{20}{64}$$
$$P(4\text{ heads}) = {}_6C_4 \cdot \tfrac{1}{64} = \tfrac{15}{64}$$
$$P(5\text{ heads}) = {}_6C_5 \cdot \tfrac{1}{64} = \tfrac{6}{64}$$
$$P(6\text{ heads}) = {}_6C_6 \cdot \tfrac{1}{64} = \tfrac{1}{64}$$

3. There are 435 members of the House of Representatives, and thus 435 − 365 = 70. There are at least 70 pairs, or 140 members who share their birthday with another member.

5. That the blank card had been removed and had been replaced by a second X card.

REVIEW EXERCISES FOR CHAP. 9

1. So that we could use the fraction-of-the-time interpretation.

3. So that we can divide the outcomes so as to classify them.

5. Among other things, that the choice of a card is truly random.

7. That all chips remaining are equally likely.

9. Assuming that the set remains on and that only complete shows are viewed, the tree-diagram method applies, and the total is 49.

11.

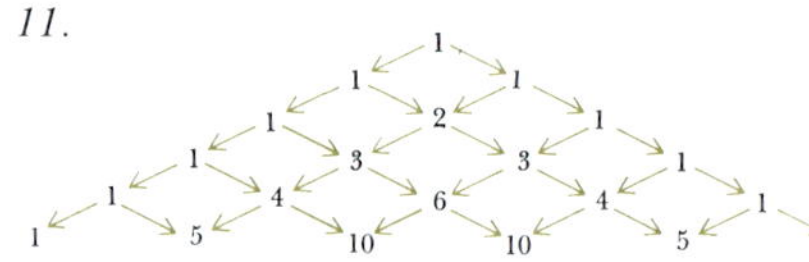

Each entry is the sum of the two with arrows pointing to it.

13. (*a*) $_{52}C_4 = \frac{52}{4} \times \frac{51}{3} \times \frac{50}{2} \times \frac{49}{1}$
$= 13 \times 17 \times 25 \times 49$

(*b*) $\dfrac{1}{_{52}C_4}$

15. $\{\{a,b,c\},\{d,e\},\emptyset,\{j,k\},\{f,g,h,i\}\}$
They are also disjoint subsets of *A*.

SECTION 10-1

1. *Outcome*	*Probability*
1	$\frac{1}{6}$
2	$\frac{1}{6}$
3	$\frac{1}{6}$
4	$\frac{1}{6}$
5	$\frac{1}{6}$
6	$\frac{1}{6}$

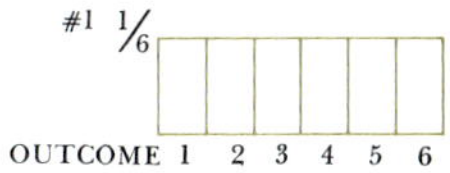

3. So that we can draw the parallels to the fraction-of-the-time definition of probability.

5. Because it describes many distributions of information or data.

7. So that we can classify outcomes, and so that the probability always adds up to 1.

SECTION 10-2

1. $EV = \frac{1}{15{,}000} \times \$1000 - \$1.05 \times \frac{14{,}999}{15{,}000}$
$= \$0.065 - \$1.05 = -\$0.98$ expected loss

3. $EV = 35 \times \frac{1}{38} - 1 \times \frac{37}{38}$
$= -\frac{2}{38} = \frac{-1}{19}$ expected loss
therefore not a fair event.

5. *Error class* Accepting an incorrect assumption.

Error class Rejecting a correct assumption.

SECTION 10-3

1. See "Statistical Abstract of the United States" issued by the United States Government for over 1,000,000 examples of descriptive statistics. These include:
(*a*) Areas of selected natural lakes.
(*b*) Production of United States coins by denomination, 1940 to 1965.
(*c*) Forest fires, number and area burned.

3. Those which estimate location and size and those which estimate distribution or spread.

5. The estimated median is 80; there is no middle value. This is difficult to estimate because there is an even number of entries and no clear-cut middle one.

7. No, it should get closer to the results of Exercise 6.

9. The expected value can be considered the average loss over a large number of trials.

11. The probability of snake eyes $= \frac{1}{36}$. Therefore $100 \times \frac{1}{36} = 2.778$ is the expected number.

SECTION 10-4

1. $\mu = 11 \times \frac{1}{2} = 5.5$

$\sigma = \sqrt{11 \times \frac{1}{2} \times \frac{1}{2}} = 1.658$

$Z_1 = \dfrac{3.5 - 5.5}{1.658} = -1.206$

$Z_2 = \dfrac{4.5 - 5.5}{1.658} = \dfrac{-1.0}{1.658} = -0.603$

Area $Z_1 = 0.3861$
Area $Z_2 = 0.2268$
Area Z_1 − Area $Z_2 = 0.1593$

3. Because a smoothed curve was used in place of the steps of the actual case.

5. Direct methods involve calculating a number of probabilities and adding them up. Calculating a normal-curve estimate is simpler.

7. Statistical decision making involves drawing conclusions about actual quantities based on the observations of symbolic models.

9. Under identical conditions of weather sometimes one kind occurs, and other times other kinds.

11. In Exercise 10, even if the data existed and the theory worked out corresponding advances, general technology would have been needed to put them to use.

REVIEW EXERCISES FOR CHAPTER 10

1. There are methods that revolve around concepts of purely random outcomes.

3. So that each possible outcome falls into one and only one classification, so that the fraction-of-the-time definition can be used.

5. If, on the average, a loss occurs on each bet, we have an outcome of "well we lose a little on each sale but make it up in volume."

7. For one thing, how many divers were asked the question before they got three out of four who preferred their air.

9. Insurance is a bet. For example, in simple life insurance you bet that you will die, and the company bets that you will not. Because they hold the "stakes" and can invest them, they can afford a fair bet.

11. The mode is 80.

INDEX